**Unmanned System Technologies**

Springer's Unmanned System Technologies (UST) book series publishes the latest developments in unmanned vehicles and platforms in a timely manner, with the highest of quality, and written and edited by leaders in the field. The aim is to provide an effective platform to global researchers in the field to exchange their research findings and ideas. The series covers all the main branches of unmanned systems and technologies, both theoretical and applied, including but not limited to:

- Unmanned aerial vehicles, unmanned ground vehicles and unmanned ships, and all unmanned systems related research in:
- Robotics Design
- Artificial Intelligence
- Guidance, Navigation and Control
- Signal Processing
- Circuit and Systems
- Mechatronics
- Big Data
- Intelligent Computing and Communication
- Advanced Materials and Engineering

The publication types of the series are monographs, professional books, graduate textbooks, and edited volumes.

Ke Zhang · Wanglei Cheng · Qingyi Liu · Bin Jiang

# Fixed-Time/Prescribed-Time Fault-Tolerant Control for Multiple Unmanned Systems

Springer

Ke Zhang
College of Automation Engineering
Nanjing University of Aeronautics and Astronautics
Nanjing, Jiangsu, China

Wanglei Cheng
College of Automation Engineering
Nanjing University of Aeronautics and Astronautics
Nanjing, Jiangsu, China

Qingyi Liu
College of Automation Engineering
Nanjing University of Aeronautics and Astronautics
Nanjing, Jiangsu, China

Bin Jiang
College of Automation Engineering
Nanjing University of Aeronautics and Astronautics
Nanjing, Jiangsu, China

ISSN 2523-3734 ISSN 2523-3742 (electronic)
Unmanned System Technologies
ISBN 978-981-95-2912-4 ISBN 978-981-95-2913-1 (eBook)
https://doi.org/10.1007/978-981-95-2913-1



This Springer imprint is published by the registered company Springer Nature Singapore Pte Ltd.
The registered company address is: 152 Beach Road, #21-01/04 Gateway East, Singapore 189721, Singapore

# Preface

Multiple unmanned systems, especially quadrotor unmanned aerial vehicles (UAVs), fixed-wing UAVs, unmanned helicopters, and UAVs-unmanned ground vehicles (UAVs-UGVs) system, have received an increasing amount of attention. Compared with single unmanned vehicle, multiple unmanned systems can realize information exchange through interactive network and cooperate with each other to take the place of humans to complete repetitive and arduous tasks in many areas, such as target surveillance and tracking, intelligence reconnaissance, search and rescue missions. However, due to the complex composition of the actuator, sensor, or other system parts, faults occur more frequently in multiple unmanned systems than those in one unmanned system. In the framework of multiple unmanned systems, the faults in one or more systems may be unevenly disseminated among neighboring agents by using a communication graph. Consequentially, the performance of the whole system will be affected. Hence, it is necessary to research fault-tolerant control (FTC) methods for multiple unmanned systems to enhance safety, reliability, and maintainability. On the other hand, convergence rate is an important performance index in the investigation of FTC problem, and the existing FTC methods with asymptotic stability are not feasible in many practical situations. To pursue fast convergence, finite-time FTC has drawn great attention due to its accuracy and robustness. However, the resultant convergence time of the finite-time FTC relies on initial states, and may grow unboundedly as the initial states are large enough. Hence, fixed-time/prescribed-time stability is proposed to correct this flaw, which guarantees that the convergence settling time is upper bounded by a constant and independent of initial states. Meanwhile, the actual system has the characteristics of dynamic instability, parameter uncertainty, strong coupling. Therefore, designing the fixed-time/prescribed-time FTC is challenging.

Fixed-time/prescribed-time FTC for multiple unmanned systems is an emerging research direction aimed at improving the safety and maneuverability of practical systems in the fields of quadrotor UAVs, fixed-wing UAVs, unmanned helicopter, and UAVs-UGVs system. However, due to the collaboration between multiple unmanned systems under practical environment confronts many difficulties, there are no more research reports of fixed-time/prescribed-time FTC, and the existing problems that should be addressed to improve the system safety are as follows

(1) Many studies in the FTC research area of multiple unmanned systems merely deal with the asymptotic convergence regardless of the requirement of the convergence time, hence, the rapid response and high steady-state tracking precision should be considered with fixed-time convergence or prescribed-time convergence.
(2) The existing results have not fully considered the impact of uncertainties on design of FTC strategy. However, the uncertainties of external disturbances, the dynamic and time-varying nature of topologies, and the interactivity of system structure can affect the quality of task completion, thus requiring targeted research.
(3) The existing research results are mostly focused on homogeneous multiple unmanned systems, and there are very few research results on FTC of heterogeneous multiple unmanned systems represented by UAVs-UGVs formation system. It is imperative to carry out relevant work to meet the practical requirements.

Based on the aforementioned discussions, this monograph focuses on the fixed-time/prescribed-time FTC problems for multiple unmanned systems. The main contributions of this monograph are summarized as the following aspects.

**Part I: Sliding Mode-Based Fixed-Time FTC for Multiple Unmanned Systems**

(1) For multiple quadrotor UAVs with actuator faults, the fixed-time fault-tolerant formation control is proposed to keep the safe distancing of UAVs when cooperating through an intensive formation. Based on an augmented fixed-time observer estimating actuator faults and disturbances, a distributed fixed-time sliding-mode controller that integrates the gradient of repulsive potential function into the sliding manifold is designed to achieve leader-follower formation control and collision avoidance simultaneously.
(2) The fixed-time collision-free FTC problem is studied under actuator faults for quadrotor UAVs. A fixed-time observer is proposed to estimate and compensate actuator faults in translational and rotational dynamics. Then, a fixed-time command filter is designed to ensure that the tracking error of each quadrotor UAV is practical fixed-time stable. In addition, an integral sliding manifold is designed based on the formation error, and an artificial potential field (APF) function-based fixed-time formation controller with collision avoidance is developed to guarantee the safety and maneuverability of the whole formation system.
(3) For UAVs-UGVs with actuator faults, a fixed-time fault-tolerant law with prescribed performance is presented. A performance function is designed to guarantee that formation error converges to a prescribed arbitrarily small region within a known time. Then, two novel distributed and decentralized adaptive FTC laws are proposed, which not only achieves stable formation tracking with fixed-time prescribed behavioral metrics but also ensures that the formation errors are uniformly bounded within a settling time.

**Part II: Fixed-Time FTC for Multiple Unmanned Systems with Switching Topologies**

(1) The FTC for fixed-wing UAVs in longitudinal formation system under switching topologies with actuator fault, input saturation and model uncertainty is investigated based on hierarchical control structure. The upper layer is divided into virtual speed formation system and virtual height formation system, and the lower system utilizes the output of the upper system to design a fixed-time anti-saturation fault-tolerant tracking controller, which ensures that the tracking error converges to the region near zero within a fixed time.
(2) The novel continuous fixed-time fault-tolerant controller is proposed to eliminate the chattering problem of fixed-time controller for UAVs-UGVs formation system with actuator faults and model uncertainties under fixed and switching topologies. A fixed-time observer is constructed to estimate the actuator faults and uncertainties, and then a continuous fixed-time formation tracking control scheme is proposed based on high-order sliding-mode and backstepping technologies.

**Part III: Prescribed-Time FTC for Multiple Unmanned Systems**

(1) For multiple unmanned helicopters subject to actuator faults, three high-order fully actuated subsystems are derived for design convenience. Then, a prescribed-time fault estimation observer is developed to approximate the subsystem states and lumped faults, and we incorporate the high-order fully actuated system (HOFAS) approach into the prescribed-time fault-tolerant formation control, to eliminate the influence of faults and preserve the target geometric pattern. In this manner, the prescribed-time property and the higher tracking accuracy can be accomplished.
(2) The neural network observer-based prescribed-time fault-tolerant formation strategy is designed for UAVs-UGVs under actuator faults, matching and mismatched disturbances to offer both chattering elimination and satisfactory tracking performance. A neural network observer is employed to approximate the unknown actuator faults and lumped disturbances. Then, a prescribed-time FTC strategy is constructed, which not only guarantees that the desired formation can be realized in an user-assignable finite time but also ensures that the control action can be smooth.
(3) For heterogeneous UAVs-UGVs, a prescribed-time optimal FTC algorithm based on hierarchical control, game theory, and prescribed-time method is proposed under actuator faults. In virtual layer, a quadratic performance index function is designed and an optimal control strategy is designed to achieve consensus tracking. In actual control layer, a prescribed-time fault-tolerant tracking controller is designed based on the optimal signal, sliding-mode and adaptive technologies, which ensures that the global convergence time is completely independent of any other parameters.

This monograph presents rigorous analysis and design process of the fixedtime/prescribed-time FTC issues for quadrotor UAVs, fixed-wing UAVs, unmanned helicopters, and UAVs-UGVs systems. The designed methods can adjust the structure and parameters of the system controller in a timely manner, ensuring the safety of the unmanned systems and restoring critical control performance as soon as possible. Therefore, these research results will enrich the development of FTC for multiple unmanned systems, and have important theoretical significance and practical application value for ensuring the safety and control of complex control systems.

Nanjing, China
July 2025

Ke Zhang
Wanglei Cheng
Qingyi Liu
Bin Jiang

# Acknowledgements

This monograph was supported in part by National Natural Science Foundation of China under Grants (62173180, 62020106003, 62503221), the Natural Science Foundation of Jiangsu Province of China under Grant (BZ2024037, BK20253024, BK20251366), and the Postdoctoral Fellowship Program of China Postdoctoral Science Foundation under Grant (GZB20240974). The authors would like to thank the students Guangrun Liu, Zhunyuan Ye, Tianxiao Ma, Bo Meng, and Yuan Lu for their contributions and efforts to this monograph.

# Contents

# Acronyms

| | |
|---|---|
| APF | Artificial Potential Field |
| CF | Command Filtering |
| DOF | Degree-of-Freedom |
| FTC | Fault-Tolerant Control |
| FTPF | Finite-Time Performance Function |
| HITL | Hardware-in-the-Loop |
| HJB | Hamilton-Jacobi-Bellman |
| HOFAS | High-Order Fully Actuated System |
| HOSMC | High-Order Sliding Mode Control |
| ITSM | Integral Terminal Sliding Mode |
| MASs | Multiagent Systems |
| PID | Proportional-Integral-Derivative |
| RBFNN | Radial Basis Function Neural Network |
| UAVs | Unmanned Aerial Vehicles |
| UAVs-UGVs | UAVs-Unmanned Ground Vehicles |
| UUB | Uniformly Ultimately Bounded |

# Chapter 1
# Introductions

## 1.1 Background

To date, the multiple unmanned systems have received an increasing amount of attention because they can take the place of humans to complete repetitive and arduous tasks in many areas, such as target surveillance and tracking, intelligence reconnaissance, search and rescue missions, and telecommunication relay [13, 15, 19, 49, 51]. With the increasing complexity of mission scenarios, the application of a single unmanned system in military and civilian fields is becoming increasingly limited. Therefore, a great deal of the scholars have directed their energies to research on multiple unmanned systems.

The concept of formation control for multiple unmanned systems has been presented according to the cooperative hunting methods of dolphins and wolves. The formation control of multiple unmanned systems can be divided into leader-following [21, 70, 93], behavioral approaches [1, 37], and virtual structure [2, 76] based on the basic principle of control tactics. As the number and complexity of actuators, sensors, and components in multiple unmanned systems increase, the overall formation system becomes more susceptible to faults, potentially leading to significant performance degradation and compromise flight assignments.

Different from the faults in a single unmanned system, which primarily impact its own operational states, faults in one unmanned system within a formation team affect not only the stability of the faulty system but also disrupt neighboring unmanned systems through information transmission among them. Hence, the fault-tolerant control (FTC) should be developed to stabilize the faulty unmanned systems and the formation team based on the system states and the estimated information received from neighboring unmanned systems, thus increasing the design challenge.

Practically, a faster convergence time/rate is also a significant performance indicator of formation flying in carrying out tasks. Apart from the asymptotic FTC, the finite-time FTC can restrain the system states around the origin within a limited time, which thus makes the settling time to be constrained by a computable constant.



K. Zhang et al., *Fixed-Time/Prescribed-Time Fault-Tolerant Control for Multiple Unmanned Systems*, Unmanned System Technologies,
https://doi.org/10.1007/978-981-95-2913-1_1

However, the settling time is usually relevant to the initial conditions, which indicates that a longer estimation or control convergence time/rate might be achieved when the values of the initial conditions are large. To conquer this drawback, the fixed-time FTC was realized, whose settling-time is correlated with the designed parameters only. Furthermore, to predetermined the convergence time/rate in advance, the prescribed-time FTC was garnered much attentions. From the existing literature, there are still few available works integrating the fixed-time/predefined-time property into the FTC strategy, designing a fixed-time/prescribed-time fault estimation observer and then compensating the redundancy within a faster convergence, which is the main motivation of this monograph.

This monograph will present systematical and comprehensive descriptions on the fixed-time/predefined-time FTC for multiple unmanned systems in the presence of diverse faults. In the following sections, several multiple unmanned systems, FTC and fixed-time/predefined-time FTC are analyzed in detail.

## 1.2 Multiple Unmanned Systems

Multiple unmanned systems have good environmental adaptability and autonomous capabilities, which can complement each other in tactics and equipment to improve combat efficiency and the success rate of completing tasks. Therefore, they can complete complex tasks that cannot be completed by a single unmanned system. As a result, studying the multiple unmanned systems has important theoretical significance and application value. In this section, we mainly concentrate on the background of quadrotor UAVs, fixed-wing UAVs, unmanned helicopters, and UAVs-UGVs.

### *1.2.1 Quadrotor UAVs*

A quadrotor UAV is actuated by four motors mounted on an airframe. Its maneuvering is achieved by alternating the rotor speed to generate torque and thrust, which grants the advantages of vertical take-off and landing as well as simple assembly and configuration, paving the way for its popular application in various scenarios from aerial photographing to cargo delivering. However, the unique mechanics of quadrotor UAVs also limit the weight-lifting ability of a single UAV. Because, as the weight of a quadrotor UAV increases, the size of its motor should also expand for adequate lift, while the response speed of a motor is inverse to its size, bringing difficulty in the design of motors and controllers.

To undertake more complex missions, multiple quadrotors can be organized into a formation to dispatch together. The formation of quadrotor UAVs is endowed with the advantages of payload-carrying [5] and multi-tasking [22], to maximize such superiority, global formation consensus must be guaranteed, and various control schemes such as optimal control [7], adaptive control [26], state-feedback control [6], and output-feedback control [58, 71] have been raised accordingly to pursue various control objectives. A distributed optimal control method based on linear quadratic regulator was proposed in [7] for the leader-follower structure. To deal with disturbances, an adaptive control scheme was adopted in [26]. To confront the complexity of quadrotors' dynamics, data-driven mechanisms such as reinforcement learning and neural networks, were adopted in [4]. To tackle the influence of communication delay, robust formation controller was designed in [40] to restrain time delay as well as nonlinearity and parametric perturbation. In the absence of critical sensors, output-feedback with auxiliary dynamics was constructed in [58] to overcome the shortness of velocity and angular velocity sensors. In [86], event-triggered technique was employed to secure the formation tracking of quadrotor UAVs under replay attacks. It is worth noticing that the prevailing solutions toward quadrotor formation have all employed a cascaded framework in which the translational and rotational movements of a quadrotor UAV are separately considered, thus combining the consensus control of multiagent systems (MASs) with the features of quadrotor dynamics.

### 1.2.2 Fixed-Wing UAVs

The fixed-wing UAV is an aircraft that typically uses wings to provide lift, rather than relying on rotors (as multirotor UAVs do) to stay aloft. Its propulsion usually comes from propellers or jet engines. Although rotor-UAVs are more widely employed in various fields at present, fixed-wing UAVs also have incomparable advantages over rotor-wing UAVs. For example, fixed-wing UAVs have longer flight durations, faster speeds, and can cover greater distances with less energy loss [8, 30]. Therefore, these advantages enable fixed-wing UAVs to perform missions that rotor-UAVs cannot accomplish. From the perspective of the UAV model, the complex and highly coupled nonlinear characteristics inherent in the dynamics model of fixed-wing UAVs make it very difficult to design effective control strategies. Furthermore, the flight state and aerodynamic environment can also affect the control performance of UAVs.

Due to the increasing demands on the level of mission completion, a single UAV will inevitably face the dilemma of insufficient capabilities in both military and civilian applications. Fortunately, UAV formation can utilize a communication network to integrate multiple UAVs into an unified system, thereby compensating for and extending the performance of a single UAV. Recently, leader-following control has attracted considerable attention from numerous scholars due to its simpler structure and easier implementation, better stability and robustness, and greater advantages in mission allocation. This has led to a large number of significant studies emerging in the field. For example, in [27], the problem of formation control for fixed-wing UAVs under unknown wind disturbances, velocity and overload constraints was addressed. Based on locally sampled relative position information, a distributed control scheme with an inner-outer loop structure and a self-triggered sampling mechanism was proposed. In [101], a fully distributed model reference robust adaptive controller was designed. It employed an adaptive projection formation operator and a novel high-frequency robust term to ensure the consensus asymptotic stability of the formation system under the influence of uncertainties and disturbances. In [82], the construction of an adaptive formation controller for fixed-wing UAVs with structural parameter uncertainties (i.e., mass and inertia uncertainties) and the issue of software-in-the-loop implementation were investigated.

### *1.2.3 Unmanned Helicopters*

As an important development subfield of aircrafts, the unmanned helicopter mainly obtains upward lift through the main rotor, and changes its attitude angle by changing the pitch of the main rotor and the tail rotor, so as to perform maneuverable flight at low altitude (several meters over the ground), low speed (starting from hover), and invariant nose direction, especially vertical takeoff and landing over small areas. Owing to these unique flight characteristics, the unmanned helicopter has been extensively utilized in military aspects for various tasks, like enemy reconnaissance, targets attack, weapons delivery and logistics support. Additionally, its civil applications include short-distance transport, medical rescue and precision agriculture. With the increase of task complexity and number, it is difficult for a single unmanned helicopter to complete the task, which requires multiple unmanned helicopters formed as a formation to efficiently execute cooperative tasks.

Nevertheless, the unmanned helicopter is a complex system with strong coupled, under-actuated and nonlinear characteristics, which makes the formation control research more challenging. Therefore, compared with quadrotor UAVs and fixed-wing UAVs, there are relatively few research results on formation control of unmanned helicopters. Based on spherical abstraction of the state space, a hybrid supervisory control scheme was proposed in [28, 29] to realize the three-dimensional leader–follower formation control objective of unmanned helicopters. In [31], an intelligent leader-following consensus formation control method based on the backstepping technology was presented for a team of unmanned helicopters, which could online estimate system uncertainties by recurrent neural networks. In the framework of hierarchical design, the authors in [64] employed the nonsingular terminal sliding mode method to design formation controller and attitude tracking controller for unmanned helicopters. Additionally, the authors in [65] provided a fast formation tracking control protocol for unmanned helicopters, enabling them to form the desired formation mode in the finite time. To avoid the case that the unmanned helicopter may collide with its neighbors before stabilizing the control input, the authors in [68] extended these aforesaid results into a formation control without collisions under a repulsive APF. It is worth noting that the aforementioned research results can meet the general needs of formation control, but they fall short in terms of robust formation mobility and flexibility. Therefore, it is necessary to employ different control strategies according to different flight performance and control characteristics of unmanned helicopters to achieve more efficient formation flight purposes, which needs to be further studied.

### 1.2.4 UAVs-UGVs

The heterogeneous UAVs-UGVs formation systems have satisfactory payload and task configuration capabilities (height, range, duration, etc), as well as numerous command, control and data collection functions [16]. In particular, the complementary group of UAVs and UGVs combines their advantages, such as satisfactory payload and task configuration capabilities, and localization capabilities, which provides the better performance. Moreover, through the communication, cooperation, scheduling and control between different agents to analyze and operate the structure, function and behavior characteristics of the systems, the UAVs-UGVs formation system can realize information exchange through interactive network and cooperate with each other to complete the more complex and multivariate tasks [59]. Therefore, the strategy of UAVs-UGVs cooperation has been the research hotspot in the field of cooperative formation control, and has widely used in military and civilian domains, such as tracking and tracing, cooperative surveillance, radiation search, resource exploration and target localization.

Cooperative formation control of heterogeneous UAVs-UGVs formation system is an emerging research direction. Collaboration means that the movement goal of all agents is to improve the overall formation performance of the system. Due to the external environment and task switching have strong randomness, and the practical system is inevitably subject to model uncertainties and disturbances, the UAVs-UGVs formation confronts many challenges. Over the past few years, very few results have been reported on this issue. For example, based on centralized information interaction method, the authors in [53] designed formation controllers for UAV to track the reference trajectory provided by the leader-UGV and completed linear or nonlinear formation tasks. However, the centralized methods rely on a perfect interactive network and prone to fail due to connection faults or occurring faults in the central controller. To overcome this problem, decentralized formation controllers are widely used in air-ground formation problems [61]. In the past decades, inspired by the consensus theory, distributed formation control methods using the local interaction information have received extensive attention [27, 42]. Hence, it is necessary to study the distributed formation control problem of UAVs-UGVs system, and new alternative and effective distributed schemes should be developed.

Based on the the aforementioned discussions, it can be concluded that the the multiple unmanned systems have been extensively studied by many researchers and some important results were presented. However, the above control systems in theory analysis are assumed to be fault-free. With the passage of time and continuous changes in internal and external environments, the multiple unmanned systems inevitably experience state gradients or mutations, leading to faults in actuators, sensors, and communication network. These faults can range from reducing operation quality to causing major accidents. Therefore, it is necessary to design fault-tolerant mechanisms for health management, to timely detect, predict, and handle faults, thereby ensuring the safety and reliability of the whole unmanned system.

## 1.3 FTC

At the same time, the standards for the reliability and performance of multiple unmanned systems are increasing. Regrettably, despite any technological advances, the appearance of faults is inevitable. This is primarily attributable to the fact that the multiple unmanned systems embed a variety of subsystems, sensors, actuators and components that are susceptible to faults. Figure 1.1 shows an example of a multiple unmanned system with faults. In addition, unforeseeable conditions and events can occur in their operating environment. To accommodate faults in these realities, passive and active FTC methodologies are frequently applied to the cooperative control for multiple unmanned systems [80]. According to different types of FTC, a list of reviewed publications is provided in Fig. 1.2. In the passive strategy, fault values are regarded as bounded disturbances and it is an extension of traditional robust control [14, 66]. The active one contains both fault estimation and reconfigurable control, and fault estimation can provide real-time fault values, which are used to design the reconfigurable control protocol [38, 100].

Passive FTC does not depend on detection and acquire fault redundancy, only designing a fixed controller to against presumed faults. Up to now, several passive FTC srategies have been proposed for single unmanned system. For example, an event-triggered adaptive fuzzy fixed-time FTC scheme was presented in [88] for a tailless flying-wing UAV. An adaptive fault-tolerant controller was designed in [72] to achieve asymptotic tracking control for unmanned underwater vehicles by employing robust integral of the sign of error feedback method. In [63] an adaptive sliding-mode-based control allocation scheme was proposed for helicopter to

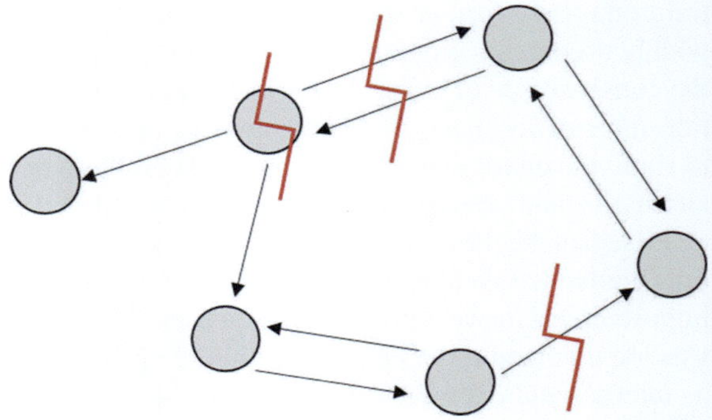

**Fig. 1.1** An example of a multiple unmanned system with faults

| Type of FTC | FTC schemes | References |
|---|---|---|
| Passive FTC | Distributed adaptive FTC | [3][72][88][91] |
| | Sliding-mode-based FTC | [23][63] |
| | Intelligent FTC | [33][43][98] |
| Active FTC | Observer-based FTC | [12][39][56][93] |
| | Adaptive FTC | [17][69][73][102] |
| | Data-driven FTC | [83][95] |

**Fig. 1.2** Classication of FTC results

accommodate simultaneous actuator faults. As for multiple unmanned system, the authors in [91] proposed a distributed adaptive FTC scheme for multiple trailing fixed-wing UAVs against actuator faults and wake vortices, by using the sliding-mode estimator, dynamic surface control architecture, neural networks, and disturbance observers. As for the unmanned helicopters, the authors in [3, 99] focused on an adaptive passive FTC strategy that ensures the tracking accuracy and flight stability of a single unmanned helicopter, and the authors in [81] developed a passive FTC-based cooperative control algorithm to synchronize the attitudes of multiple three degree-of-freedom (DOF) unmanned helicopters. In [98], a passive FTC policy was designed for the team of heterogeneous UAVs-UGVs formation system via reinforcement learning but without knowledge of the vehicle dynamics. In [43], the authos proposed a neural-network-based adaptive passive FTC strategy for heterogeneous UAVs-UGVs formation system.

In terms of the fault diagnosis technique that can detect, identify and estimate faults, active FTC is devised to reconfigure the control scheme or modify controller parameters to eliminate the affects of faults based on fault estimation observers [56, 93], adaptive approximators [17, 102], and data-driven methods [83, 95]. For single unmanned system, an active FTC scheme was constructed in [73] for a helicopter with actuator/sensor faults. In [55], an active FTC system was proposed for spacecraft attitude control in the presence of actuator faults, fault estimation errors, and control input constraints. In [69], an active FTC scheme was proposed to achieve trajectory tracking for unmanned underwater vehicle subject to sensor faults and multiple uncertainties. In [52], an active FTC was designed for UAV to rely on uncertain information provided by the fault detection and identification algorithm. In [96], a fuzzy adaptive observer based active FTC scheme was developed for unmanned surface vehicle with unmeasurable states, sideslip angle, and steering machine fault.

As for multiple unmanned system, the decentralized fault estimation and distributed active FTC hierarchy was developed in [38] for multiple 3-DOF unmanned helicopters. In [77], an active FTC method of fault diagnosis, estimation, and FTC for UAVs-UGVs attitude system was designed under actuator time-varying faults and external disturbances. In [39], an unknown input observer-based fault estimation and distributed active FTC approach were developed for multiple unmanned surface vehicles with aperiodic denial-of-service attacks.

Based on the above analysis, it can be concluded that some effective schemes have been developed in FTC for multiple unmanned systems to finish specific tasks, such as the UAVs, UAVs-UGVs, unmanned surface vehilces, and unmanned helicopters. Moreover, the methods involved are also extensive, including adaptive control technology, sliding-mode control technology, intelligent control technology, and observer-based technology.

## 1.4 Fixed-Time FTC

From a practical standpoint, the settling time is an important performance specification of a time-critical system, which characterizes the corresponding convergent speed. However, the convergence time derived from the above-mentioned FTC schemes escapes to infinity. Recently, finite-time FTC has received considerable research attention for multiple unmanned systems [9, 33, 41]. However, the convergence time of associated algorithms is usually dependent on the initial conditions of the system. Sufficiently large or even unavailable initial states in some cases result to realize the finite convergence time. Hence, fixed-time FTC method was proposed in [48] to correct this flaw, which can ensure that the settling time is upper bounded by a constant independent of initial conditions.

Based on the fixed-time theorem, several important FTC results are proposed for single unmanned system. The authors in [88] proposed an event-triggered adaptive fuzzy fixed-time FTC scheme for a tailless flying-wing UAV under limited communication resources, actuator faults, external disturbances, and system uncertainties. A distributed fixed-time FTC methodology for six-DOF fixed-wing UAV was designed in [45] for velocity and attitude channels. In [62], a fixed-time FTC based on the backstepping control was designed for robot manipulators to enhance the tracking performance of the system. A fixed-time FTC protocol was proposed in [36] for autonomous ground vehicles to follow reference paths and velocities with preset performance constraints. An adaptive fault-tolerant approach with a fixed-time sliding mode was proposed in [94] for trajectory tracking of uncertain robot manipulators with actuator effectiveness faults. A fixed-time FTC scheme was proposed in [87] for a hypersonic gliding vehicle to counteract actuator faults and model uncertainties. A fuzzy adaptive FTC strategy was proposed in [46] to solve the fast fixed-time constrained tracking problem for hypersonic flight vehicles. A fixed-time attitude tracking maneuver was designed in [10] of rigid spacecraft in the presence of actuator faults, external disturbances, and inertia uncertainties. In [23], a novel fixed-time sliding mode FTC method was developed for a rigid spacecraft subject to external disturbance, actuator faults, and input saturation. In [89], a fixed-time adaptive fuzzy FTC method was designed for tailless aircraft attitude tracking problem in the absence of angular velocity measurements.

As for multiple unmanned systems, preliminary research work of theoretical research for fixed-time FTC has been carried out. For example, an adaptive event-triggered fixed-time FTC method was proposed in [25] for a category of homogeneous MASs. In [50], a fixed-time sliding mode consensus FTC strategy was designed for a class of nonlinear homogeneous MASs with actuator faults. In [78], an adaptive fuzzy fixed-time FTC scheme was proposed of homogeneous MASs suffering from nonaffine faults. In the application field of multiple unmanned systems, a distributed fixed-time FTC methodology was proposed in [92] for multiple six-DOF fixed-wing UAVs subject to actuator faults and full-state constraints. A novel fixed-time FTC scheme was proposed in [47] for fixed-wing UAVs with uncertainties, communication link and actuator faults. In [90], a fixed-time intelligent FTC scheme with prescribed performance was proposed for multiple unmanned airships against actuator faults.

The above analysis shows that the fixed-time FTC has received considerable research attention in single unmanned system. However, there have not been many achievements in fixed-time FTC of multiple unmanned systems based on the interactive information. Moreover, due to the existence of the heterogeneity, the results of fixed-time FTC for heterogeneous multiple unmanned systems are even rarer. Since agents often have different dynamics on many practical occasions, it is of practical interest to extend it to more common dynamic systems, especially, heterogeneous multiple unmanned systems. Therefore, sustained efforts are needed to conduct in-depth research on fixed-time FTC for multiple unmanned systems.

## 1.5 Prescribed-Time FTC

Although the aforementioned results overcome a variety of fixed-time FTC issues in certain ways, the methods aforementioned have a certain limitation, i.e., the convergence time is not able to be predefined in advance. For this purpose, the prescribed-time convergence has garnered much attentions [20, 84, 85]. Compared with the finite/fixed-time control methods where the finite convergence time is determined by system initial condition or a number of design parameters, the prescribed-time method renders the finite convergence time fully independent of initial condition and any other parameter, thus it can be uniformly prespecified. Moreover, the prescribed-time control is based on regular feedback of local system states, thus it generates smooth control action everywhere. Owing to this superior performance, prescribed-time convergence reveals unique advantages in synchronous stabilization and cooperative maneuvering, like the faster obstacle avoidance [54], (bipartite) consensus control [32, 79], formation tracking [35, 97], and the containment control [24, 74].

For systems with faults, there have not been many research results on prescribed-time FTC. As for single unmanned system, an adaptive prescribed-time control method was proposed in [75] for a class of nonlinear telerobotic system with actuator faults and position error constraints. The authors in [60] developed an improved iterative learning observer to handle the unknown faults, and a prescribed-time active FTC scheme was presented for dynamic positioning vessel. A prescribed-time FTC was proposed in [44] for micro-electro-mechanical system using type-2 fuzzy logic system. Moreover, a prescribed-time FTC strategy was proposed in [57] for a strict-feedback nonlinear system with sensor faults and unknown control directions. In [34], an adaptive practical prescribed-time FTC method was proposed for autonomous underwater vehicle under tracking error constraints and velocity constraints. In [12], a prescribed-time FTC scheme with prescribed performance was proposed for quadrotor UAV to track the desired trajectory under actuator faults.

As for multiple unmanned systems, a prescribed-time fuzzy adaptive consensus control method was proposed in [67] for homogeneous MASs with sensor faults and dead-zone input. The authors in [12] proposed a prescribed-time tracking FTC scheme for quadrotor UAVs under actuator faults based on extended state observer. In [18], a novel prescribed-time FTC method was proposed for vehicle platoon under actuator faults and leader-follower link faults. In [11], a novel distributed prescribed-time consensus tracking control strategy is presented for heterogeneous nonlinear MASs under deception attacks and actuator faults.

Recently, the prescribed-time FTC for multiple unmanned system has become an emerging popular topic. However, based on the above analysis, it can be concluded that the prescribed-time FTC of multiple unmanned systems is still in the initial stage of research and the problem of reconstructing the prescribed-time controller under faults has not been effectively addressed. Especially in the field of heterogeneous multiple unmanned systems represented by UAVs-UGVs, implementing prescribed-time FTC is more challenging. The existing results of prescribed-time/fixed-time FTC for multiple unmanned systems can be found in Fig. 1.3. We can find that there are relatively few reports on prescribed-time/fixed-time FTC for multiple unmanned systems. Therefore, it is imperative to carry out research on prescribed-time/fixed-time FTC for multiple unmanned systems.

## 1.6 Organization of the Monograph

This monograph consists of 10 chapters. This chapter presents the background and the descriptions of multiple unmanned systems, as well as the descriptions of FTC and fixed-time/prescribed-time FTC. The whole organization diagram of this monograph is shown in Fig. 1.4.

Chapters 2, 3 and 4 are about the sliding mode-based fixed-time FTC for multiple unmanned systems. Specifically, Chap. 2 investigates the command filter-based fixed-time fault-tolerant formation control of multiple quadrotor UAVs under actuator faults. Chapter 3 aims to propose a novel fixed-time collision-free FTC scheme

| Type of FTC | Main contents | References |
|---|---|---|
| Fixed-Time FTC | Formation tracking for fixed-wing UAVs | [45][88] |
| | Synchronization tracking for multiple unmanned airships | [90] |
| | Consensus tracking for multiagent systems | [25][50] |
| Prescribed-Time FTC | Formation tracking for quadrotor UAVs | [12] |
| | Consensus tracking for vehicle platoon | [18] |
| | Consensus tracking for multiagent systems | [11][67] |

**Fig. 1.3** Existing results of prescribed-time/fixed-time FTC for multiple unmanned systems

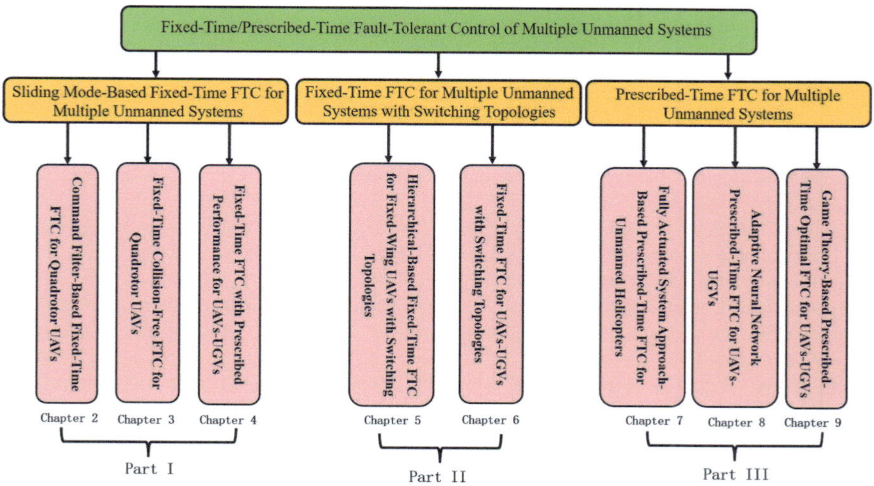


**Fig. 1.4** Organization diagram of this monograph

for quadrotor UAVs, which considers the collision avoidance among UAVs when faults occur, and the convenience of engineering application. Chapter 4 proposes a fixed-time fault-tolerant formation control strategy for UAVs-UGVs system with prescribed performance, which not only achieves stable formation tracking but also ensures that the formation errors are uniformly bounded.

Chapters 5 and 6 are about the fixed-time FTC for multiple unmanned systems with switching topologies. Specifically, Chap. 5 solves the fault-tolerant formation control problem for multiple fixed-wing UAVs with actuator faults under switching communication topologies. Chapter 6 proposes a novel continuous fixed-time fault-tolerant formation control framework for UAVs-UGVs system with actuator faults and model uncertainties under fixed and switching topologies.

Chapters 7, 8, and 9 are about the prescribed-time FTC for multiple unmanned systems. Specifically, Chap. 7 investigates the fully actuated system approach-based prescribed-time FTC for unmanned helicopters. An adaptive neural network prescribed-time formation controller is proposed for a specific heterogeneous UAVs-UGVs system in Chap. 8. Then, a prescribed-time optimal formation control strategy is designed for UAVs-UGVs system in Chap. 9 based on game theory. Finally, Chap. 10 gives the conclusions and future directions of MASs. Therefore, this monograph presents the fixed-time/prescribed-time FTC design for homogeneous and heterogeneous multiple unmanned systems against diverse presumed faults, to improve the reliability and safety during the flight.

## References


1. Balch T, Arkin RC (1998) Behavior-based formation control for multirobot teams. IEEE Trans Tobotics Autom 14(6):926–939
2. Beard RW, Lawton J, Hadaegh FY (2001) A coordination architecture for spacecraft formation control. IEEE Trans Control Syst Technol 9(6):777–790
3. Chen M, Yan K, Wu Q (2022) Multiapproximator-based fault-tolerant tracking control for unmanned autonomous helicopter with input saturation. IEEE Trans Syst Man Cybern Syst 52(9):5710–5722
4. Cheng M, Liu H, Wen G, Lü J, Lewis FL (2023) Data-driven time-varying formation-containment control for a heterogeneous air-ground vehicle team subject to active leaders and switching topologies. Automatica 153:111029
5. Doakhan M, Kabganian M, Azimi A (2023) Cooperative payload transportation with real-time formation control of multi-quadrotors in the presence of uncertainty. J Franklin Inst 360(2):1284–1307
6. Dong Z, Wang X, Zhang X, Hu M, Dinh TN (2023) Global exponential synchronization of discrete-time high-order switched neural networks and its application to multi-channel audio encryption. Nonlinear Anal Hybrid Syst 47:101291
7. Du H, Zhu W, Wen G, Duan Z, Lu J (2019) Distributed formation control of multiple quadrotor aircraft based on nonsmooth consensus algorithms. IEEE Trans Cybern 49(1):342–353
8. Elijah T, Jamisola RS, Tjiparuro Z, Namoshe M (2021) A review on control and maneuvering of cooperative fixed-wing drones. Int J Dyn Control 9:1332–1349
9. Fei Y, Shi P, Lim CP, Yuan X (2023) Finite-time observer-based formation tracking with application to omnidirectional robots. IEEE Trans Industr Electron 70(10):10598–10606
10. Gao J, Fu Z, Zhang S (2019) Adaptive fixed-time attitude tracking control for rigid spacecraft with actuator faults. IEEE Trans Industr Electron 66(9):7141–7149
11. Gao Y, Zhou W, Niu B, Kao Y, Wang H, Sun N (2024) Distributed prescribed-time consensus tracking for heterogeneous nonlinear multi-agent systems under deception attacks and actuator faults. IEEE Trans Autom Sci Eng 21(4):6920–6929
12. Gong W, Li B, Ahn CK, Yang Y (2023) Prescribed-time extended state observer and prescribed performance control of quadrotor UAVs against actuator faults. Aerosp Sci Technol 138:108322
13. Gu J, Su T, Wang Q, Du X, Guizani M (2018) Multiple moving targets surveillance based on a cooperative network for multi-UAV. IEEE Commun Mag 56(4):82–89

14. Guo X-G, Tan D-C, Ahn CK, Wang J-L (2022) Fully distributed adaptive fault-tolerant sliding-mode control for nonlinear leader-following multiagent systems with ANASs and IQCs. IEEE Trans Cybern 52(5):2763–2774
15. Guo Y, Yin S, Hao J (2020) Joint placement and resources optimization for multi-user UAV-relaying systems with underlaid cellular networks. IEEE Trans Veh Technol 69(10):12374–12377
16. Duan H, Liu S (2010) Unmanned air/ground vehicles heterogeneous cooperative techniques: current status and prospects. Scie China Technol Sci 53(5):1349–1355
17. Han J, Liu X, Wei X, Hu X (2023) Adaptive adjustable dimension observer based fault estimation for switched fuzzy systems with unmeasurable premise variables. Fuzzy Sets Syst 452:149–167
18. Han J, Zhang J, He C, Lv C, Li C, Ji Y, Hou X (2023) Prescribed-time performance recovery fault tolerant control of platoon with nominal constraints guarantee. IEEE Trans Intell Transp Syst 24(3):3501–3513
19. Hu C, Zhang Z, Yang N, Shin H-S, Tsourdos A (2019) Fuzzy multiobjective cooperative surveillance of multiple UAVs based on distributed predictive control for unknown ground moving target in urban environment. Aerosp Sci Technol 84:329–338
20. Hua C-C, Li H, Li K, Ning P (2024) Adaptive prescribed-time stabilization of uncertain nonlinear systems with unknown control directions. IEEE Trans Autom Control 69(6):3968–3974
21. Hua Y, Dong X, Han L, Li Q, Ren Z (2018) Finite-time time-varying formation tracking for high-order multiagent systems with mismatched disturbances. IEEE Trans Syst Man Cybern Syst 50(10):3795–3803
22. Ivanov D, Kalyaev I, Kapustyan S (2015) Formation task in a group of quadrotors. In: Robot intelligence technology and applications, vol 3. Springer, pp 183–191
23. Jiang B, Hu Q, Friswell MI (2016) Fixed-time attitude control for rigid spacecraft with actuator saturation and faults. IEEE Trans Control Syst Technol 24(5):1892–1898
24. Jiang Y, Lv J, Wang C, Kao Y, Wang F (2024) Prescribed-time disturbance observer-based fully distributed prescribed-time containment control of multiagent systems. IEEE Trans Circuits Syst II Express Briefs 71(4):2044–2048
25. Jin D, Xiang Z (2024) Adaptive event-triggered fixed-time fault-tolerant consensus control for a class of multiagent systems. IEEE Trans Syst Man Cybern Syst 54(8):4853–4862
26. Jin X-Z, Che W-W, Wu Z-G, Deng C (2022) Robust adaptive general formation control of a class of networked quadrotor aircraft. IEEE Trans Syst Man Cybern Syst 52(12):7714–7726
27. Jin Z, Bai L, Wang Z, Zhang P (2024) Self-triggered distributed formation control of fixed-wing unmanned aerial vehicles subject to velocity and overload constraints. IEEE Trans Autom Sci Eng 21(3):4082–4093
28. Karimoddini A, Lin H, Chen BM, Lee TH (2013) Hybrid three-dimensional formation control for unmanned helicopters. Automatica 49(2):424–433
29. Karimoddini A, Lin H, Chen BM, Lee TH (2013) A bumpless hybrid supervisory control algorithm for the formation of unmanned helicopters. Mechatronics 23(6):677–688
30. Keller J, Thakur D, Likhachev M, Gallier J, Kumar V (2016) Coordinated path planning for fixed-wing UAS conducting persistent surveillance missions. IEEE Trans Autom Sci Eng 14(1):17–24
31. Kuo CW, Tsai CC, Lee CT (2021) Intelligent leader-following consensus formation control using recurrent neural networks for small-size unmanned helicopters. IEEE Trans Syst Man Cybern Syst 51(2):1288–1301
32. Li J, Xiang X, Dong D, Yang S (2023) Prescribed time observer based trajectory tracking control of autonomous underwater vehicle with tracking error constraints. Ocean Eng 274:114018
33. Li Y, Feng K, Li K (2024) Finite-time fuzzy adaptive dynamic event-triggered formation tracking control for USVs with actuator faults and multiple constraints. IEEE Trans Industr Inf 20(4):5285–5296

34. Li Y, He J, Shen H, Zhang W, Li Y (2023) Adaptive practical prescribed-time fault-tolerant control for autonomous underwater vehicles trajectory tracking. Ocean Eng 277:114263
35. Li Y, Zheng X, Li K (2024) Prescribed-time adaptive intelligent formation controller for nonlinear multiagent systems based on time-domain mapping. IEEE Trans Artif Intell 5(4):1778–1790
36. Liang Z, Wang Z, Zhao J, Wong PK, Yang Z, Ding Z (2023) Fixed-time and fault-tolerant path-following control for autonomous vehicles with unknown parameters subject to prescribed performance. IEEE Trans Syst Man Cybern Syst 53(4):2363–2373
37. Lin J-L, Hwang K-S, Wang Y-L (2014) A simple scheme for formation control based on weighted behavior learning. IEEE Trans Neural Netw Learn Syst 25(6):1033–1044
38. Liu C, Jiang B, Patton RJ, Zhang K (2019) Hierarchical-structure-based fault estimation and fault-tolerant control for multiagent systems. IEEE Trans Control Netw Syst 6(2):586–597
39. Liu C, Jiang B, Wang X, Zhang Y, Xie S (2024) Event-based distributed secure control of unmanned surface vehicles with DoS attacks. IEEE Trans Syst Man Cybern Syst 54(4):2159–2170
40. Liu H, Ma T, Lewis FL, Wan Y (2020) Robust formation trajectory tracking control for multiple quadrotors with communication delays. IEEE Trans Control Syst Technol 28(6):2633–2640
41. Liu K, Yang P, Wang R, Jiao L, Li T, Zhang J (2023) Observer-based adaptive fuzzy finite-time attitude control for quadrotor UAVs. IEEE Trans Aerosp Electron Syst 59(6):8637–8654
42. Liu P, Hao Y, Wang Q (2024) Distributed formation-containment control of networked mobile robots using the udwadia–kalaba approach. IEEE Trans Netw Sci Eng 11(1):848–857
43. Liu S, Jiang B, Mao Z, Zhang Y (2024) Neural-network-based adaptive fault-tolerant cooperative control of heterogeneous multiagent systems with multiple faults and DoS attacks. IEEE Trans Neural Netw Learn Syst 35(5):6273–6285
44. Luo S, Song Y, Deng G, Zhang J, Ouakad HM (2024) Prescribed-time fault-tolerant control of the FO decoupled dual-mass MEMS gyro with deferred constraints-design and implementation. IEEE Trans Syst Man Cybern Syst 54(10):6372–6384
45. Lv M, Ahn CK, Zhang B, Fu A (2023) Fixed-time antisaturation cooperative control for networked fixed-wing unmanned aerial vehicles considering actuator failures. IEEE Trans Aerosp Electron Syst 59(6):8812–8825
46. Lv M, Li Y, Wan L, Dai J, Chang J (2022) Fast nonsingular fixed-time fuzzy fault-tolerant control for HFVs with guaranteed time-varying flight state constraints. IEEE Trans Fuzzy Syst 30(11):4555–4567
47. Meng B, Zhang K, Jiang B (2023) Fixed-time optimal fault-tolerant formation control with prescribed performance for fixed-wing UAVs under dual faults. IEEE Trans Signal Inf Process over Netw 9:875–887
48. Polyakov A (2012) Nonlinear feedback design for fixed-time stabilization of linear control systems. IEEE Trans Autom Control 57(8):2106–2110
49. Qin Z, Wang H, Wei Z, Qu Y, Xiong F, Dai H, Wu T (2021) Task selection and scheduling in UAV-enabled mec for reconnaissance with time-varying priorities. IEEE Internet Things J 8(24):17290–17307
50. Ren H, Ma H, Li H, Wang Z (2023) Adaptive fixed-time control of nonlinear mass with actuator faults. IEEE/CAA J Automatica Sinica 10(5):1252–1262
51. Ribeiro RG, Cota LP, Euzébio TA, Ramírez JA, Guimarães FG (2021) Unmanned-aerial-vehicle routing problem with mobile charging stations for assisting search and rescue missions in postdisaster scenarios. IEEE Trans Syst Man Cybern Syst 52(11):6682–6696
52. Rudin K, Ducard GJJ, Siegwart RY (2020) Active fault-tolerant control with imperfect fault detection information: Applications to UAVs. IEEE Trans Aerosp Electron Syst 56(4):2792–2805
53. Santana LV, Brandao AS, Sarcinelli-Filho M (2016) Heterogeneous leader-follower formation based on kinematic models. In: International conference on unmanned aircraft systems (ICUAS), pp 342–346
54. Shakouri A, Assadian N (2022) Prescribed-time control for perturbed euler-lagrange systems with obstacle avoidance. IEEE Trans Autom Control 67(7):3754–3761

55. Shen Q, Yue C, Goh CH, Wang D (2019) Active fault-tolerant control system design for spacecraft attitude maneuvers with actuator saturation and faults. IEEE Trans Industr Electron 66(5):3763–3772
56. Su X, Liu X, Song Y-D (2018) Fault-tolerant control of multiarea power systems via a sliding-mode observer technique. IEEE/ASME Trans Mechatron 23(1):38–47
57. Sun W, Gao Y, Su S-F, Zhao X, Xie X (2024) Prescribed-time control for nonlinear systems with sensor faults and unknown control directions. IEEE Trans Autom Sci Eng 21(3):4495–4503
58. Sun Y, Xia K, Zou Y, Fu Q, He X (2022) Distributed output-feedback formation tracking control for clustered quadrotors. IEEE Trans Aerosp Electron Syst 58(3):1894–1905
59. Tanner HG, Christodoulakis DK (2007) Decentralized cooperative control of heterogeneous vehicle groups. Robot Auton Syst 55(11):811–823
60. Tuo Y, Li J, Peng Z, Wang S (2024) Prescribed-time active fault-tolerant control of dynamic positioning vessels based on improved iterative learning observer. Ocean Eng 311:118786
61. Vallejo-Alarcón M, Castro-Linares R, Velasco-Villa M (2015) Unicycle-type robot & quadrotor leader-follower formation backstepping control. IFAC-PapersOnLine 48(19):51–56
62. Van M, Sun Y, Mcllvanna S, Nguyen M-N, Khyam MO, Ceglarek D (2023) Adaptive fuzzy fault tolerant control for robot manipulators with fixed-time convergence. IEEE Trans Fuzzy Syst 31(9):3210–3219
63. Wang B, Zhang Y (2018) An adaptive fault-tolerant sliding mode control allocation scheme for multirotor helicopter subject to simultaneous actuator faults. IEEE Trans Industr Electron 65(5):4227–4236
64. Wang D, Zong Q, Tian B, Shao S, Zhang X, Zhao X (2018) Neural network disturbance observer-based distributed finite-time formation tracking control for multiple unmanned helicopters. ISA Trans 73:208–226
65. Wang D, Zong Q, Tian B, Wang F, Dou L (2018) Finite-time fully distributed formation reconfiguration control for UAV helicopters. Int J Robust Nonlinear Control 28(18):5943–5961
66. Wang H, Dong J (2024) Robust fault-tolerant formation maneuver control for multiagent systems with mismatched disturbances. IEEE Trans Syst Man Cybern Syst 54(6):3762–3773
67. Wang J, Liu J, Li Y, Chen CLP, Liu Z, Li F (2024) Prescribed time fuzzy adaptive consensus control for multiagent systems with dead-zone input and sensor faults. IEEE Trans Autom Sci Eng 21(3):4016–4027
68. Wang J, Wang D, Wang C, Deng F (2020) Robust formation control for unmanned helicopters with collision avoidance. J Franklin Inst 357(17):11997–12018
69. Wang X (2020) Active fault tolerant control for unmanned underwater vehicle with sensor faults. IEEE Trans Instrum Meas 69(12):9485–9495
70. Wang S, Wang H, Che W, Li Q (2025) Distributed leader-following formation control of networked mobile robots via global orientation estimation. IEEE Trans Netw Sci Eng 12(3):2135–2150
71. Wang X, Park JH, Liu H, Zhang X (2021) Cooperative output-feedback secure control of distributed linear cyber-physical systems resist intermittent DoS attacks. IEEE Trans Cybern 51(10):4924–4933
72. Wang X, Wang Q, Su Y, Ouyang Y, Sun C (2024) Adaptive sensor-fault tolerant control of unmanned underwater vehicles with input saturation. IEEE/CAA J Automatica Sinica 11(4):907–918
73. Wang X, Wang Y, Zhang Z, Wang X, Patton R (2023) Sensor fault tolerant control for a 3-DOF helicopter considering detectability loss. IEEE Trans Circuits Syst I Regul Pap 70(10):1112–4125
74. Wang Y, Song Y, Hill DJ, Krstic M (2018) Prescribed-time consensus and containment control of networked multiagent systems. IEEE Trans Cybern 49(4):1138–1147
75. Wang Z, Liang B, Sun Y, Zhang T (2020) Adaptive fault-tolerant prescribed-time control for teleoperation systems with position error constraints. IEEE Trans Industr Inf 16(7):4889–4899

76. Wu Y, Zuo Z, Han Q, Wang Y, Yang H (2023) Formation control of wheeled mobile robots with multiple virtual leaders under communication failures. IEEE Trans Control Syst Technol 31(1):295–305
77. Zhao X, Li X, Zhang Z (2022) Active fault-tolerant control scheme for unmanned air-ground attitude system with time-varying delay faults. Appl Sci 12:5882
78. Xin B, Cheng S, Wang Q, Chen J, Deng F (2023) Fixed-time prescribed performance consensus control for multiagent systems with nonaffine faults. IEEE Trans Fuzzy Syst 31(10):3433–3446
79. Xu C, Wu B, Zhang Y (2021) Distributed prescribed-time attitude cooperative control for multiple spacecraft. Aerosp Sci Technol 113:106699
80. Yang H, Han Q-L, Ge X, Ding L, Xu Y, Jiang B, Zhou D (2020) Fault-tolerant cooperative control of multiagent systems: A survey of trends and methodologies. IEEE Trans Industr Inf 16(1):4–17
81. Yang H, Jiang B, Liu HH, Yang H, Zhang Q (2019) Attitude synchronization for multiple 3-DOF helicopters with actuator faults. IEEE/ASME Trans Mechatron 24(2):597–608
82. Yang J, Wang X, Baldi S, Singh S, Farì S (2019) A software-in-the-loop implementation of adaptive formation control for fixed-wing UAVs. IEEE/CAA J Automatica Sinica 6(5):1230–1239
83. Yang Y, Yue D (2019) Observer-based decentralized adaptive NNs fault-tolerant control of a class of large-scale uncertain nonlinear systems with actuator failures. IEEE Trans Syst Man Cybern Syst 49(3):528–542
84. Ye H, Song Y (2023) Prescribed-time control for linear systems in canonical form via nonlinear feedback. IEEE Trans Syst Man Cybern Syst 53(2):1126–1135
85. Ye H, Song Y (2023) Prescribed-time tracking control of MIMO nonlinear systems under non-vanishing uncertainties. IEEE Trans Autom Control 68(6):3664–3671
86. Yin T, Gu Z, Xie X (2023) Observer-based event-triggered sliding mode control for secure formation tracking of multi-UAV systems. IEEE Trans Netw Sci Eng 10(2):887–898
87. Yu X, Li P, Zhang Y (2018) The design of fixed-time observer and finite-time fault-tolerant control for hypersonic gliding vehicles. IEEE Trans Industr Electron 65(5):4135–4144
88. Yu Z, Li Y, Lv M, Pei B, Fu A (2024) Event-triggered adaptive fuzzy fault-tolerant attitude control for tailless flying-wing UAV with fixed-time convergence. IEEE Trans Veh Technol 73(4):4858–4869
89. Yu Z, Li Y, Lv M, Pei B, Xu W, Bu X (2024) Fixed-time adaptive fuzzy fault-tolerant attitude control for tailless aircraft without angular velocity measurements. IEEE Trans Fuzzy Syst 32(4):2458–2470
90. Yu Z, Zhang Y, Jiang B, Su C-Y, Fu J, Jin Y, Chai T (2022) Enhanced recurrent fuzzy neural fault-tolerant synchronization tracking control of multiple unmanned airships via fractional calculus and fixed-time prescribed performance function. IEEE Trans Fuzzy Syst 30(10):4515–4529
91. Yu Z, Zhang Y, Jiang B, Yu X, Fu J, Jin Y, Chai T (2020) Distributed adaptive fault-tolerant close formation flight control of multiple trailing fixed-wing UAVs. ISA Trans 106:181–199
92. Zhang B, Sun X, Lv M, Liu S, Li L (2022) Distributed adaptive fixed-time fault-tolerant control for multiple 6-DOF UAVs with full-state constraints guarantee. IEEE Syst J 16(3):4792–4803
93. Zhang K, Jiang B, Shi P (2016) Adjustable parameter-based distributed fault estimation observer design for multiagent systems with directed graphs. IEEE Trans Cybern 47(2):306–314
94. Zhang L, Liu H, Tang D, Hou Y, Wang Y (2022) Adaptive fixed-time fault-tolerant tracking control and its application for robot manipulators. IEEE Trans Industr Electron 69(3):2956–2966
95. Zhang S, Yang P, Kong L, Chen W, Fu Q, Peng K (2021) Neural networks-based fault tolerant control of a robot via fast terminal sliding mode. IEEE Trans Syst Man Cybern Syst 51(7):4091–4101
96. Zhang X, Xu X, Li J, Zhang Z, Zhang Z, Brunauer G (2024) Observer-based fuzzy adaptive fault-tolerant control for NSC system of USV with sideslip angle and steering machine fault. IEEE Trans Intell Veh 9(1):372–382

97. Zhang Y, Chadli M, Xiang Z (2023) Prescribed-time formation control for a class of multiagent systems via fuzzy reinforcement learning. IEEE Trans Fuzzy Syst 31(12):4195–4204
98. Zhao W, Liu H, Valavanis KP, Lewis FL (2022) Fault-tolerant formation control for heterogeneous vehicles via reinforcement learning. IEEE Trans Aerosp Electron Syst 58(4):2796–2806
99. Zhao W, Xia Y, Zhai D-H, Cui B (2023) Adaptive event-triggered coordination control of unknown autonomous underwater vehicles under communication link faults. Automatica 158:111277
100. Zhao X, Zong Q, Tian B, Liu W (2020) Integrated fault estimation and fault-tolerant tracking control for Lipschitz nonlinear multiagent systems. IEEE Trans Cybern 50(2):678–688
101. Zhi Y, Liu L, Guan B, Wang B, Cheng Z, Fan H (2022) Distributed robust adaptive formation control of fixed-wing UAVs with unknown uncertainties and disturbances. Aerosp Sci Technol 126:107600
102. Zuo Z, Zhang J, Wang Y (2015) Adaptive fault-tolerant tracking control for linear and Lipschitz nonlinear multi-agent systems. IEEE Trans Industr Electron 62(6):3923–3931

# Part I
# Sliding Mode-Based Fixed-Time FTC for Multiple Unmanned Systems

# Chapter 2
# Command-Filter-Based Fixed-Time FTC for Quadrotor UAVs

## 2.1 Introduction

Due to the repeated differentiations of virtual control inputs required in traditional backstepping methods, which often lead to the so-called "explosion of complexity" problem, the command-filter-based control has emerged as an effective technique to overcome this problem. The core idea of command-filter-based control is to pass the desired virtual control signal through a low-pass filter that approximates its derivative, thereby avoiding the direct computation of high-order derivations. These filtered signals are then used in the recursive design of controllers, enabling a simpler and more practical implementation of backstepping steps. Therefore, the command filter mechanism facilitates better integration with adaptive or robust control components, making it a highly flexible tool in nonlinear control design [11, 22].

On the other hand, in contrast to conventional finite-time control strategies where the convergence time depends on the initial state, the fixed-time control guarantees that system states reach to the equilibrium within a fixed time upper bound, regardless of the initial states [3, 13, 19]. This property is particularly desirable in time-critical applications, such as cooperative UAV missions or fault recovery scenarios, where predictable and bounded transient behavior is essential. On this foundation, this chapter investigates the command-filter-based fixed-time FTC problem of quadrotor UAV formations, which inherently exhibit strong nonlinearities, underactuation and potential actuator faults. Specifically, the design of a fault-tolerant outer-loop controller based on command-filtered backstepping and an integral sliding mode surface are presented. Then, an inner-loop fault-tolerant attitude controller leveraging piecewise terminal sliding mode and fixed-time sliding mode observer. Finally, comprehensive numerical simulations are conducted to validate the effectiveness and performance of the proposed approach. The main contributions of this chapter are summarized as follows



K. Zhang et al., *Fixed-Time/Prescribed-Time Fault-Tolerant Control for Multiple Unmanned Systems*, Unmanned System Technologies,
https://doi.org/10.1007/978-981-95-2913-1_2

(1) This chapter focuses on the fault-tolerant formation control of quadrotor UAVs under actuator faults. A fixed-time convergent fault-tolerant controller is developed to ensure that tracking errors converge within a predefined time regardless of initial conditions, while maintaining strong robustness.

(2) A virtual leader mechanism is introduced into the control framework, and command filter techniques are employed to overcome the "explosion of complexity" commonly encountered in backstepping designs, thereby enhancing system stability and practical implementation feasibility.

(3) An integral sliding mode-based approach is adopted to improve the tracking accuracy and robustness in the presence of actuator faults and external disturbances, effectively mitigating large state deviations. A complete control scheme is proposed including formation control objectives, command filter design, outer-loop sliding mode control and observer-based fault estimation. Extensive numerical simulations are conducted to demonstrate the effectiveness and superiority of the proposed approach.

## 2.2 Problem Formulation and Preliminaries

### *2.2.1 Problem Formulation*

Consider the actuator faults during flight, the dynamical model of the $i$th quadrotor UAV can be described as

$$\begin{cases} \dot{X}_i = V_i, \\ \dot{V}_i = u_i^F + f_i + d_{Xi}. \end{cases} \tag{2.1}$$

where the vector $X_i = [x_i, y_i, z_i]^{\mathrm{T}}$ is the position; $V_i = [\dot{x}_i, \dot{y}_i, \dot{z}_i]^{\mathrm{T}}$ is the velocity; $d_{xi} = [d_{xi}, d_{yi}, d_{zi}]^{\mathrm{T}}$ is the unknown external disturbance of the position system; $f_i = \left[-\frac{\zeta_{xi}\dot{x}_i}{m_i}, -\frac{\zeta_{yi}\dot{y}_i}{m_i}, -\frac{\zeta_{zi}\dot{z}_i}{m_i}\right]^{\mathrm{T}}$ represents the nonlinear terms; $u_i^F$ denotes the applied control signal under actuator faults, and the following holds

$$u_i^F = \rho_{Ti} u_{ci} + u_{bi} \tag{2.2}$$

$$u_{ci} = \begin{bmatrix} \frac{T_{di}}{m_i}(\cos\phi_i \sin\theta_i \cos\psi_i + \sin\phi_i \sin\psi_i) \\ \frac{T_{di}}{m_i}(\cos\phi_i \sin\theta_i \sin\psi_i - \sin\phi_i \cos\psi_i) \\ \frac{T_{di}}{m_i}\cos\phi_i \cos\theta_i \end{bmatrix} \tag{2.3}$$

In the equation, $u_{ci}$ represents the command control signal, $\rho_{Ti}$ denotes the total lift efficiency factor, and it satisfies $\rho_{mi} \le \rho_{Ti} \le 1$ Additionally, $u_{bi} = [u_{bi1}, u_{bi2}, u_{bi3}]^{\mathrm{T}}$ represents the unknown additive fault.

**Assumption 2.1** ([23]) The nonlinear term $f_i$ satisfies $\|f_i\| \leq f_m$, and the unknown external airflow disturbance is bounded by $|d_{Xi}| \leq d_m$. The unknown additive fault $u_{bi}$ is bounded, and there exists a known constant such that $\|u_{bi}\| \leq u_{bm}$.

**Assumption 2.2** ([20]) Consider a formation system Eq. (2.1) composed of multiple quadrotor UAVs. The control objective is to design a fault-tolerant controller such that, under any initial state, the desired formation trajectory is achieved within a fixed time

$$\begin{aligned} \lim_{t \to T_1} (\Theta_i - \Theta_{di}) = 0, \quad i = 1, 2, \cdots, N \\ \lim_{t \to T_1 + T_2} (X_i - X_0 - h_i) = 0, \quad i = 1, 2, \cdots, N \end{aligned} \tag{2.4}$$

where $T_1$ and $T_2$ are the fixed time to be calculated in the subsequent part. $X_0$ represents the trajectory of the virtual leader, and $h_i$ denotes the desired relative position of the $i$th follower with respect to the leader's desired signal. In current formation research, rigid graph theory is widely applied, and undirected graphs are commonly used. This chapter also employs an undirected graph, indicating that the communication topology among the quadrotor UAVs is bidirectional.

### 2.2.2 *Preliminaries*

Consider the following nonlinear system

$$\dot{x} = f(t, x), \quad x(0) = x_0 \tag{2.5}$$

where $x \in R^n$ represents the system state vector, $f(t, x) : R^+ \times R^n \to R^n$ denotes a nonlinear mapping, and $x_0$ signifies the initial state of the system. Based on the described system, we first present the definition of fixed-time convergence.

**Definition 2.1** ([2]) If the origin of Eq. (2.5) satisfies the following conditions, then the origin of the system is said to be globally fixed-time convergent

(i) The origin of the system is globally finite-time convergent.
(ii) The system Eq. (2.5) has a limited-time collection time function $T(x_0)$ with bounds, which indicates that there exists a constant $T_{\max}$ such that for any initial state $x_0$, it holds that $T(x_0) \leq T_{\max}$.

**Lemma 2.1** ([12]) *For the system Eq. (2.5), if there exists a continuous positive function $V(x) : R^n \to R$ such that the following inequality holds, then the origin of the system is globally asymptotically stable*

$$\dot{V}(x) + \alpha V^p(x) + \beta V^q(x) \leq 0 \tag{2.6}$$

*In the above,* $\alpha > 0$, $\beta > 0$, $p > 1$, $0 < q < 1$, *and the convergence time function* $T$ *satisfies*

$$T \leq T_{\max} = \frac{1}{\alpha(p-1)} + \frac{1}{\beta(1-q)} \tag{2.7}$$

*Furthermore, if* $p = 1 + \tau$ *and* $q = 1 - \tau$, *the convergence time* $T$ *can have a less conservative upper bound* $\bar{T}_{\max}$ *as follows*

$$T \leq \bar{T}_{\max} = \frac{\pi}{2\tau\sqrt{\alpha\beta}} \tag{2.8}$$

**Lemma 2.2** ([2]) *For* $x_1, x_2, \cdots, x_N \geq 0$, *the following inequalities hold*

$$\begin{cases} \sum_{i=1}^{N} x_i^a \geq \left(\sum_{i=1}^{N} x_i\right)^a, & 0 < a < 1 \\ \left(\sum_{i=1}^{N} x_i\right)^a \geq \sum_{i=1}^{N} x_i^a \geq N^{1-a}\left(\sum_{i=1}^{N} x_i\right)^a, & a > 1 \end{cases} \tag{2.9}$$

**Lemma 2.3** ([24]) *For* $y \in \mathbb{R}^n$ *and* $a > b > 0$, *the following inequality holds*

$$\|y\|_a \leq \|y\|_b \leq n^{\frac{1}{b}-\frac{1}{a}}\|y\|_a \tag{2.10}$$

## 2.3 Main Results

### 2.3.1 Command-Filter-Based Fixed-Time FTC for Position Subsystems

To achieve fast and accurate trajectory tracking in multi-UAV formation systems, especially in the presence of uncertainties and actuator faults, a command-filter-based fixed-time FTC framework is proposed. This section focuses on the design of an outer-loop controller for the position subsystem under nominal conditions, where the system dynamics are assumed to be free of nonlinear perturbations, external disturbances, and faults. By formulating the nominal dynamics and defining the formation tracking errors with respect to a virtual leader, the control objective is to drive these errors to zero within a fixed time. Fixed-time stability theory, which guarantees convergence within a predefined time regardless of initial states, offers a significant advantage over conventional finite-time methods [4, 12]. The integration of CF into the backstepping framework avoids the need for complex signal differentiation, alleviating the derivative explosion problem and reducing computational complexity [17]. These features enhance both the responsiveness and robustness of the controller, thereby laying a solid foundation for fixed-time FTC design in coordinated UAV formations [5].

For a nominal system without nonlinear terms, external disturbances, and actuator faults, the dynamics can be described as

$$\begin{cases} \dot{X}_i = V_i \\ \dot{V}_i = u_{ci} \end{cases} \tag{2.11}$$

Define the formation position tracking error for each quadrotor UAV is $e_{Xi} = X_i - X_0 - h_i$, and the velocity error is $e_{Vi} = V_i - V_0 - \dot{h}_i$, where $V_0$ denotes the velocity of the virtual leader. The nominal position error system is then expressed as

$$\begin{cases} \dot{e}_{Xi} = e_{Vi} \\ \dot{e}_{Vi} = u_{ci} - \alpha_0 - \dot{h}_i \end{cases} \tag{2.12}$$

where $\alpha_0 = \dot{V}_0$ is the acceleration of the virtual leader. Using the backstepping method, a virtual velocity $V_i^*$ for the $i$th follower is introduced to achieve fixed-time formation tracking.

Define the error between the actual velocity and the virtual velocity is

$$e_i = V_i - V_i^* \tag{2.13}$$

Subsequently, the nominal system Eq. (2.11) can be rewritten in terms of $X_i$ and $e_i$

$$\dot{X}_i = e_i + V_i^* \dot{e}_i = u_i^* \tag{2.14}$$

where $u_i^* = u_{ci} - V_i^*$ is referred as the virtual control signal. The virtual control signal $u_i^*$ and the virtual velocity $V_i^*$ are designed as follows

$$u_i^* = -\alpha_1 \mathrm{sig}^{1-\tau}(e_i) - \beta_1 \mathrm{sig}^{1+\tau}(e_i) \tag{2.15}$$

$$V_i^* = -\alpha_2 \mathrm{sig}^{1-\tau}(\xi_i) - \beta_2 \mathrm{sig}^{1+\tau}(\xi_i) + \dot{h}_i + V_0 \tag{2.16}$$

$$\xi_i = \sum_{j=1}^{N} a_{ij} \left[(X_i - h_i) - (X_j - h_j)\right] + b_i (X_i - X_0 - h_i) \tag{2.17}$$

where $\alpha_1 > 0$, $\beta_1 > 0$, $\alpha_2 > 0$, $\beta_2 > 0$, $0 < \tau < 1$, and $\xi_i$ represents the formation tracking neighbor error.

Hence, the distributed formation controller $u_{ci}$ is designed as

$$u_{ci} = u_i^* + \dot{V}_i^* \tag{2.18}$$

According to [19], it is straightforward to prove that the position tracking error $e_{Xi}$ will converge to zero within a fixed time. However, this control law includes the differential term of the virtual velocity, and Eq. (2.16) contains power terms less than 1, which may lead to singularity. To avoid this phenomenon, the CF technique can be considered to generate the differential terms required in the backstepping method. Next, define the compensated position tracking error signal $\varepsilon_i$

$$\varepsilon_i = e_{Xi} - \mu_i \tag{2.19}$$

where $\mu_i$ is the compensation signal

$$\dot{\mu}_i = -k_{\mu i}\mu_i - k_1\mathrm{sig}^{1-\tau}(\mu_i) - k_2\mathrm{sig}^{1+\tau}(\mu_i) - k_{\omega i}\mu_i + \omega_{\mu i} \tag{2.20}$$

where $k_{\mu i} \geq \frac{1}{2}, k_1 > 0, k_2 > 0$, and $0 < \tau < 1$. The compensation signal $\mu_i$ is introduced to compensate for the filtering error $\omega_{\mu i}$ generated by the command filter designed below. The parameter $k_{\omega i}$ satisfies the following inequality

$$k_{\omega i} \geq \frac{\frac{1}{2}\left\|\omega_{\mu i}\right\|^2 + \lambda_{\max}(L+B)\left[-k_1\left|\varepsilon_i^{\mathrm{T}}\right|\mathrm{sig}^{1-\tau}(\mu_i) - k_2\left|\varepsilon_i^{\mathrm{T}}\right|\mathrm{sig}^{1+\tau}(\mu_i)\right]}{\|\mu_i\|^2 - \lambda_{\max}(L+B)\left|\varepsilon_i^{\mathrm{T}}\mu_i\right|} \tag{2.21}$$

The filtering error $\omega_{\mu i}$ is defined as $\omega_{\mu i} = \bar{V}_i^* - V_i^c$, where $\bar{V}_i^*$ is the designed virtual velocity and $V_i^c$ is the pseudo-control signal

$$V_i^c = -k_{\mu i}\mu_i - k_3\mathrm{sig}^{1-\tau}(\xi_i) - k_4\mathrm{sig}^{1+\tau}(\xi_i) + V_0 + \dot{h}_i \tag{2.22}$$

where $k_3 > 0, k_4 > 0$, and $\xi_i$ is the newly defined compensation formation neighbor error as

$$\xi_i = \sum_{j=1}^{N} a_{ij}\left[(X_i - h_i - \mu_i) - (X_j - h_j - \mu_j)\right] + b_i(X_i - X_0 - h_i - \mu_i) \tag{2.23}$$

To avoid direct differentiation, a CF is used to generate the virtual velocity $\bar{V}_i^*$ and the virtual velocity derivative term $\dot{\bar{V}}_i^*$. The equations are as follows

$$\begin{cases} \dot{f}_{i1} = \omega_n f_{i2} \\ \dot{f}_{i2} = 2\zeta\omega_n f_{i2} - \omega_n(f_{i1} - V_i^c) \end{cases} \tag{2.24}$$

where $f_{i1}, f_{i2}$ are the states of the filter, $\bar{V}_i^* = f_{i1}, \dot{\bar{V}}_i^* = \omega_n f_{i2}$ are the outputs of the filter, $\zeta, \omega_n$ are suitable parameters. The actual velocity $V_i$ and the virtual velocity $\bar{V}_i^*$ have an error $\bar{e}_i = V_i - \bar{V}_i^*$, and the virtual control signal $\bar{u}_i^*$ is designed as

$$\bar{u}_i^* = -k_5\mathrm{sig}^{1-\tau}(\bar{e}_i) - k_6\mathrm{sig}^{1+\tau}(\bar{e}_i) \tag{2.25}$$

where $k_5 > 0, k_6 > 0$. The nominal control $u_{ni}$ is designed as

$$u_{ni} = \bar{u}_i^* + \dot{\bar{V}}_i^* \tag{2.26}$$

The integral sliding mode surface is designed as

$$s_i = \gamma \left( V_i - V_i(t_0) - \int_{t_0}^{t} u_{ni} \mathrm{d}v \right) \tag{2.27}$$

where $\gamma > 0$ is a positive parameter, $V_i(t_0)$ is the initial state of $V_i$. The compensation term $u_{comi}$ is designed as

$$\begin{aligned} u_{comi} &= -\sigma_i \operatorname{sign}(s_i) - s_i^3 \\ \sigma_i &= \frac{(1-\rho_m)\,\|u_{ni}\| + u_{bm} + f_m + d_m + \delta}{\rho_m} \end{aligned} \tag{2.28}$$

where $\rho_m$ is the lower bound of actuator efficiency loss faults, $u_{bm}$ is the upper bound of bias faults, $f_m$ is the upper bound of nonlinear terms, $d_m$ is the upper bound of disturbances, and $\delta > 0$ is a constant. The fault-tolerant controller for the position loop of the $i$th quadrotor UAV is designed as

$$u_{ci} = u_{ni} + u_{comi} \tag{2.29}$$

where $u_{ni}$ is the nominal control term for the healthy system, and $u_{comi}$ is the designed compensation term to suppress the lumped uncertainties of the system.

**Theorem 2.1** *Consider the quadrotor UAV formation system (2.1) with actuator faults, uncertainties, and external disturbances. If the distributed position control law (2.29) is applied, the follower position tracking error will converge to zero within a fixed time $T_4$.*

**Remark 2.1** The position controller designed in this chapter for the quadrotor formation system is fully distributed, requiring no global information about the communication topology. However, this control scheme requires the upper bounds of the lumped uncertainties and actuator faults. The function sign(·) commonly used in sliding mode control often introduces chattering issues. To mitigate this, saturation functions are typically employed. In this chapter, the following function replaces the sign function

$$\operatorname{signod}(x) = \frac{2}{1+\exp(-20x)} - 1 \tag{2.30}$$

In the formation control of quadrotor UAVs, the robustness and fault-tolerant capability of the attitude subsystem are critical, particularly in the presence of actuator faults and external disturbances [16]. To ensure system convergence within a fixed time independent of initial conditions, this study proposes a sliding mode-based fixed-time FTC strategy for the attitude subsystem. Sliding mode control is widely adopted for nonlinear systems due to its strong robustness [15], while fixed-time stability further guarantees rapid and predictable system response [12].

This section designs an integral sliding mode surface for the attitude subsystem and rigorously proves the fixed-time convergence of sliding variables through Lyapunov analysis. Specifically, by examining the dynamics of the sliding mode surface and incorporating fixed-time control theory, we derive stability conditions ensuring that the system state converges within a predetermined time. This proof not only validates the effectiveness of the control law but also establishes a theoretical foundation for subsequent FTC design.

The following proof begins by constructing a Lyapunov function, followed by an analysis of its time derivative to determine the convergence conditions of the sliding mode surface. This ensures that the system achieves stability within a fixed time, as required for reliable and high-performance control in practical applications.

***Proof*** First, we need to prove that the integral sliding mode surface $s_i$ can converge to zero [18]. Construct the following Lyapunov function

$$V_1 = \frac{1}{2}\sum_{i=1}^{N} s_i^{\mathrm{T}} s_i \tag{2.31}$$

Taking the time-derivative of $V_1$, it derives that

$$\dot{V}_1 = \sum_{i=1}^{N} \gamma s_i^{\mathrm{T}} \left[\rho_i (u_{ni} + u_{comi}) + u_{bi} + f_i + d_{Xi} - u_{ni}\right] \tag{2.32}$$

Substituting the control laws Eqs. (2.26) and (2.28), it can be obtained that

$$\begin{aligned}
\dot{V}_1 &= \sum_{i=1}^{N} -\gamma \rho_{is_i}^{\mathrm{T}} \left(\sigma_i \operatorname{sign}(s_i) + s_i^3\right) + \gamma s_i^{\mathrm{T}} (\rho_i - I_3)\, u_{ni} + \gamma s_i^{\mathrm{T}} (u_{bi} + f_i + d_{Xi}) \\
&\le \sum_{i=1}^{N} \left[ -\gamma \rho_m \sum_{k=1}^{3} s_{ik}^4 - \gamma \left(u_{bm} + f_m + d_m + \delta\right) \sum_{k=1}^{3} |s_{ik}| + \gamma s_i^{\mathrm{T}} (u_{bi} + f_i + d_{Xi}) \right] \\
&\le -\gamma \rho_m \sum_{i=1}^{N}\sum_{k=1}^{3} s_{ik}^4 - \gamma \delta \sum_{i=1}^{N}\sum_{k=1}^{3} |s_{ik}| \\
&\le -\frac{4}{3N} \gamma \rho_m V_1^2 - \sqrt{2}\gamma \delta V_1^{\frac{1}{2}}
\end{aligned} \tag{2.33}$$

According to Lemma 2.1, the integral sliding mode surface satisfies $s_i = 0$ and $\dot{s}_i = 0$ within the following time $T_1$

$$T_1 \le \frac{3N}{4\gamma \rho_m} + \frac{\sqrt{2}}{\gamma \delta} \tag{2.34}$$

Once the integral sliding mode surface reaches zero, system Eq. (2.1) can be further transformed into

$$
\begin{aligned}
\dot{X}_i &= \overline{e}_i + \overline{V}_i^* \\
\dot{\overline{e}}_i &= \overline{u}_i^*
\end{aligned}
\tag{2.35}
$$

Next, we need to prove that the actual velocity $V_i$ can track the virtual velocity $\overline{V}_i^*$. Construct the following Lyapunov function

$$
V_2 = \frac{1}{2}\sum_{i=1}^{N} \overline{e}_i^{\mathrm{T}} \overline{e}_i \tag{2.36}
$$

Taking the derivative of $V_2$, we can obtain

$$
\begin{aligned}
\dot{V}_2 &= \sum_{i=1}^{N} \overline{e}_i^{\mathrm{T}} \left[-k_5 \mathrm{sig}^{1-\tau}\left(\overline{e}_i\right) - k_6 \mathrm{sig}^{1+\tau}\left(\overline{e}_i\right)\right] \\
&= \sum_{i=1}^{N} \left[-k_5 \left(\overline{e}_i^{\mathrm{T}} \overline{e}_i\right)^{\frac{2-\tau}{2}} - 3^{-\frac{\tau}{2}} k_6 \left(\overline{e}_i^{\mathrm{T}} \overline{e}_i\right)^{\frac{2+\tau}{2}}\right] \\
&\le -k_5 \left(\sum_{i=1}^{N} \overline{e}_i^{\mathrm{T}} \overline{e}_i\right)^{\frac{2-\tau}{2}} - 3^{-\frac{\tau}{2}} N^{-\frac{\tau}{2}} k_6 \left(\sum_{i=1}^{N} \overline{e}_i^{\mathrm{T}} \overline{e}_i\right)^{\frac{2+\tau}{2}} \\
&= -2^{\frac{2-\tau}{2}} k_5 V_2^{\frac{2-\tau}{2}} - 2^{\frac{2+\tau}{2}} 3^{\frac{\tau}{2}} N^{-\frac{\tau}{2}} k_6 V_2^{\frac{2+\tau}{2}}
\end{aligned}
\tag{2.37}
$$

According to Lemma 2.1, the error $\overline{c_i}$ between the actual velocity $V_i$ and the virtual velocity $\overline{V_i}$ will converge to zero within the time $T_2$.

$$
T_2 \le \frac{\pi}{2\tau\sqrt{3^{-\frac{\tau}{2}} N^{-\frac{\tau}{2}} k_5 k_6}} \tag{2.38}
$$

Finally, the compensation position tracking error signal $\varepsilon_i$ and the compensation signal $\mu_i$ converge to zero are analyzed. By constructing the Lyapunov function

$$
V_3 = \frac{1}{2}\varepsilon^{\mathrm{T}}(H \otimes I_3)\varepsilon + \frac{1}{2}\mu^{\mathrm{T}}\mu \tag{2.39}
$$

where $H = L + B$, $\varepsilon = \left[\varepsilon_1^{\mathrm{T}}, \varepsilon_2^{\mathrm{T}}, \ldots, \varepsilon_N^{\mathrm{T}}\right]^{\mathrm{T}}$, $\mu = \left[\mu_1^{\mathrm{T}}, \mu_2^{\mathrm{T}}, \ldots, \mu_N^{\mathrm{T}}\right]^{\mathrm{T}}$. Taking the time-derivative of $V_3$, we get

$$
\dot{V}_3 = \varepsilon^{\mathrm{T}}(H \otimes I_3)(\dot{\varepsilon}_X - \dot{\mu}) + \sum_{i=1}^{N} \mu_i^{\mathrm{T}} \dot{\mu}_i \tag{2.40}
$$

where $\varepsilon_X = \left[\varepsilon_{X1}^{\mathrm{T}}, \varepsilon_{X2}^{\mathrm{T}}, \ldots, \varepsilon_{XN}^{\mathrm{T}}\right]^{\mathrm{T}}$. From Eq. (2.19), it follows that

$$\begin{aligned}\dot{\varepsilon}_X - \dot{\mu} =& \bar{V}^* - I_3 \otimes V_0 - \dot{h} - \dot{\mu} \\ =& \omega_\mu + V^c - I_3 \otimes V_0 - \dot{h} - \dot{\mu} \\ =& -k_3 \mathrm{sig}^{1-\tau}[(H \otimes I_3)\,\varepsilon] - k_4 \mathrm{sig}^{1+\tau}[(H \otimes I_3)\,\varepsilon] \\ &+ k_1 \mathrm{sig}^{1-\tau}(\mu) + k_2 \mathrm{sig}^{1+\tau}(\mu) + k_\omega \mu\end{aligned} \tag{2.41}$$

where $\bar{V}^* = \left[\bar{V}_1^{*\mathrm{T}}, \bar{V}_2^{*\mathrm{T}}, \ldots, \bar{V}_N^{*\mathrm{T}}\right]^{\mathrm{T}}$, $h = \left[h_1^{\mathrm{T}}, h_2^{\mathrm{T}}, \ldots, h_N^{\mathrm{T}}\right]^{\mathrm{T}}$, $\omega_\mu = \left[\omega_{\mu 1}^{\mathrm{T}}, \omega_{\mu 2}^{\mathrm{T}}, \ldots, \omega_{\mu N}^{\mathrm{T}}\right]^{\mathrm{T}}$, $V^c = \left[V_1^{c\mathrm{T}}, V_2^{c\mathrm{T}}, \ldots, V_N^{c\mathrm{T}}\right]^{\mathrm{T}}$.

By substituting Eq. (2.41) into Eq. (2.40), the derivative of $V_3$ is given by

$$\begin{aligned}\dot{V}_3 \leq& \varepsilon^{\mathrm{T}}(H \otimes I_3)\big[-k_3 \mathrm{sig}^{1-\tau}[(H \otimes I_3)\,\varepsilon] - k_4 \mathrm{sig}^{1+\tau}[(H \otimes I_3)\,\varepsilon] \\ &+ k_1 \mathrm{sig}^{1-\tau}(\mu) + k_2 \mathrm{sig}^{1+\tau}(\mu) + k_\omega \mu\big] + \sum_{i=1}^{N}[-k_{\mu i}\mu_i^{\mathrm{T}}\mu_i \\ &- k_1 \mu_i^{\mathrm{T}} \mathrm{sig}^{1-\tau}(\mu_i) - k_2 \mu_i^{\mathrm{T}} \mathrm{sig}^{1+\tau}(\mu_i) - k_{\omega i}\mu_i^{\mathrm{T}}\mu_i + \mu_i^{\mathrm{T}}\omega_{\mu i}]\end{aligned} \tag{2.42}$$

Using the inequality of parameter $k_{\omega i}$, and combining the Lemmas 2.2 and 2.3, it can be further simplified to

$$\begin{aligned}\dot{V}_3 \leq& -k_3 \left\|(H \otimes I_3)\,\varepsilon\right\|_{2-\tau}^{2-\tau} - k_4 \left\|(H \otimes I_3)\,\varepsilon\right\|_{2+\tau}^{2+\tau} - k_1 \sum_{i=1}^{N} \|\mu_i\|_{2-\tau}^{2-\tau} - k_2 \sum_{i=1}^{N} \|\mu_i\|_{2+\tau}^{2+\tau} \\ \leq& -\left(k_3 \left\|(H \otimes I_3)\,\varepsilon\right\|_2^2 + k_1 \sum_{i=1}^{N} \|\mu_i\|_2^2\right)^{\frac{2-\tau}{2}} \\ &- (3N)^{\frac{\tau}{2}}\,(N+1)^{-\frac{\tau}{2}} \left(k_4 \left\|(H \otimes I_3)\,\varepsilon\right\|_2^2 + k_2 \sum_{i=1}^{N} \|\mu_i\|_2^2\right)^{\frac{2+\tau}{2}} \\ \leq& -\alpha V_3^{\frac{2-\tau}{2}} - \beta V_3^{\frac{2+\tau}{2}}\end{aligned} \tag{2.43}$$

where $\alpha$ and $\beta$ are given as

$$\begin{aligned}\alpha &= \min\left\{(2\lambda_{\min}(H)k_3)^{2-\tau/2}, (2k_1)^{2-\tau/2}\right\} \\ \beta &= \min\left\{(2\lambda_{\min}(H)k_4)^{2+\tau/2}, (2k_2)^{2+\tau/2}\right\} \times (3N)^{\frac{\tau}{2}}(N+1)^{\frac{\tau}{2}}.\end{aligned}$$

According to the Lemma 2.1, it can be concluded that after time $T_3$, the compensation position tracking error $\varepsilon = 0$ and the compensation signal $\mu = 0$.

$$T_3 \leq \frac{\pi}{\tau\sqrt{\alpha\beta}} \tag{2.44}$$

Furthermore, from Eq. (2.19), the formation tracking position error $e_{Xi} = 0$ at this time. In summary, the position error of the entire system converges to zero within the fixed time $T_4 = T_1 + T_2 + T_3$.

### 2.3.2 Sliding Mode-Based Fixed-Time FTC for Attitude Subsystems

According to the unit quaternion differential equation [14], considering actuator faults and external disturbances, the attitude dynamics can be described as

$$\begin{cases} \dot{q}_0 = -\dfrac{1}{2} q_v^{\mathrm{T}} \omega \\ \dot{q}_v = \dfrac{1}{2} \left( q_v^{\times} + q_0 I_3 \right) \omega \\ J\dot{\omega} = -\omega^{\times} J\omega + \rho M_c + d_M \end{cases} \tag{2.45}$$

Extracting the commanded control torque $M_c$, we can obtain that

$$J\dot{\omega} = -\omega^{\times} J\omega + M_c + (\rho - I_3) M_c + d_M \tag{2.46}$$

where $D_M = (\rho - I_3)\, M_c + d_M$ can be regarded as the total uncertainty.

By considering the attitude tracking problem, it is transformed into an error system. The system of equations is given by

$$\begin{cases} \dot{q}_{e0} = -\dfrac{1}{2} q_{ev}^{\mathrm{T}} e_{\omega} \\ \dot{q}_{ev} = \dfrac{1}{2} \left( q_{ev}^{\times} + q_{e0} I_3 \right) e_{\omega} \\ J\dot{e}_{\omega} = -\omega^{\times} J\omega + J e_{\omega}^{\times} C\omega_d - JC\dot{\omega}_d + M_c + (\rho - I_3) M_c + d_M \end{cases} \tag{2.47}$$

**Assumption 2.3** ([10]) The external disturbance torque $d_M$ is unknown and bounded, $\|d_M\| \leq \overline{d}_M$, and the total uncertainty $D_M$ has a constant upper bound $L_1$ for its time derivative.

Accurate estimation of system uncertainties is crucial for robust control of dynamic systems. This work presents a fixed-time sliding mode observer that guarantees convergence within a predetermined time, independent of initial conditions. The proposed design overcomes limitations of conventional sliding mode observers by incorporating nonlinear feedback terms and carefully selected gain parameters [9]. Theoretical analysis demonstrates the observer's ability to precisely estimate disturbances while maintaining robustness against uncertainties. The developed method provides explicit convergence time bounds and parameter selection criteria, offering practical advantages for control applications requiring predictable performance. This chapter details the observer structure, stability proof, and implementation aspects through rigorous mathematical treatment.

Design the following sliding mode observer to estimate the total uncertainty $D_M$

$$\begin{cases} \dot{\eta}_1 = -k_9 \dfrac{\lambda_1}{\|\lambda_1\|^{\frac{1}{2}}} - k_{10}\|\lambda_1\|^{\kappa-1} + \eta_2 - J^{-1}\omega^{\times} J\omega + J^{-1}M_c \\ \dot{\eta}_2 = -k_{11}\dfrac{\lambda_1}{\|\lambda_1\|} \end{cases} \tag{2.48}$$

where $\eta_1$, $\eta_2$ represent the states of the observer, the variable $\lambda_1 = \eta_1 - \omega$, and the observer parameters satisfy

$$\kappa > 1, k_9 > \sqrt{2k_{11}}, k_{10} > 0, k_{11} > 4\|J^{-1}\|L_1 \tag{2.49}$$

The total uncertainty $D_M$ can be estimated within a fixed time by the state $J\eta_2$, hence the estimation error will converge to zero within $T_5$.

$$T_5 \le \left( \frac{1}{k_{10}(\kappa - 1)\}^{\kappa-1}} + \frac{2(\sqrt{2\vartheta})^{\frac{1}{2}}}{k_9} \right) \left( 1 + \frac{k_{11} + L_1}{(k_{11} - L_1)\left(1 - \frac{\sqrt{2k_{11}}}{k_9}\right)} \right) \tag{2.50}$$

where $\vartheta > 0$.

***Proof*** Define the observer error $e_\eta = \eta_2 - J^{-1}D_M$, then the equations for the variable $\lambda_1$ and the observer error $e_\eta$ are obtained

$$\begin{cases} \dot{\lambda}_1 = -k_9 \dfrac{\lambda_1}{\|\lambda_1\|^2} - k_{10}\|\lambda_1\|^{\kappa-1} + e_\eta \\ \dot{e}_\eta = -k_{11}\dfrac{\lambda_1}{\|\lambda_1\|} - J^{-1}\dot{D}_M \end{cases} \tag{2.51}$$

According to [1], the observer error $e_\eta$ can converge to zero within a fixed time $T_5$. Consider the characteristics of attitude kinematics, the terminal sliding mode surface for the attitude control system is introduced as follows

$$s_2 = e_\omega + s_a \tag{2.52}$$

where $s_a = [s_{a1}, s_{a2}, s_{a3}]^{\mathrm{T}}$ is defined as

$$
\begin{aligned}
s_{am} &= \begin{cases} k_7 \text{sig}^{g_1}(q_{evm}) + k_8 \text{sig}^{g_2}(q_{evm}), & \bar{s}_m = 0 \text{ or } \bar{s}_m \neq 0, \\ l_1 q_{evm} + l_2 \text{sig}^2(q_{evm}), & \bar{s}_m \neq 0, \end{cases} \\
l_1 &= k_7 (2 - g_1) \varepsilon^{g_1 - 1} + k_8 (2 - g_2) \varepsilon^{g_2 - 1} \\
l_2 &= k_7 (g_1 - 1) \varepsilon^{g_1 - 2} + k_8 (g_2 - 1) \varepsilon^{g_2 - 2}
\end{aligned} \tag{2.53}
$$

$$
\bar{s}_m = e_{\omega m} + k_7 \text{sig}^{g_1}(q_{evm}) + k_8 \text{sig}^{g_2}(q_{evm}) \tag{2.54}
$$

where $m = 1, 2, 3$, $k_7 > 0$, $k_8 > 0$, $0 < g_1 < 1$, $g_2 > 1$, and $\varepsilon > 0$ is a very small positive constant. The switching part $s_a$ is introduced to avoid the singularity in the differentiation of the sliding surface. The selection of parameters $l_1$ and $l_2$ ensures that $s_a$ is continuous and its derivative is also continuous. By choosing a very small positive constant $\varepsilon$, the differentiation of $s_a$ is given by

$$
\dot{s}_a = \begin{cases} k_7 g_1 \text{diag}\left(|q_{evm}|^{g_1 - 1}\right) \dot{q}_{ev} + k_8 g_2 \text{diag}\left(|q_{evm}|^{g_2 - 1}\right) \dot{q}_{ev}, & \bar{s}_m = 0 \text{ or } \bar{s}_m \neq 0, \\ |q_{evm}| > \varepsilon \\ l_1 \dot{q}_{ev} + 2 l_2 \text{diag}(|q_{evm}|) \dot{q}_{ev}, \quad \bar{s}_m \neq 0, \ |q_{evm}| \leq \varepsilon \end{cases} \tag{2.55}
$$

Design the attitude fault-tolerant controller $M_c$ as

$$
\begin{aligned}
M_c &= \omega^x J \omega - J e_\omega^x C \omega_d + J C \dot{\omega}_d - J W(q_e) - J \alpha_3 \text{sig}^{P_1}(s_2) - J \beta_3 \text{sig}^{P_2}(s_2) - J \eta_2 \\
W(q_e) &= \begin{cases} k_7 g_1 \text{diag}\left(|q_{evm}|^{g_1 - 1}\right) \dot{q}_{ev} + k_8 g_2 \text{diag}\left(|q_{evm}|^{g_2 - 1}\right) \dot{q}_{ev}, & \bar{s}_m = 0 or \\ \bar{s}_m \neq 0, \ |q_{evm}| > \varepsilon \\ l_1 \dot{q}_{ev} + 2 l_2 \text{diag}(|q_{evm}|) \dot{q}_{ev}, \quad \bar{s}_m \neq 0, \ |q_{evm}| \leq \varepsilon \end{cases}
\end{aligned} \tag{2.56}
$$

where $0 < p_1 < 1$ and $p_2 > 1$.

**Theorem 2.2** *Consider the quadrotor UAV attitude system (2.44) with actuator faults and external disturbance torques. If the FTC law (2.55) is applied, the attitude tracking error $q_{ev}$ and angular velocity error $e_\omega$ will converge to zero within a fixed time $T_g$.*

***Proof*** Construct the following Lyapunov function

$$
V_4 = \frac{1}{2} s_2^{\mathrm{T}} s_2 \tag{2.57}
$$

By taking the derivative of $V_4$, we can get

$$
\begin{aligned}
\dot{V}_4 &= s_2^{\mathrm{T}} (\dot{e}_\omega + \dot{s}_a) \\
&= s_2^{\mathrm{T}} \left[ J^{-1} \left( -\omega^\times J \omega + J e_\omega^\times C \omega_d - J C \dot{\omega}_d + M_c + (\rho - I_3) M_c + d_M \right) + \dot{s}_a \right]
\end{aligned} \tag{2.58}
$$

Substituting the torque controller, it has

$$\begin{aligned}\dot{V}_4 &= s_2^{\mathrm{T}}\left[J^{-1}\left(-\omega^{\times} J\omega + J e_{\omega}^{\times} C\omega_d - JC\dot{\omega}_d + M_c + (\rho - I_3)\, M_c + d_M\right) + \dot{s}_a\right] \\ &= s_2^{\mathrm{T}} J^{-1}\left(-J\alpha_3 \mathrm{sig}^{p_1}(s_2) - J\beta_3 \mathrm{sig}^{p_2}(s_2) + (\rho - I_3)\, M_c + d_M - J\eta_2\right) \\ &= s_2^{\mathrm{T}}\left(-\alpha_3 \mathrm{sig}^{p_1}(s_2) - \beta_3 \mathrm{sig}^{p_2}(s_2) - e_{\eta}\right)\end{aligned} \tag{2.59}$$

According to the fixed-time observer, the state $J\eta_2$ can accurately estimate the uncertainty $D_M$ within the fixed time $T_5$, i.e., $e_{\eta} = 0$. Therefore, we have

$$\begin{aligned}\dot{V}_4 &= -\alpha_3 s_2^{\mathrm{T}} \mathrm{sig}^{p_1}(s_2) - \beta_3 s_2^{\mathrm{T}} \mathrm{sig}^{p_2}(s_2) \\ &= -\alpha_3 \sum_{m=1}^{3} |s_{2m}|^{p_1+1} - \beta_3 \sum_{m=1}^{3} |s_{2m}|^{p_2+1} \\ &\le -\alpha_3 \left(\sum_{m=1}^{3} s_{2m}^2\right)^{\frac{p_1+1}{2}} - 3^{\frac{1-p_2}{2}} \beta_3 \left(\sum_{m=1}^{3} s_{2m}^2\right)^{\frac{p_2+1}{2}} \\ &= -2^{\frac{p_1+1}{2}} \alpha_3 V_4^{\frac{p_1+1}{2}} - 3^{\frac{1-p_2}{2}} 2^{\frac{p_2+1}{2}} \beta_3 V_4^{\frac{p_2+1}{2}}\end{aligned} \tag{2.60}$$

According to Lemma 2.1, after time $T_5$, the sliding mode surface $s_2$ converges to 0.

$$T_5 \le \frac{2}{3^{\frac{-p_2+1}{2}} 2^{\frac{p_2+1}{2}} \beta_3 (p_2 - 1)} + \frac{2}{2^{\frac{p_1+1}{2}} \alpha_3 (1 - p_1)} \tag{2.61}$$

By choosing a sufficiently small positive parameter $\varepsilon$, we can make $s_a = 0$, and we have

$$e_{\omega} = -k_7 \mathrm{sig}^{g_1}(q_{ev}) - k_8 \mathrm{sig}^{g_2}(q_{ev}) \tag{2.62}$$

Select the following Lyapunov function

$$V_5 = q_{ev}^{\mathrm{T}} q_{ev} + (1 - q_{e0})^2 \tag{2.63}$$

By taking the derivative of $V_5$, we get

$$\begin{aligned}\dot{V}_5 &= -2\dot{q}_{e0} \\ &= -k_7 q_{ev}^{\mathrm{T}} \mathrm{sig}^{g_1}(q_{ev}) - k_8 q_{ev}^{\mathrm{T}} \mathrm{sig}^{g_2}(q_{ev}) \\ &= -k_7 \sum_{i=1}^{3} \left(q_{evi}^2\right)^{\frac{g_1+1}{2}} - k_8 \sum_{i=1}^{3} \left(q_{evi}^2\right)^{\frac{g_2+1}{2}} \\ &\le -k_7 \left(1 - q_{e0}^2\right)^{\frac{g_1+1}{2}} - 3^{1-\frac{g_2+1}{2}} k_8 \left(1 - q_{e0}^2\right)^{\frac{g_2+1}{2}}\end{aligned} \tag{2.64}$$

**Case 1.** $q_{e0}(t_0) \geq 0$
Since $0 \leq q_{e0} \leq 1$, it follows that $1 - q_{e0} \leq 1 - q_{e0}^2$. Continuing the simplification

$$\begin{aligned}\dot{V}_5 &\leq -k_7 (1 - q_{e0})^{\frac{g_1+1}{2}} - 3^{\frac{1-g_2}{2}} k_8 (1 - q_{e0})^{\frac{g_2+1}{2}} \\ &= -2^{-\frac{g_1+1}{2}} k_7 V_5^{\frac{g_1+1}{2}} - 2^{-\frac{g_2+1}{2}} \times 3^{\frac{1-g_2}{2}} k_8 V_5^{\frac{g_2+1}{2}}\end{aligned} \tag{2.65}$$

Based on Lemma 2.1, after an additional time $T_6$, the error quaternion $q_{ev}$ converges to zero, and $q_{e0} \in [0, 1]$ converges to 1.

$$T_6 \leq \frac{2^{\frac{g_2+3}{2}}}{3^{\frac{1-g_2}{2}} k_8 (g_2 - 1)} + \frac{2^{\frac{g_1+3}{2}}}{k_7 (1 - g_1)} \tag{2.66}$$

**Case 2.** $-1 < q_{e0}(t_0) < 0$
Choose the Lyapunov function

$$V_6 = |q_{e0}| \tag{2.67}$$

Taking its derivative, we obtain

$$\begin{aligned}\dot{V}_6 &= -\dot{q}_{e0} \\ &= -\frac{1}{2} k_7 q_{ev}^{\mathrm{T}} \mathrm{sig}^{g_1}(q_{ev}) - \frac{1}{2} k_8 q_{ev}^{\mathrm{T}} \mathrm{sig}^{g_2}(q_{ev}) \\ &= -\frac{1}{2} k_7 \sum_{i=1}^{3} \left(q_{evi}^2\right)^{\frac{g_1+1}{2}} - \frac{1}{2} k_8 \sum_{i=1}^{3} \left(q_{evi}^2\right)^{\frac{g_2+1}{2}} \\ &\leq -\frac{1}{2} k_7 \left(1 - |q_{e0}|^2\right)^{\frac{g_1+1}{2}} \\ &\leq -\frac{1}{2} k_7 \left(1 - V_6^2\right)^{\frac{g_1+1}{2}}\end{aligned} \tag{2.68}$$

According to [6], after time $T_7$, $q_{e0}$ will reach 0. Then, following Case 1, after an additional time $T_6$, $q_{e0}$ will converge from 0 to 1. From the analysis of the two cases, $q_{e0}$ will always reach 1 after time $T_7 + T_6$, and $q_{ev}$ will always converge to 0. By Eq. (2.58), the angular velocity error $e_\omega$ will also converge to 0. In summary, after time $T_8 = T_5 + T_6 + T_7$, the error quaternion $q_{ev}$ and the angular velocity error $e_\omega$ converge to zero, and $q_{e0}$ converges to 0. Thus, the tracking control objective for the attitude is achieved.

### *2.3.3 Simulation Results*

This section verifies the fault-tolerant formation control algorithm designed in this chapter through numerical simulations. The Simulink - based 6 - DOF rigid body model is used as the system object. The formation consists of a virtual leader and three quadrotor UAVs, i.e., $N = 3$. Some parameters of the quadrotor UAVs are different, while others are the same $c_i = 0.1$, $L_i = 0.2\text{m}$. For the three UAVs, their masses and inertia matrices are given as follows

$$\begin{aligned}
m_1 &= 1.20\,\text{kg}, \quad J_1 = \text{diag}\{0.030, 0.030, 0.036\}\,\text{kg} \cdot \text{m}^2 \\
m_2 &= 1.25\,\text{kg}, \quad J_2 = \text{diag}\{0.020, 0.024, 0.036\}\,\text{kg} \cdot \text{m}^2 \\
m_3 &= 1.50\,\text{kg}, \quad J_3 = \text{diag}\{0.030, 0.030, 0.036\}\,\text{kg} \cdot \text{m}^2
\end{aligned}$$

The desired trajectory of the virtual leader is defined as

$$P_0 = \left[5\sin(0.2t), 5\cos(0.2t), 2.0 + 0.2t\right]^{\text{T}}\,\text{m}$$

with the desired yaw angle $\psi_{id} = 0\,\text{rad}$. The desired relative positions in the formation control are

$$\begin{aligned}
h_1 &= \left[\sin(t/5), \cos(t/5), -1\right]^{\text{T}}\,\text{m}, \\
h_2 &= \left[0, 0, -2\right]^{\text{T}}\,\text{m}, \\
h_3 &= \left[-\sin(t/5), -\cos(t/5), -1\right]^{\text{T}}\,\text{m}.
\end{aligned}$$

Figure 2.1 shows the undirected topology graph of the formation system.

Based on the undirected communication topology graph, the weighted adjacency matrix $A$, the leader matrix $B$, and the in-degree matrix $D$ are given as

$$A = \begin{bmatrix} 0 & 1 & 0 \\ 1 & 0 & 1 \\ 0 & 1 & 0 \end{bmatrix}, \quad B = \begin{bmatrix} 1 & 0 & 0 \\ 0 & 1 & 0 \\ 0 & 0 & 1 \end{bmatrix}, \quad D = \begin{bmatrix} 1 & 0 & 0 \\ 0 & 2 & 0 \\ 0 & 0 & 1 \end{bmatrix}$$

The Laplacian matrix is

$$L = D - A = \begin{bmatrix} 1 & -1 & 0 \\ -1 & 2 & -1 \\ 0 & -1 & 1 \end{bmatrix}$$

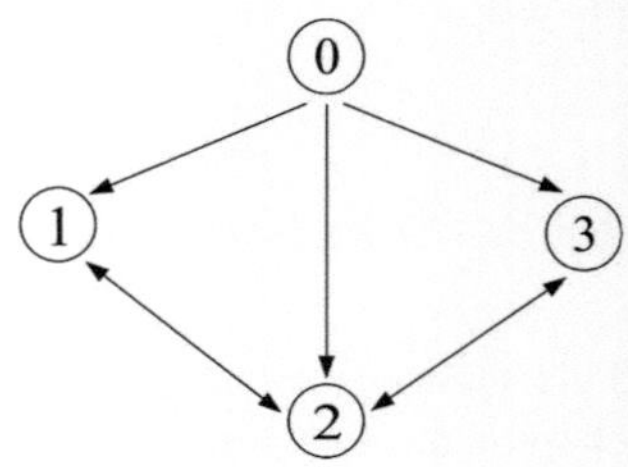


**Fig. 2.1** Undirected topology graph of quadrotor formation

The initial position and velocity vectors are

$$
\begin{aligned}
p_1(t_0) &= [0.5, 6.4, 1.8]^{\mathrm{T}},\ v_1(t_0) = [0.8, 0.1, 0.1]^{\mathrm{T}} \\
p_2(t_0) &= [-0.9, 6.2, 0.8]^{\mathrm{T}},\ v_2(t_0) = [0.8, 0.1, 0.1]^{\mathrm{T}} \\
p_3(t_0) &= [-0.5, 6, -0.1]^{\mathrm{T}},\ v_3(t_0) = [0.8, 0.1, 0.1]^{\mathrm{T}}
\end{aligned}
$$

The initial attitude angles and initial angular velocity vectors are

$$
\begin{aligned}
\phi_1(t_0) &= 0,\ \theta_1(t_0) = 0,\ \psi_1(t_0) = 0,\ \omega_1(t_0) = [0, 0, 0]^{\mathrm{T}} \\
\phi_2(t_0) &= 0,\ \theta_2(t_0) = 0,\ \psi_2(t_0) = \frac{\pi}{3},\ \omega_2(t_0) = [0, 0, 0]^{\mathrm{T}} \\
\phi_3(t_0) &= 0,\ \theta_3(t_0) = 0,\ \psi_3(t_0) = 0,\ \omega_3(t_0) = [0, 0, 0]^{\mathrm{T}}
\end{aligned}
$$

The aerodynamic coefficients for the quadrotor are $\zeta_{xi} = \zeta_{yi} = \zeta_{zi} = 0.02$, and the aerodynamic disturbances are $d_{xi} = d_{yi} = d_{zi} = 0.03\sin(0.5t)$ N. The disturbance torque is $d_M = [0.001, 0.001, 0.001]^{\mathrm{T}}$ N · m.

The actuator faults for the quadrotor formation during flight are defined as follows

$$
\rho_2 = \begin{cases} I_4, t \le 45s \\ \operatorname{diag}\left\{ \begin{array}{l} 0.85 + 0.15\exp\left(-10(t-30)\right), 0.85 + 0.15\exp\left(-10(t-30)\right), \\ 0.85 + 0.15\exp\left(-10(t-30)\right), 0.85 + 0.15\exp\left(-10(t-30)\right) \end{array} \right\} I_4, \\ t > 45s \end{cases}
$$

$$
\rho_1 = \begin{cases} I_4, t \le 30s \\ \operatorname{diag}\left\{0.75 + 0.25\exp\left(-10(t-30)\right), 0.75 + 0.25\exp\left(-10(t-30)\right), 1, 1\right\} I_4, \\ t > 30s \end{cases}
$$

$$
\begin{aligned}
\rho_3 &= I_4 \\
u_{bi} &= 0.01\sin(0.5t)
\end{aligned}
$$

In this scenario, the third quadrotor does not experience any actuator faults during flight, while the second quadrotor has a fault at 30s, and the first quadrotor has a fault at 45s. The minimum efficiency factor $\rho_m$ is set to 0.5 for the simulation.

For the position subsystem, the formation controller parameters are selected as $k_1 = k_2 = k_3 = k_4 = 1$, $k_5 = k_6 = 0.4$, $k_{\mu i} = 4$, $\tau = 0.5$, $\omega_n = 15$, $\xi = 0.5$, $\gamma = \operatorname{diag}\{0.05, 0.05, 1\}$, $u_{bm} = 0.01$, $f_m = 0.06$, $d_m = 0.03$, $\delta = 0.01$.

For the attitude subsystem, the controller parameters are selected as $k_7 = k_8 = 3$, $k_9 = 5$, $k_{10} = 2$, $k_{11} = 4$, $k = 1.5$, $g_1 = 0.5$, $g_2 = 1.5$, $\alpha_3 = \beta_3 = 3$.

Figures 2.2, 2.3 and 2.4 show the formation error response curves of the three quadrotor UAVs. It can be observed that each quadrotor UAV can quickly and accurately track the desired trajectory, maintaining the position error within a small neighborhood. Under the influence of the actuator fault of the second quadrotor UAV at

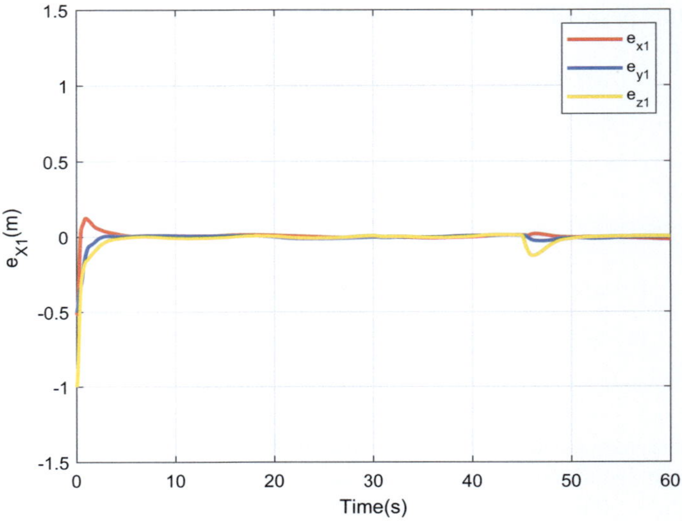


**Fig. 2.2** The position error of quadrotor 1

30s, the tracking error initially increases but then rapidly converges to near zero, maintaining good tracking performance. The simulation results demonstrate that the formation controller has excellent fault-tolerant capability and robustness.

As shown in Fig. 2.2, quadrotor 1 maintains excellent tracking precision throughout the entire operation period, with all three position error components ($e_{x1}$, $e_{y1}$, $e_{z1}$) converging to and remaining within $\pm 0.5$ m after the initial transient phase. The error curves exhibit smooth convergence characteristics, indicating well-tuned control parameters.

Figure 2.3 reveals the system response to actuator fault in quadrotor 2 at $t = 30$s. Prior to the fault occurrence, the tracking errors remain bounded within $\pm 1.0$ m. Following the actuator fault, we observe a temporary deviation where the x-position error ($e_{x2}$) peaks at approximately 1.5 m. However, the controller demonstrates remarkable recovery capability, bringing all error components back to the pre-fault level within 5 s. This rapid recovery highlights the effectiveness of the FTC scheme.

The consistency of the formation control is further confirmed in Fig. 2.4, which shows the error trajectories of quadrotor 3. Despite the fault in quadrotor 2, quadrotor 3 maintains stable tracking with errors bounded within $\pm 0.8$ m, demonstrating the decentralized nature of the control architecture. The experimental results exhibit three key characteristics: all quadrotors achieve initial convergence within 10 s with steady-state errors satisfying design specifications ($\pm 0.5$ m nominal and $\pm 1.0$ m during faults), and cross-coupling effects are effectively minimized as evidenced by the z-axis tracking errors remaining below $\pm 0.3$ m, the system demonstrates precise trajectory tracking under normal conditions while maintaining rapid fault recovery capability without destabilizing the formation. These results collectively verify the

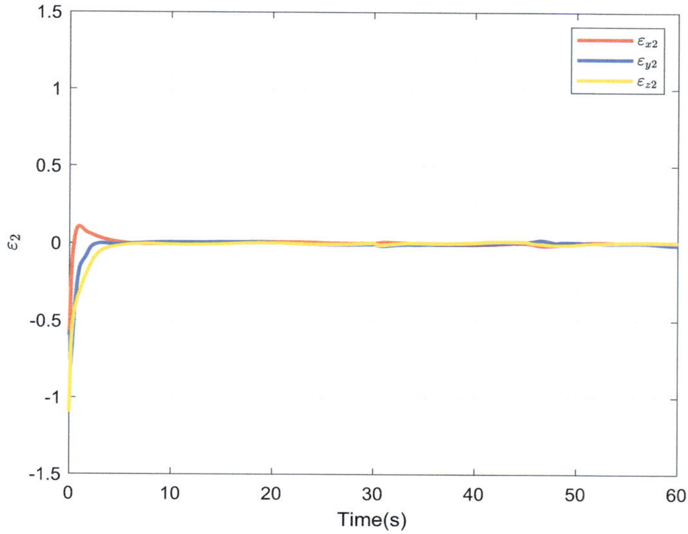


**Fig. 2.3** The position error of quadrotor 2

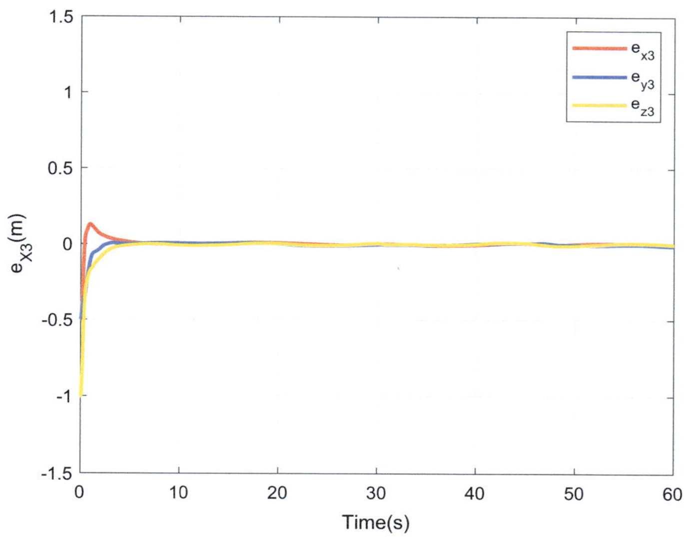


**Fig. 2.4** The position error of quadrotor 3

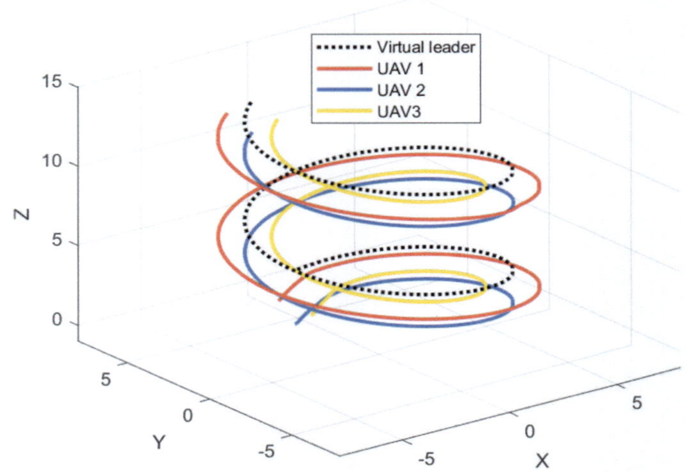


**Fig. 2.5** The actual flight trajectory of the quadrotor formation

proposed strategy's decentralized robustness against individual quadrotor faults and its ability to preserve formation integrity during fault events.

The desired trajectory for the quadrotor UAV formation system is as follows: the three quadrotors perform uniform circular motions with different radii in the $xoy$ plane and ascend uniformly along the $z$-axis at the same speed. Figure 2.5 shows the desired and actual motion trajectories of the entire system in three-dimensional space. The figure demonstrates that, despite the presence of actuator faults and airflow disturbances, each quadrotor in the system can form the desired formation.

As shown in Fig. 2.6, when viewed from the $z$-axis direction in the $xoy$ plane, the three quadrotor UAVs perform uniform circular motions with different radii around the origin.The concentric circular pattern confirms the controller's ability to simultaneously regulate both radial and tangential motion components.

Figure 2.7 shows the error quaternion response curve of the second quadrotor. After the fault occurs at 30s, the error quaternion deviates but quickly converges, demonstrating the effectiveness of the attitude fault-tolerant controller.

Figure 2.8 shows the desired output lift forces of the four rotors of the second quadrotor UAV. At 30s, an actuator fault occurs, but due to the design of the fault-tolerant controller, the desired lift forces increase to achieve compensation. At 45s, influenced by the fault of the first quadrotor, the desired lift forces exhibit rapid fluctuations but quickly converge back to their original amplitudes.

Figure 2.9 shows the actual output lift force curves of the four rotors of the second quadrotor. At 30s, motors 1 and 2 experience efficiency loss faults, causing the actual lift forces to drop rapidly but quickly recover to their original amplitudes. At 45s, the actual lift forces are affected by the neighboring fault, resulting in a spike.

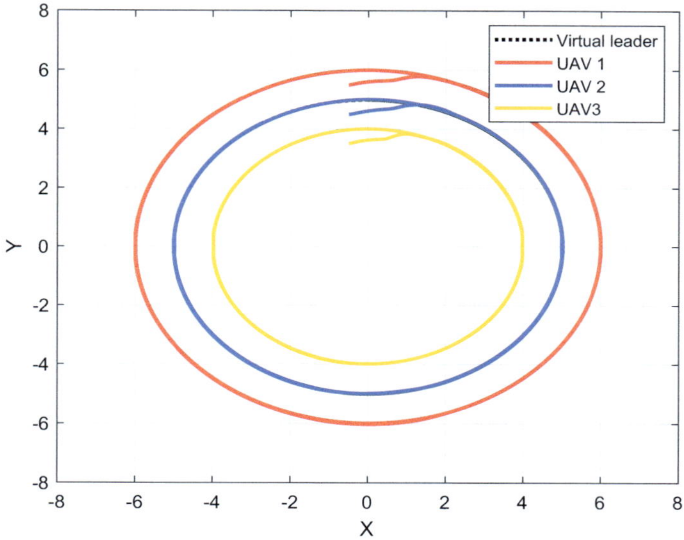


**Fig. 2.6** The $xoy$ plane

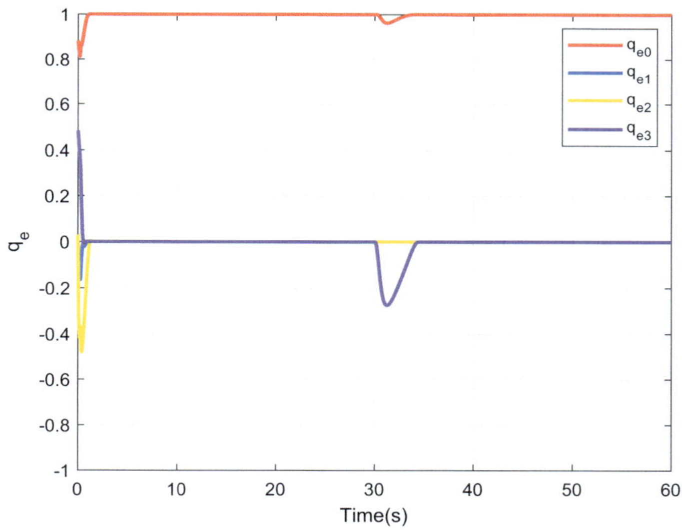


**Fig. 2.7** The error quaternion curve of quadrotor 2

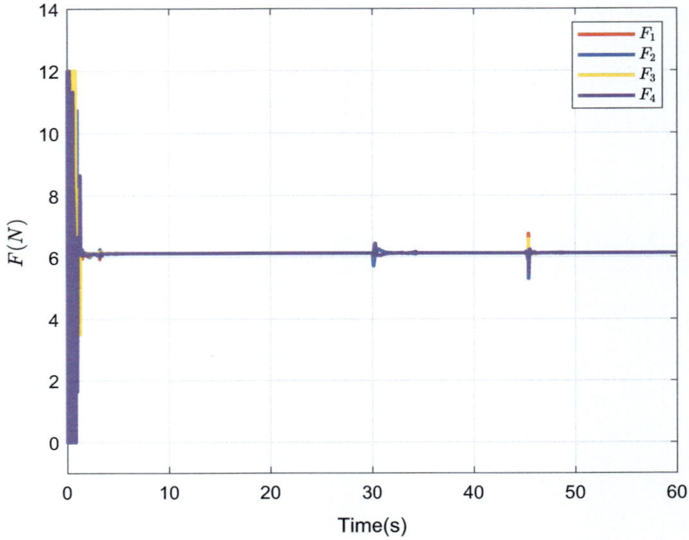


**Fig. 2.8** The expected lift force for each rotor of quadrotor 2

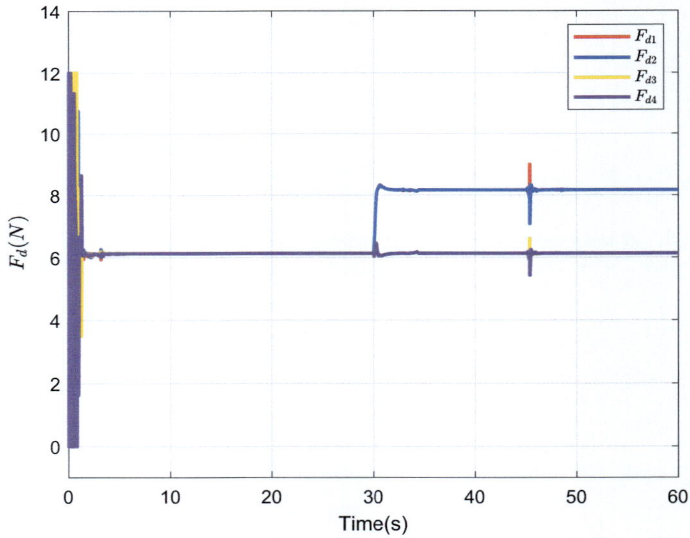


**Fig. 2.9** The actual lift force of each rotor of the quadrotor 2

## 2.4 Conclusions

This chapter investigates the fault-tolerant formation control problem for multiple quadrotor UAVs under an undirected communication topology, with a focus on meeting stricter requirements on convergence time and ensuring fixed-time stability of the system. For the position loop, a nominal controller is designed by incorporating a command-filtering technique to circumvent the derivative singularity issues commonly encountered in fixed-time control theory. Additionally, a robust integral sliding mode control scheme is proposed to enhance the system's robustness against disturbances and uncertainties. In the attitude loop, a piecewise sliding mode function is introduced to eliminate potential singular terms in the controller, while a sliding mode observer is employed to estimate and compensate for the aggregated uncertainties in real time. The stability of the overall system is rigorously analyzed and proven through theoretical derivations, ensuring that the proposed control scheme achieves fixed-time convergence. Numerical simulations are conducted to validate the effectiveness of the designed controller, demonstrating its superior theoretical performance and practical applicability. The results confirm that the proposed approach offers significant advantages in both theoretical rigor and real-world implementation.

## References


1. Basin M, Bharath Panathula C, Shtessel Y (2017) Multivariable continuous fixed-time second-order sliding mode control: design and convergence time estimation. IET Control Theory Appl 11(8):1104–1111
2. Chen Q, Xie S, He X (2020) Neural-network-based adaptive singularity-free fixed-time attitude tracking control for spacecrafts. IEEE Trans Cybern 51(10):5032–5045
3. Cunha JPV, Hsu L, Costa RR, Lizarralde F (2003) Output-feedback model-reference sliding mode control of uncertain multivariable systems. IEEE Trans Autom Control 48(12):2245–2250
4. Gong W, Li B, Yang Y, Ban H, Xiao B (2019) Fixed-time integral-type sliding mode control for the quadrotor UAV attitude stabilization under actuator failures. Aerosp Sci Technol 95:105444
5. Huang Y, Jia Y (2017) Fixed-time consensus tracking control for second-order multi-agent systems with bounded input uncertainties via nfftsm. IET Control Theory Appl 11(16):2900–2909
6. Jiang B, Hu Q, Friswell MI (2016) Fixed-time attitude control for rigid spacecraft with actuator saturation and faults. IEEE Trans Control Syst Technol 24(5):1892–1898
7. Liang X, Wang Q, Hu C, Dong C (2020) Fixed-time observer based fault tolerant attitude control for reusable launch vehicle with actuator faults. Aerosp Sci Technol 107:106314
8. Liu Y, Ma J, Tu H (2018) Robust command filtered adaptive backstepping control for a quadrotor aircraft. J Control Sci Eng 2018(1):1854648
9. Long H, Guo T, Zhao J (2021) Adaptive disturbance observer-based novel fixed-time nonsingular terminal sliding-mode control for a class of dof nonlinear systems. IEEE Trans Industr Inf 18(9):5905–5914
10. Ma Y, Jiang B, Wang J, Gong J (2023) Adaptive fault-tolerant formation control for heterogeneous UAVs-UGVs systems with multiple actuator faults. IEEE Trans Aerosp Electron Syst 59(5):6705–6716

11. Mi W, Luo L, Zhong S (2023) Fixed-time consensus tracking for multi-agent systems with a nonholomonic dynamics. IEEE Trans Autom Control 68(2):1161–1168
12. Polyakov A (2011) Nonlinear feedback design for fixed-time stabilization of linear control systems. IEEE Trans Autom Control 57(8):2106–2110
13. Sarrafan N, Zarei J (2021) Bounded observer-based consensus algorithm for robust finite-time tracking control of multiple nonholonomic chained-form systems. IEEE Trans Autom Control 66(10):4933–4938
14. Shao X, Hu Q, Shi Y, Jiang B (2018) Fault-tolerant prescribed performance attitude tracking control for spacecraft under input saturation. IEEE Trans Control Syst Technol 28(2):574–582
15. Shtessel Y, Edwards C, Fridman L, Levant A et al (2014) Sliding mode control and observation. Springer
16. Wang B, Shen Y, Li N, Zhang Y, Gao Z (2023) An adaptive sliding mode fault-tolerant control of a quadrotor unmanned aerial vehicle with actuator faults and model uncertainties. Int J Robust Nonlinear Control 33(17):10182–10198
17. Wang Y, Chen W, Zhang S, Zhu J, Cao L (2018) Command-filtered incremental backstepping controller for small unmanned aerial vehicles. J Guid Control Dyn 41(4):954–967
18. Wang Z, Su Y, Zhang L (2020) Fixed-time fault-tolerant attitude tracking control for rigid spacecraft. J Dyn Syst Meas Contr 142(2):024502
19. Wei X, Yu W, Wang H, Yao Y, Mei F (2018) An observer-based fixed-time consensus control for second-order multi-agent systems with disturbances. IEEE Trans Circuits Syst II Express Briefs 66(2):247–251
20. Wu X, Lu S, Wei C, Liao Y (2021) Observer-based fault-tolerant attitude tracking control for rigid spacecraft with actuator saturation and faults. Acta Astronaut 178:824–834
21. Yang Y, Yue D (2018) Adaptive decentralized fault-tolerant tracking control of a class of uncertain large-scale nonlinear systems with actuator faults. Trans Inst Meas Control 40(3):831–842
22. Yu J, Chen B, Yu H, Lin C, Zhao L (2018) Neural networks-based command filtering control of nonlinear systems with uncertain disturbance. Inf Sci 426:50–60
23. Lu Y, Liu W (2023) Disturbance observer-based adaptive fuzzy fault-tolerant tracking control for uncertain nonlinear systems with input delay and actuator faults. Int J Control Autom Syst 21(3):820–828
24. Liu J, Wu Y, Sun M, Sun C (2022) Fixed-time cooperative tracking for delayed disturbed multi-agent systems under dynamic event-triggered control. IEEE/CAA J Automatica Sinica 9(5):930–933

# Chapter 3
# Fixed-Time Collision-Free FTC for Quadrotor UAVs

## 3.1 Introduction

Collision prevention in quadrotor UAV formations is essential for operational safety, particularly when high-density configurations must traverse restricted spaces. Unanticipated actuator faults inducing abrupt maneuvers can further compromise the situation. Various collision avoidance methodologies have been explored, including APF, model predictive control strategies, learning-based techniques, game-theoretic approaches, barrier Lyapunov function methods, and damped Hooke's law implementations. Previous studies have addressed this challenge through distinct frameworks. The authors in [8, 24] utilized APF by modeling inter-agent interactions via distance-dependent attractive and repulsive potentials. The authors in [1] integrated minimum separation requirements as state constraints within a model predictive control paradigm. Meanwhile, the reference [2] employed iterative learning algorithms to resolve collision scenarios, whereas the reference [13] formulated the problem as a strategic game to derive Nash-equilibrium solutions.

To ensure practical engineering applicability, the control strategy must be both feasible for implementation and supported by realistic verification methods to facilitate iterative development. Many hardware-based studies, such as those in [12, 20], require a complete overhaul of the UAV's existing control architecture when deploying new algorithms. This process often proves time-consuming and costly, primarily due to the challenges associated with tuning an entirely new controller from scratch. A more cost-effective solution has been demonstrated in [7, 25], where fixed-time FTC modules were integrated into baseline controllers to improve the safety of quadrotor UAVs. This method is particularly advantageous when multiple UAVs with standard controllers require safety upgrades. Furthermore, hardware-in-the-loop (HITL) testing, as performed in [15], offers a high-fidelity yet economical validation approach, enabling the development of multi-UAV formation control protocols in a more efficient and budget-friendly manner.

 
K. Zhang et al., *Fixed-Time/Prescribed-Time Fault-Tolerant Control for Multiple Unmanned Systems*, Unmanned System Technologies,
https://doi.org/10.1007/978-981-95-2913-1_3

The contributions of this chapter are as follows

(1) Enhanced dynamic response via fixed-time control theory: While prior studies on quadrotor formation control achieved asymptotic or finite-time convergence [18, 21], the fixed-time approach adopted here ensures faster formation stabilization. Unlike existing fixed-time works in [3, 10, 14] that simplified agents as second-order systems, this work incorporates the complete dynamics of quadrotor UAVs–including position, attitude, and motor dynamics–while accounting for actuator faults and external disturbances, aspects overlooked in [4, 22]. The proposed scheme thus combines rapid fixed-time convergence with full dynamic consideration.

(2) Addressing a critical operational scenario: Actuator faults compromise both UAV stability and formation integrity by inducing positional drift, necessitating integrated collision avoidance. Prior fault-tolerant formation studies in [3, 11] ignored inter-agent safety margins, while collision avoidance works in [9, 23] neglected fault impacts. The developed distributed hierarchical control decouples these coupled risks, simultaneously enabling FTC and collision prevention–a dual-layer safety mechanism for formations.

(3) Hardware-ready implementation framework: Unlike most formation control studies relying solely on numerical validation [4, 11, 19, 22], this work deploys algorithms on flight controllers and verifies them via HITL testing for practical fidelity. Additionally, the solution is standardized for PX4 compatibility, which is omitted in [7, 20, 25]. In comparison, this chapter bridges the gap between theoretical design and considered systems.

## 3.2 Problem Formulation and Preliminaries

### *3.2.1 Problem Formulation*

The positional and rotational dynamics of the $i$th quadrotor UAV are systematically characterized, with each subsystem decomposed into its kinematic and dynamic models. The positional dynamics can be mathematically represented as

$$\text{Kinematic:} \quad \dot{p}_i = v_i \tag{3.1}$$

$$\text{Dynamic:} \quad \dot{v}_i = \left(G + R_b^e F_i^b + F_{di}\right)/m_i \tag{3.2}$$

Let $p_i = [x_i, y_i, z_i]^{\mathrm{T}}$ and $v_i = [v_{ix}, v_{iy}, v_{iz}]^{\mathrm{T}}$, which stand for the position and velocity in the earth frame (North-East-Down, NED). Here, $G = [0, 0, m_i g]^{\mathrm{T}}$ with $m_i$ indicating the UAV's mass and $g$ the gravitational acceleration. The rotation matrix from the body frame to the earth frame, defined by Euler angles, is denoted by $R_b^e$ [7]. Under the body frame (Forward-Right-Down, FRD), the joint force is expressed as $F_i^b = [0, 0, -T_i]^{\mathrm{T}}$, where $T_i$ corresponds to the total thrust of the quadrotor UAV. Meanwhile, $F_{di}$ refers to the aerodynamic disturbance arising from air drag or wind gusts.

The attitude subsystem is modeled through the following formulation

$$\text{Kinematic:} \quad \dot{\Lambda}_i = M\omega_i \tag{3.3}$$

$$\text{Dynamic:} \quad \dot{\omega}_i = J_i^{-1}(-\omega_i^{\times} J_i\omega_i + \tau_i + \tau_{di}) \tag{3.4}$$

Let $\Lambda_i = [\phi_i, \theta_i, \psi_i]^{\mathrm{T}}$ signify the Euler angle, with $\omega_i = [p_i, q_i, r_i]^{\mathrm{T}}$ indicating the angular velocity in the body frame. The inertial matrix is $J_i = \mathrm{diag}\left\{\mathrm{I}_{\mathrm{i,xx}}, \mathrm{I}_{\mathrm{i,yy}}, \mathrm{I}_{\mathrm{i,zz}}\right\}$, while $\tau_i = \left[\tau_{i,\phi}, \tau_{i,\theta}, \tau_{i,\psi}\right]^{\mathrm{T}}$ represents the control torque. Additionally, $\tau_{di}$ refers to the external disturbance in rotational dynamics. The following matrix $M$ serves as the transformation matrix that converts the derivative of Euler angles into angular velocity

$$M = \begin{bmatrix} 1 & \tan\theta_i \sin\phi_i & \tan\theta_i \cos\phi_i \\ 0 & \cos\phi_i & -\sin\phi_i \\ 0 & \sin\phi_i / \cos\theta_i & \cos\phi_i / \cos\theta_i \end{bmatrix} \tag{3.5}$$

The expressions for the total thrust and control torque are given by $[\tau_i, T_i]^{\mathrm{T}} = E\left[w_{1i}^2, w_{2i}^2, w_{3i}^2, w_{4i}^2\right]^{\mathrm{T}}$, in which matrix $E$ represents the control effectiveness matrix of a quadrotor UAV as

$$E = \begin{bmatrix} -\frac{\sqrt{2}}{4}dC_t & \frac{\sqrt{2}}{4}dC_t & \frac{\sqrt{2}}{4}dC_t & -\frac{\sqrt{2}}{4}dC_t \\ \frac{\sqrt{2}}{4}dC_t & -\frac{\sqrt{2}}{4}dC_t & \frac{\sqrt{2}}{4}dC_t & -\frac{\sqrt{2}}{4}dC_t \\ C_m & C_m & -C_m & -C_m \\ C_t & C_t & C_t & C_t \end{bmatrix} \tag{3.6}$$

In this formulation, $d$ represents the distance between opposing motors, with $C_t$ denoting the thrust coefficient and $C_m$ the torque coefficient. The spinning speed of each motor is denoted by $w_{ji}$ where $j$ ranges from 1 to 4.

Taking the motor response model into account, the motor dynamics can be expressed as

$$\dot{w}_{ji} = \frac{1}{\varepsilon}(w_{ji\_d} - w_{ji}) \tag{3.7}$$

$$w_{ji\_d} = C_r\left(w_{ji\_c}\right) + W_b \tag{3.8}$$

In this context, $\varepsilon$ represents the time constant of the motors, while $w_{ji\_d}$ signifies the desired motor speed. Here, $C_r$ and $W_b$ denote the slope and intercept of the motor response curve, respectively, and $w_{ji\_c}$ stands for the duty cycle.

**Remark 3.1** The actual control input for a quadrotor is the motor duty cycle, though control torque and thrust are commonly selected as the control inputs for the position and attitude subsystems to facilitate system analysis. Once the required control torque and thrust are computed, they are distributed to derive the motor duty cycles, and actual torque and thrust are generated via the motor control effectiveness.

The rotational velocity of motors experiencing partial effectiveness reduction can be expressed as $w_{ji} = \rho_j \overline{w}_{ji}$, with $\rho_j$ representing the remaining speed ratio for individual motors, where $\rho_j \in (0, 1)$, while $\overline{w}_{ji}$ corresponds to the nominal motor velocity without faults. By constructing $P = \text{diag}\{\rho_1, \rho_2, \rho_3, \rho_4\}$, the resulting torque and aggregate thrust generation in the presence of actuator faults becomes expressible as

$$\begin{aligned} [\tau_i, T_i]^{\text{T}} &= EP^2 \left[\overline{w}_{1i}^2, \overline{w}_{2i}^2, \overline{w}_{3i}^2, \overline{w}_{4i}^2\right]^{\text{T}} = EP^2E^{-1}[\tau_{ui}, T_{ui}]^{\text{T}} \\ &= [\tau_{ui}, T_{ui}]^{\text{T}} + \underbrace{\left(EP^2E^{-1} - I\right)[\tau_{ui}, T_{ui}]^{\text{T}}}_{\Xi_i} \end{aligned} \tag{3.9}$$

The vector $[\tau_{ui}, T_{ui}]^{\text{T}}$ represents the ideal torque and thrust output associated with $\left[\overline{w}_{1i}^2, \overline{w}_{2i}^2, \overline{w}_{3i}^2, \overline{w}_{4i}^2\right]^{\text{T}}$ as determined by the control system, while $\Xi_i$ characterizes the disturbance effects induced by actuator faults.

**Remark 3.2** This mathematical formulation converts motor-related multiplicative faults into an aggregated additive uncertainty within the torque and thrust control framework, ensuring the fault compensation mechanism remains effective for both multiplicative and additive fault scenarios.

Drawing upon Eqs. (3.9), (3.2), and (3.4), the subsequent formulations can be further expanded as follows

$$\dot{v}_i = \left(G + R_b^e F_{ui}^b + R_b^e \Xi_{Ti} + F_{di}\right)/m_i \tag{3.10}$$

$$\dot{\omega}_i = J_i^{-1}(-\omega_i^{\times} J_i \omega_i + \tau_i + \Xi_{\tau i} + \tau_{di}) \tag{3.11}$$

where $F_{ui}^b = [0, 0, -T_{ui}]^{\text{T}}$. The influence of actuator faults on the position subsystem is characterized by $\Xi_{Ti} = [0, 0, -[0, 0, 0, 1]\Xi_i]^{\text{T}}$, whereas $\Xi_{\tau i} = [I_{3\times3}, 0_{3\times1}]\Xi_i$ signifies the impact on the attitude subsystem. Here, the first three elements of the term $\Xi_i$ are considered in these definitions (The first 3 elements of the term $\Xi_i$).

For achieving independent control of position and attitude dynamics, the reference acceleration is formulated as $a_{ui} = \left(G + R_b^e F_{ui}^b\right)/m_i$. From this relation, both the required thrust magnitude $T_{ui}$ and target orientation angles $[\phi_{di}, \theta_{di}]^{\text{T}}$ become determinable (Note that in practical implementations, yaw rate typically serves as the controlled variable while yaw orientation is obtained through direct measurement).

$$\begin{cases} T_{ui} = m_i \sqrt{a_{ui,x}^2 + a_{ui,y}^2 + (g - a_{ui,z})^2} \\ \phi_{di} = \arcsin\left(m_i(a_{ui,y}\cos(\psi_i) - a_{ui,x}\sin(\psi_i))/T_{ui}\right) \\ \theta_{di} = \arctan\left(a_{ui,x}\cos(\psi_i) + a_{ui,y}\sin(\psi_i)/(a_{ui,z} - g)\right) \end{cases} \tag{3.12}$$

The desired acceleration enables independent control between position and attitude dynamics, allowing Eqs. (3.10) and (3.11) to be reformulated as

$$\dot{v}_i = a_{ui} + d_i \tag{3.13}$$

$$\dot{\omega}_i = h(\omega_i) + J_i^{-1}\tau_{ui} + f_i \tag{3.14}$$

Here, $d_i = \left(R_b^e \Xi_{Ti} + F_{di}\right)/m_i$ represents the combined disturbance arising from actuator faults and aerodynamic effects, while $h(\omega_i) = J_i^{-1}(-\omega_i^{\times} J_i \omega_i)$ describes the gyroscopic dynamics. The term $f_i = J_i^{-1}(\Xi_{\tau i} + \tau_{di})$ quantifies the total torque disturbance resulting from both actuator faults and environmental disturbances.

**Assumption 3.1** ([25]) The disturbance terms $d_i$ and $f_i$ associated with system faults, along with their time derivatives, possess finite but unknown bounds satisfying $\|d_i\| \leq D_1$, $\|\dot{d}_i\| \leq D_2$, $\|f_i\| \leq L_1$, and $\|\dot{f}_i\| \leq L_2$.

**Remark 3.3** While the fault-induced disturbances $d_i$ and $f_i$ incorporate control torque and thrust components, these actuation signals originate from the motor dynamics (Eq. 3.7) that exhibit first-order filter characteristics. This inherent dynamic behavior guarantees smooth transitions in torque and thrust outputs, thereby preserving the bounded nature of $\dot{d}_i$ and $\dot{f}_i$ without disruption from control actuation signals.

### 3.2.2 Preliminaries

The networked formation of multiple UAVs can be represented by a graph $G = (N, E, A)$, where $N = \{v_1, v_2, \ldots, v_n\}$ corresponds to the collection of nodes, $E \subseteq N \times N$ describes the interconnection links, and $A = [a_{ij}] \in \mathbb{R}^{n \times n}$ represents the adjacency matrix. The element $a_{ij}$ quantifies the communication strength between distinct nodes, satisfying $a_{ij} > 0$ exclusively when $(v_i, v_j) \in E$ with $i \neq j$. The graph exhibits undirected characteristics when the adjacency matrix $A$ is symmetric. The degree matrix $D = \text{diag}\{d_1, d_2, \ldots, d_n\}$ is constructed with $d_i = \sum_{j=1}^{n} a_{ij}$, and the graph connectivity is guaranteed when $D$ possesses full rank. The Laplacian matrix $L = D - A$ captures the overall connectivity pattern. The parameter $b_i$ indicates the communication link quality with a reference leader, where $b_i > 0$ signifies an active connection for UAV $i$, while $b_i = 0$ denotes no connection. The collective leader-follower interactions are characterized by the diagonal matrix $B = \text{diag}\{b_1, b_2, \ldots, b_n\}$.

**Lemma 3.1** ([26]) *For a multi-UAV formation with undirected and connected communication topology $G$, if there exists at least one quadrotor UAV capable of receiving information from the leader, then the composite matrix $H = L + B$ maintains symmetry and positive definiteness.*

**Lemma 3.2** ([5]) *For a system possessing a continuous radially unbounded Lyapunov function $\Gamma(x) : \mathbb{R}^n \to \mathbb{R}_+ \cup \{0\}$ satisfying*

$$\dot{\Gamma}(x) \leq -\alpha\Gamma^{\mu}(x) - \beta\Gamma^{\lambda}(x) + \Delta$$

*with parameters* $\mu > 1$, $0 < \lambda < 1$, $\alpha, \beta > 0$, *and* $0 < \Delta < \infty$, *the origin achieves practical fixed-time stability. The convergence time is rigorously bounded by*

$$T_{\max} = \frac{1}{\alpha\chi(\mu - 1)} + \frac{1}{\beta\chi(1 - \lambda)}, \quad 0 < \chi < 1$$

*with the residual set characterized by*

$$x \in \left\{ \Gamma(x) \leq \min \left\{ \left( \frac{\Delta}{\alpha(1 - \chi)} \right)^{\frac{1}{\mu}}, \left( \frac{\Delta}{\beta(1 - \chi)} \right)^{\frac{1}{\lambda}} \right\} \right\}$$

**Control Objective**: In a distributed leader-follower network topology, each follower UAV $i$ must achieve fixed-time convergence of its formation tracking error $e_i = p_i - \Delta_i - p_d$, where $\Delta_i$ represents the prescribed formation geometry and $p_d$ denotes the leader's position. This convergence must be guaranteed despite actuator faults in individual quadrotor UAVs, while maintaining collision-free conditions throughout the formation.

## 3.3 Main Results

### 3.3.1 *Fixed-Time Fault-Disturbance Estimation Observer*

For actuator fault estimation and compensation, an augmented fixed-time observer from [17] with measurement noise suppression capability is implemented across both translational and rotational dynamics.

Taking sensor noise into account, the dynamics described by Eqs. (3.13) and (3.14) can be unified through the following formulation

$$\begin{cases} x_{ij\_1m} = x_{ij\_1} + n_{ij} \\ \dot{x}_{ij\_1} = f\left(x_{ij\_1}\right) + g\left(u_{ij}\right) + \Upsilon_{ij} \end{cases} \tag{3.15}$$

where $x_{ij\_1m} (j = 1, 2)$ represents the measured states affected by sensor noise, with $x_{i1\_1} = \omega_i$ and $x_{i2\_1} = v_i$. The terms $n_{i1}$ and $n_{i2}$ correspond to zero-mean high-frequency time-varying noise in the angular velocity and velocity measurements, respectively. The nonlinear functions are defined as $f\left(x_{i1\_1}\right) = h(\omega_i)$ and $f\left(x_{i2\_1}\right) = 0$, while the input-related terms are $g\left(u_{i1}\right) = J_i^{-1}\tau_{ui}$ and $g\left(u_{i2}\right) = a_{ui}$. Additionally, $\Upsilon_{i1} = f_i$ and $\Upsilon_{i2} = d_i$ denote the respective disturbance terms.

To mitigate sensor noise effects, we introduce an auxiliary variable $x_{ij\_0} = \int_0^t x_{ij\_1m}\, \mathrm{d}t$, which yields the decomposition $x_{ij\_0} = \int_0^t \left(x_{ij\_1} + n_{ij}\right) \mathrm{d}t = \int_0^t x_{ij\_1}\, \mathrm{d}t + o$, where $\|o\| \to 0$ asymptotically. The fault-disturbance estimation observer is then constructed as

$$\begin{cases} \dot{z}_{ij\_0} = z_{ij\_1} + k_{m0}\mathrm{sig}^{\alpha_f}(x_{ij\_0} - z_{ij\_0}) + k_{m0}\mathrm{sig}^{\beta_f}(x_{ij\_0} - z_{ij\_0}) \\ \dot{z}_{ij\_1} = f(z_{ij\_1}) + g\left(u_{ij}\right) + k_{m1}\mathrm{sig}^{2\alpha_f - 1}(x_{ij\_0} - z_{ij\_0}) \\ \qquad + k_{m1}\mathrm{sig}^{2\beta_f - 1}(x_{ij\_0} - z_{ij\_0}) + z_{ij\_2} \\ \dot{z}_{ij\_2} = k_{m2}\mathrm{sig}^{3\alpha_f - 2}(x_{ij\_0} - z_{ij\_0}) + k_{m2}\mathrm{sig}^{3\beta_f - 2}(x_{ij\_0} - z_{ij\_0}) \end{cases} \tag{3.16}$$

where $z_{ij\_\kappa}$ $(\kappa = 0, 1)$ estimates the states $x_{ij\_\kappa}$ $(\kappa = 0, 1)$, and $z_{ij\_3}$ estimates the uncertainty $\Upsilon_{ij}$. The parameters $\alpha_f \in (1 - \delta_\alpha, 1)$ and $\beta_f \in (1, 1 + \delta_\beta)$ are bounded by small positive constants $\delta_\alpha > 0$ and $\delta_\beta > 0$. The gain $k_{m\kappa}$ $(\kappa = 0, 1, 2)$ is selected to ensure the matrix $W$ is Hurwitz.

$$W = \begin{bmatrix} -k_{m0} & 1 & 0 \\ -k_{m1} & 0 & 1 \\ -k_{m2} & 0 & 0 \end{bmatrix} \tag{3.17}$$

where the nonlinear operator $\mathrm{sig}^\eta(\zeta)$ is defined for a vector $\zeta = [\zeta_1, \zeta_2, \ldots, \zeta_n]^\mathrm{T}$ as $\mathrm{sig}^\eta(\zeta) = \left[\mathrm{sign}(\zeta_1)|\zeta_1|^\eta,\ \mathrm{sign}(\zeta_2)|\zeta_2|^\eta,\ \cdots,\ \mathrm{sign}(\zeta_n)|\zeta_n|^\eta\right]^\mathrm{T}$

As established in [17], under Assumption 2.1, the fault-disturbance estimation observer guarantees practical fixed-time convergence with bounded settling time $T_{1\max} = \frac{1}{p}\left(\frac{\beta_{V\infty}}{\beta_\infty} + \frac{\alpha_{V0}}{|\alpha_0|}\right)$ where $p > 0$, $\beta_{V\infty} \geq \max\{1, \beta_f, 2\beta_f - 1\}$, $\alpha_{V0} \geq \max\{1, \alpha_f, 2\alpha_f - 1\}$, $\beta_\infty = \beta_f - 1$, and $\alpha_0 = \alpha_f - 1$. The estimation errors satisfy $\|f_i - \hat{f}_i\| \leq L_{\Upsilon_f}$, $\|d_i - \hat{d}_i\| \leq L_{\Upsilon_d}$ with positive constants $L_{\Upsilon_d}$ and $L_{\Upsilon_f}$, where $\hat{f}_i$ and $\hat{d}_i$ estimate the attitude subsystem's faulty torque $f_i$ and position subsystem's lumped disturbance $d_i$, respectively.

**Remark 3.4** Compared to the fixed-time observers in [7, 25], the proposed fault-disturbance estimation observer enhances robustness against onboard electronic noise through auxiliary variable construction. In implementation, only the estimates $\hat{f}_i$ and $\hat{d}_i$ are injected into the controller. The velocity $v_i$ and angular rate $\omega_i$ required by downstream controllers are obtained directly from onboard sensors to prevent observer-induced phase lag. Furthermore, the observer's nonlinear term $f(z_{ij\_1})$ is replaced by $f(x_{ij\_1})$ during operation, as these can be computed directly from sensor measurements.

### 3.3.2 *Fixed-Time Velocity Tracking Controller Design*

The baseline attitude control system consists of two components: a proportional-integral-derivative (PID)-based angular velocity controller and a proportional attitude controller. Let $\tau_{i\text{-}base}$ denote the torque output from this baseline controller. The final control torque is then computed as $\tau_{ui} = \tau_{i\text{-}base} - \hat{f}_i$ where $\hat{f}_i$ represents the estimated fault torque from the observer.

To obtain the derivative of the desired velocity while preventing excessive acceleration commands due to potential noise and oscillations in the velocity reference, we design a fixed-time command filter as follows

$$\dot{\hat{v}}_{di} = -k_c \left( \text{sig}^{\alpha_1}(\hat{v}_{di} - v_{di}) + \text{sig}^{\beta_1}(\hat{v}_{di} - v_{di}) \right) \tag{3.18}$$

where $v_{di}$ represents the velocity command for the $i$-th quadrotor UAV, $\hat{v}_{di}$ denotes its filtered version, with design constraints $k_c > 0.5, 0 < \alpha_1 < 1$, and $\beta_1 > 1$ governing the filter's dynamic response.

**Assumption 3.2** For the velocity command's time derivative, there exists a known positive constant $L_{vd} > 0$ such that the inequality $\|\dot{v}_{di}\| \leq L_{vd}$ strictly holds, which explicitly characterizes the maximum allowable variation rate of the desired velocity profile.

**Theorem 3.1** *Under the command filter structure specified in Eq. (3.18) and subject to Assumption 3.2, the filtering error $e_{fi} = \hat{v}_{di} - v_{di}$ exhibits practical fixed-time convergence. This guarantees that the error dynamics enter and remain within an arbitrarily small vicinity of the origin in a predetermined time interval, independent of initial conditions.*

***Proof*** Choose Lyapunov candidate $V_{fi} = \frac{1}{2} e_{fi}^{\mathrm{T}} e_{fi}$.

$$\begin{aligned}
\dot{V}_{fi} &= -k_c \left( \sum_{j=1}^{3} \left| e_{fi,j} \right|^{1+\alpha_1} + \sum_{j=1}^{3} \left| e_{fi,j} \right|^{1+\beta_1} \right) - e_{fi}^{\mathrm{T}} \dot{v}_{di} \\
&\leq -k_c \left( \sum_{j=1}^{3} \left| e_{fi,j} \right|^{1+\alpha_1} + \sum_{j=1}^{3} \left| e_{fi,j} \right|^{1+\beta_1} \right) + \sum_{j=1}^{3} \left| e_{fi,j} \right| \left| \dot{v}_{di,j} \right| \\
&\leq -k_c \left( \sum_{j=1}^{3} \left| e_{fi,j} \right|^{1+\alpha_1} + \sum_{j=1}^{3} \left| e_{fi,j} \right|^{1+\beta_1} \right) + \sum_{j=1}^{3} \frac{1}{2} e_{fi,j}^2 + \sum_{j=1}^{3} \frac{1}{2} \dot{v}_{di,j}^2 \\
&\leq -\left( k_c - \frac{1}{2} \right) \sum_{j=1}^{3} 2^{\frac{1+\alpha_1}{2}} \left( \frac{1}{2} e_{fi,j}^2 \right)^{\frac{1+\alpha_1}{2}} \\
&\quad - \left( k_c - \frac{1}{2} \right) \sum_{j=1}^{3} 2^{\frac{1+\beta_1}{2}} \left( \frac{1}{2} e_{fi,j}^2 \right)^{\frac{1+\beta_1}{2}} + \frac{1}{2} \|\dot{v}_{di}\|^2
\end{aligned} \tag{3.19}$$

According to Lemma 2.2, it can be further derived that

$$\begin{aligned}
\dot{V}_{fi} \leq & \left( k_c - \frac{1}{2} \right) 2^{\frac{\alpha_1+1}{2}} V_{fi}^{\frac{\alpha_1+1}{2}} - \left( k_c - \frac{1}{2} \right) 2^{\frac{\beta_1+1}{2}} 3^{\frac{1-\beta_1}{2}} V_{fi}^{\frac{\beta_1+1}{2}} \\
& + \frac{1}{2} L_v^2 - k_{f\alpha} V_{fi}^{\frac{1+\alpha_1}{2}} - k_{f\beta} V_{fi}^{\frac{1+\beta_1}{2}} + \frac{1}{2} L_{vd}^2
\end{aligned} \tag{3.20}$$

The gain parameters $k_{f\beta}$ and $k_{f\alpha}$ are $k_{f\beta} = \left(2^{\frac{\beta_1+1}{2}} 3^{\frac{1-\beta_1}{2}} k_c - 2^{\frac{\beta_1-1}{2}} 3^{\frac{1-\beta_1}{2}}\right)$ and $k_{f\alpha} = \left(2^{\frac{\alpha_1+1}{2}} k_c - 2^{\frac{\alpha_1-1}{2}}\right)$. Following Lemma 3.2, the filter error achieves practical fixed-time convergence with a rigorously bounded settling time $T_{f\max} = \frac{2}{k_{f\beta}\chi_f(\beta_1-1)} + \frac{2}{k_{f\alpha}\chi_f(1-\alpha_1)}$, where the scaling factor $\chi_f$ satisfies $0 < \chi_f < 1$.

The filter error $e_{fi}$ is strictly bounded by a positive constant $L_f$, with its residual set characterized by $e_{fi} \in \left\{ V_{fi} \le \min\left\{ \left(\frac{0.5L_{vd}^2}{k_{f\alpha}(1-\chi_f)}\right)^{\frac{1}{\lambda_f}}, \left(\frac{0.5L_{vd}^2}{k_{f\beta}(1-\chi_f)}\right)^{\frac{1}{\mu_f}} \right\}\right\}$ where the exponents are defined as $\lambda_f = \frac{1+\alpha_1}{2}$ and $\mu_f = \frac{1+\beta_1}{2}$. This completes the proof of Theorem 3.1.

For the translational dynamics governed by Eq. (3.13), we define the velocity tracking error as $e_{vi} = v_{di} - v_i$, where $v_{di}$ denotes the desired velocity. The corresponding velocity controller is designed as

$$a_{ui} = \dot{\hat{v}}_{di} + k_{v1}\mathrm{sig}^{\alpha_2}(e_{vi}) + k_{v1}\mathrm{sig}^{\beta_2}(e_{vi}) + k_{v2}e_{vi} - \hat{d}_i \tag{3.21}$$

with control parameters satisfying $k_{v1} > 0$, $k_{v2} > 0$, $0 < \alpha_2 < 1$, and $\beta_2 > 1$.

**Theorem 3.2** *The integrated system comprising the fault-disturbance estimation observer* (3.16), *command filter* (3.18), *and velocity controller* (3.21) *ensures practical fixed-time convergence of the velocity tracking error $e_{vi}$ for each quadrotor UAV. This guarantees that all agents achieve their desired velocities within a uniformly bounded settling time, independent of initial conditions.*

***Proof*** Choose Lyapunov candidate $V_{ti} = \frac{1}{2}e_{vi}^{\mathrm{T}}e_{vi}$.

$$\begin{aligned}
\dot{V}_{ti} \le &- e_{vi}^{\mathrm{T}}\left(k_{v1}\mathrm{sig}^{\alpha_2}(e_{vi}) + k_{v1}\mathrm{sig}^{\beta_2}(e_{vi})\right) - k_{v2}e_{vi}^{\mathrm{T}}e_{vi} \\
&+ \sum_{j=1}^{3}\left|e_{vi,j}\right|\left|\dot{v}_{di,j} - \dot{\hat{v}}_{di,j} + \hat{d}_{i,j} - d_{i,j}\right| \\
\le &- k_{v1}\sum_{j=1}^{3}2^{\frac{\alpha_2+1}{2}}\left(\frac{1}{2}e_{vi,j}^2\right)^{\frac{\alpha_2+1}{2}} - k_{v1}\sum_{j=1}^{3}2^{\frac{\beta_2+1}{2}}\left(\frac{1}{2}e_{vi,j}^2\right)^{\frac{\beta_2+1}{2}} \\
&- k_{v2}\sum_{j=1}^{3}e_{vi,j}^2 + \sum_{j=1}^{3}\frac{1}{2}\left|\dot{v}_{di,j} - \dot{\hat{v}}_{di,j} + \hat{d}_{i,j} - d_{i,j}\right|^2 + \sum_{j=1}^{3}\frac{1}{2}\left|e_{vi,j}\right|^2 \\
\le &- k_{v1}\sum_{j=1}^{3}2^{\frac{\alpha_2+1}{2}}\left(\frac{1}{2}e_{vi,j}^2\right)^{\frac{\alpha_2+1}{2}} - k_{v1}\sum_{j=1}^{3}2^{\frac{\beta_2+1}{2}}\left(\frac{1}{2}e_{vi,j}^2\right)^{\frac{\beta_2+1}{2}} \\
&+ \left(\frac{1}{2} - k_{v2}\right)\sum_{j=1}^{3}\left(e_{vi,j}^2\right) + \sum_{j=1}^{3}\frac{1}{2}\left|\dot{v}_{di,j} - \dot{\hat{v}}_{di,j} + \hat{d}_{i,j} - d_{i,j}\right|^2
\end{aligned} \tag{3.22}$$

If $k_{v2} \geq 0.5$, according to Lemma 2.2, the following inequality holds

$$\dot{V}_{ti} \leq -\left(k_{v1} 2^{\frac{\alpha_2+1}{2}} (V_{ti})^{\frac{\alpha_2+1}{2}} + k_{v1} 2^{\frac{\beta_2+1}{2}} 3^{\frac{1-\beta_2}{2}} (V_{ti})^{\frac{\beta_2+1}{2}}\right) + \frac{1}{2} L_f^2 + \frac{1}{2} L_{\Upsilon d}^2 \tag{3.23}$$

Following Lemma 3.2, the velocity tracking error achieves practical fixed-time convergence with its settling time $T_v$ rigorously bounded by $T_{v\max} = \frac{2}{k_{v\beta}\chi_v(\beta_2-1)} + \frac{2}{k_{v\alpha}\chi_v(1-\alpha_2)}$, where $0 < \chi_v < 1$, $k_{v\beta} = k_{v1} 2^{\frac{\beta_2+1}{2}} 3^{\frac{1-\beta_2}{2}}$, and $k_{v\alpha} = k_{v1} 2^{\frac{\alpha_2+1}{2}}$.

If $k_{v2} < 0.5$, the following inequality holds

$$\begin{aligned}
\dot{V}_{ti} \leq & -k_{v1} \sum_{j=1}^{3} 2^{\frac{\alpha_2+1}{2}} \left(\frac{1}{2} e_{vi,j}^2\right)^{\frac{\alpha_2+1}{2}} - k_{v1} \sum_{j=1}^{3} 2^{\frac{\beta_2+1}{2}} \left(\frac{1}{2} e_{vi,j}^2\right)^{\frac{\beta_2+1}{2}} \\
& - \left(k_{v2} - \frac{1}{2}\right) \left(\sum_{j=1}^{3} 2^{\frac{\alpha_2+1}{2}} \left(\frac{1}{2} e_{vi,j}^2\right)^{\frac{\alpha_2+1}{2}} + \sum_{j=1}^{3} 2^{\frac{\beta_2+1}{2}} \left(\frac{1}{2} e_{vi,j}^2\right)^{\frac{\beta_2+1}{2}}\right) \\
& + \frac{1}{2} L_f^2 + \frac{1}{2} L_{\Upsilon d}^2 \\
\leq & -\left(k_{v1} + k_{v2} - \frac{1}{2}\right) \left(2^{\frac{\alpha_2+1}{2}} V_{ti}^{\frac{\alpha_2+1}{2}} + 2^{\frac{\beta_2+1}{2}} 3^{\frac{1-\beta_2}{2}} V_{ti}^{\frac{\beta_2+1}{2}}\right) \\
& + \frac{1}{2} L_f^2 + \frac{1}{2} L_{\Upsilon d}^2
\end{aligned} \tag{3.24}$$

The velocity tracking error likewise demonstrates practical fixed-time convergence, governed by the precisely defined gain parameters

$$\begin{aligned}
k_{v\beta} &= (k_{v1} + k_{v2} - 0.5) 2^{\frac{\beta_2+1}{2}} 3^{\frac{1-\beta_2}{2}} \\
k_{v\alpha} &= (k_{v1} + k_{v2} - 0.5) 2^{\frac{\alpha_2+1}{2}}
\end{aligned}$$

The tracking error $e_{vi}$ is strictly bounded by a positive constant $L_v$, with its residual set characterized by

$$e_{vi} \in \left\{ V_{vi} \leq \min \left\{ \left(\frac{0.5L_f^2 + 0.5L_{\Upsilon d}^2}{k_{v\alpha}(1-\chi_f)}\right)^{\frac{1}{\lambda_f}}, \left(\frac{0.5L_f^2 + 0.5L_{\Upsilon d}^2}{k_{v\beta}(1-\chi_f)}\right)^{\frac{1}{\mu_f}} \right\} \right\}$$

where the exponents are defined as $\lambda_f = \frac{1+\alpha_2}{2}$ and $\mu_f = \frac{1+\beta_2}{2}$. This completes the proof of Theorem 3.2.

These two configurations exhibit distinct characteristics in terms of convergence time and residual set size. When $k_{v2} \geq 0.5$, the system achieves accelerated convergence and reduced tracking error. However, this configuration simultaneously amplifies velocity overshoot and noise sensitivity proportional to $k_{v2}$ magnitude, necessitating careful parameter selection based on comprehensive system performance evaluation.

**Remark 3.5** While the attitude subsystem's convergence rate is not explicitly analyzed, its tracking dynamics can be effectively decoupled from position control considerations due to the inherent timescale separation between rotational and translational dynamics. Specifically, The significantly faster response of rotational dynamics permits neglecting attitude tracking time in position control design. Residual attitude tracking errors can be incorporated into the position subsystem's uncertainty estimation framework. This hierarchical decomposition allows the disturbance observer to effectively compensate for attitude-related perturbations.

### 3.3.3 *Fixed-Time Fault-Tolerant Controller Design*

The proposed control architecture ensures two fundamental properties for each quadrotor agent. Preserved individual maneuverability through decentralized control design. Established prerequisites for collision-aware formation coordination.

Let $\xi_i \in \mathbb{R}^{3\times 1}$ denote an auxiliary state vector, defined as

$$\xi_i = \sum_{j=1}^{n} a_{ij}(p_i - \Delta_{ij} - p_j) + b_i(p_i - p_d - \Delta_i) \tag{3.25}$$

The virtual leader's position is denoted by $p_d = [x_d, y_d, z_d]^{\mathrm{T}}$, while the $i$-th UAV's position is $p_i = [x_i, y_i, z_i]^{\mathrm{T}}$. The formation geometry is $\Delta_i = [\Delta_{ix}, \Delta_{iy}, \Delta_{iz}]^{\mathrm{T}}$, $\Delta_{ij} = \Delta_i - \Delta_j$ where $\Delta_i$ specifies the desired relative position between UAV $i$ and the virtual leader, and $\Delta_{ij}$ defines inter-agent spacing.

The collective formation state $\xi = [\xi_1, \xi_2, \ldots, \xi_n]^{\mathrm{T}}$ admits a compact representation $\xi = ((L + B) \otimes I_{3\times 3})\, e$ with the global tracking error $e = [e_1, e_2, \ldots, e_n]^{\mathrm{T}}$ and individual errors $e_i = p_i - \Delta_i - p_d$.

For each quadrotor UAV, the sliding mode manifold $s_i \in R^{3\times 1}$ is formulated adopting an integral structure as below, so that the sliding manifold and its derivative can maintain continuity.

$$s_{i,\kappa} = k_{arc,\kappa} \arctan(\xi_{i,\kappa}) + \int_0^t \Gamma_{i,\kappa} \left(\mathrm{sig}^{\alpha}\left(\xi_{i,\kappa}\right) + k_{\xi}\xi_{i,\kappa} + \mathrm{sig}^{\beta}\left(\xi_{i,\kappa}\right)\right) \mathrm{d}t \tag{3.26}$$

in which $1 \le i \le n$, $1 \le \kappa \le 3$, $0 < \alpha < 1$, $\beta > 1$, $k_{arc,\kappa} > 0$, $k_{\xi} > 0$, and $\Gamma_{i,\kappa} = \Gamma_{0,\kappa} + \Gamma_{APF,\kappa}$, with $\Gamma_{0,\kappa} > 0$, and $\Gamma_{APF,\kappa} \ge 0$ being the $\kappa$-th component of the repulsive force function constructed to accomplish collision avoidance.

$$\Gamma_{APF,\kappa} = \begin{cases} \left| \sum_{j=1}^{n} -Q_\kappa Grad(R, D)(p_{i,\kappa} - p_{j,\kappa}) \right|, & R_{in} \le D_{ij} \le R_{out} \\ 0, & D_{ij} > R_{out} \end{cases} \tag{3.27}$$

in which $Grad(R, D) = \frac{4(R_{out}^2 - R_{in}^2)(R_{out}^2 - D_{ij}^2)}{(D_{ij}^2 - R_{in}^2)^3}$, with $Q_\kappa > 0$ for $\kappa = 1, 2, 3$, the relative distance between UAVs is represented by $D_{ij} = \|p_i - p_j\|$, $R_{out}$ represents the maximum distance where the repulsive function remains active, and $R_{in}$ stands for the minimum safety distance. Owing to the gradient having additional linearity, the APF function utilizes the gradient of the repulsive function to achieve a stable response at diverse distances. It should be emphasized that only the amplitude of the APF function is employed, with its repulsive direction information replaced by the convergence direction of the formation error.

**Theorem 3.3** *Upon convergence of the formation error to the sliding manifold, that is $s_{i,\kappa} = 0$, the tracking error of the virtual leader e demonstrates fixed-time convergence. When the formation error converges in the vicinity of the sliding manifold, namely $|s_{i,\kappa}| \le L_{si,\kappa}$ with $L_{si,\kappa} > 0$, the tracking error e achieves practical fixed-time convergence.*

***Proof*** When $s_{i,\kappa} = 0 = \dot{s}_{i,\kappa}$, the subsequent equality is satisfied

$$\begin{cases} k_{arc,\kappa} \arctan(\xi_{i,\kappa})' = -\Gamma_{i,\kappa} \left(\mathrm{sig}^\alpha \left(\xi_{i,\kappa}\right) + \mathrm{sig}^\beta \xi_{i,\kappa} + k_\xi \xi_{i,\kappa}\right) \\ \dfrac{k_{arc,\kappa} \dot{\xi}_{i,\kappa}}{1 + \xi_{i,\kappa}^2} = -\Gamma_{i,\kappa} \left(\mathrm{sig}^\alpha \left(\xi_{i,\kappa}\right) + \mathrm{sig}^\beta \left(\xi_{i,\kappa}\right) + k_\xi \xi_{i,\kappa}\right) \\ \dot{\xi}_{i,\kappa} = -\dfrac{1 + \xi_{i,\kappa}^2}{k_{arc,\kappa}} \Gamma_{i,\kappa} \left(\mathrm{sig}^\alpha \left(\xi_{i,\kappa}\right) + \mathrm{sig}^\beta \left(\xi_{i,\kappa}\right) + k_\xi \xi_{i,\kappa}\right) \end{cases} \tag{3.28}$$

Let $K$ be defined as $K = \mathrm{diag}\{\frac{1+\xi_{i,\kappa}^2}{k_{arc,\kappa}} \Gamma_{i,\kappa}\}(\mathrm{i} = 1 \ldots \mathrm{n}, \kappa = 1 \ldots 3)$, and further derivation indicates that

$$\begin{cases} \dot{\xi} = -K \left(\mathrm{sig}^\alpha (\xi) + \mathrm{sig}^\beta (\xi) + k_\xi \xi\right) \\ (L + B) \otimes (I_{3\times3})\, \dot{e} = -K \left(\mathrm{sig}^\alpha (\xi) + \mathrm{sig}^\beta (\xi) + k_\xi \xi\right) \\ \dot{e} = -((L + B) \otimes I_{3\times3})^{-1} K \left(\mathrm{sig}^\alpha (\xi) + \mathrm{sig}^\beta (\xi) + k_\xi \xi\right) \end{cases} \tag{3.29}$$

Select the Lyapunov candidate $V_s = \frac{1}{2} e^{\mathrm{T}} He$, where $H = (L + B) \otimes I_{3\times3}$. In light of Lemma 3.1, $H$ is positive definite.

$$\dot{V}_s = \xi^{\mathrm{T}} \dot{e} = -\xi^{\mathrm{T}} H^{-1} K \left(\mathrm{sig}^\alpha (\xi) + \mathrm{sig}^\beta (\xi) + k_\xi \xi\right) \tag{3.30}$$

Given that $\frac{1+\xi_{i,\kappa}^2}{k_{arc,\kappa}}\Gamma_{i,\kappa} \geq \frac{\Gamma_{0,\kappa}}{k_{arc,\kappa}}$, it follows that $\lambda_{\min}(K) = \frac{\min \Gamma_{0,\kappa}}{\|k_{arc}\|_\infty}$ (where $\lambda_{\min}(K)$ denotes the minimum eigenvalue of $K$). By defining $k_{arc} = \left[k_{arc,1}, k_{arc,2}, k_{arc,3}\right]^{\mathrm{T}}$ and setting $k_{HK} = \lambda_{\min}(H^{-1}K) = \frac{\Gamma_0 \lambda_{\min}(H^{-1})}{\|k_{arc}\|_\infty}$, the subsequent inequality is established

$$
\begin{aligned}
\dot{V}_s \leq & -k_{KH}\left(\sum_{1\leq i\leq n, 1\leq\kappa\leq 3}\left(\xi_{i,\kappa}^2\right)\right)^{\frac{\alpha+1}{2}} - k_{KH}(3n)^{\frac{1-\beta}{2}}\left(\sum_{1\leq i\leq n, 1\leq\kappa\leq 3}\left(\xi_{i,\kappa}^2\right)\right)^{\frac{\beta+1}{2}} \\
= & -k_{KH}\left(e^{\mathrm{T}}H^{\mathrm{T}}He\right)^{\frac{\alpha+1}{2}} - k_{KH}(3n)^{\frac{1-\beta}{2}}\left(e^{\mathrm{T}}H^{\mathrm{T}}He\right)^{\frac{\beta+1}{2}} \\
& - k_{KH}(3n)^{\frac{1-\beta}{2}}\left(2\lambda_{\min}(H)\frac{1}{2}e^{\mathrm{T}}He\right)^{\frac{\beta+1}{2}} \\
\leq & -k_{KH}(2\lambda_{\min}(H))^{\frac{\alpha+1}{2}}(V_s)^{\frac{\alpha+1}{2}} - k_{KH}(3n)^{\frac{1-\beta}{2}}(2\lambda_{\min}(H))^{\frac{\beta+1}{2}}(V_s)^{\frac{\beta+1}{2}}
\end{aligned} \tag{3.31}
$$

In accordance with Lemma 2.1, the tracking error of the virtual leader achieves fixed-time convergence when the sliding mode manifold is attained. Settling time is constrained by $T_{s\max} = \frac{2}{k_{KH}(3n)^{\frac{1-\beta}{2}}(2\lambda_{\min}(H))^{\frac{\beta+1}{2}}(\beta-1)} + \frac{2}{k_{KH}(2\lambda_{\min}(H))^{\frac{\alpha+1}{2}}(1-\alpha)}$.

Given that $\left|s_{i,\kappa}\right| \leq L_{si,\kappa}$, assume that $\left|\dot{s}_{i,\kappa}\right| \leq L_{sdi,\kappa}$ with $L_{sdi,\kappa} > 0$. The derivative of Eq. (3.28) can be rewritten as $\dot{\xi}_{i,\kappa} = -\frac{1+\xi_{i,\kappa}^2}{k_{arc,\kappa}}\left(1 - \frac{\dot{s}_{i,\kappa}}{Q(\xi)}\right)Q(\xi)$, in which $Q(\xi) = \Gamma_{i,\kappa}\left(\mathrm{sig}^\alpha\left(\xi_{i,\kappa}\right) + \mathrm{sig}^\beta\left(\xi_{i,\kappa}\right) + k_\xi \xi_{i,\kappa}\right)$. Under the condition that $|Q(\xi)| > \left|\dot{s}_{i,\kappa}\right|$, $\left(1 - \frac{\dot{s}_{i,\kappa}}{Q(\xi)}\right) > 0$, the tracking error of the virtual leader is fixed-time convergent, and the proof is analogous to that under the condition when $s_{i,\kappa} = 0$.

Under the scenario where $|Q(\xi)| \leq \left|\dot{s}_{i,\kappa}\right|$, suggesting that $\xi_{i,\kappa}$ is similarly bounded by $L_{\xi i,\kappa} > 0$, and rephrase Eq. (3.28) as

$$
\dot{\xi}_{i,\kappa} = -\frac{1+\xi_{i,\kappa}^2}{k_{arc,\kappa}}\Gamma_{i,\kappa}\left(\mathrm{sig}^\alpha\left(\xi_{i,\kappa}\right) + \mathrm{sig}^\beta\left(\xi_{i,\kappa}\right) + k_\xi \xi_{i,\kappa}\right) + \frac{1+\xi_{i,\kappa}^2}{k_{arc,\kappa}}\dot{s}_{i,\kappa} \tag{3.32}
$$

and Eq. (3.30) can be rewritten as

$$
\dot{V}_s = -\xi^{\mathrm{T}}H^{-1}K\left(\mathrm{sig}^\alpha(\xi) + \mathrm{sig}^\beta(\xi) + k_\xi \xi\right) + \xi^{\mathrm{T}}H^{-1}K_L\dot{S} \tag{3.33}
$$

where $K_L = \mathrm{diag}\{\frac{1+\xi_{i,\kappa}^2}{\mathrm{k}_{\mathrm{arc},\kappa}}\}$, and $S = [s_1, s_2, \ldots, s_n]^{\mathrm{T}}$.

Let $L_\xi = \left[L_{\xi 1,1}, L_{\xi 1,2}, \ldots, L_{\xi n,3}\right]^{\mathrm{T}}$ and $L_{sd} = \left[L_{sd1,1}, L_{sd1,2}, \ldots, L_{sdn,3}\right]^{\mathrm{T}}$ be defined. Analogously to Eq. (3.31), the subsequent inequality is valid

$$\begin{aligned}\dot{V}_s \leq & -k_{KH}(2\lambda_{\min}(H))^{\frac{\alpha+1}{2}}(V_s)^{\frac{\alpha+1}{2}} - k_{KH}(3n)^{\frac{1-\beta}{2}}(2\lambda_{\min}(H))^{\frac{\beta+1}{2}}(V_s)^{\frac{\beta+1}{2}} \\ & + \lambda_{\max}\left(H^{-1}K_L\right)\left(\frac{1}{2}\left\|L_\xi\right\|^2 + \frac{1}{2}\left\|L_{sd}\right\|^2\right)\end{aligned} \tag{3.34}$$

It should be noted that as $\xi$ is bounded in this case, $\lambda_{\max}\left(H^{-1}K_L\right)$ is likewise bounded. Based on Lemma 3.2, the tracking error exhibits practical fixed-time convergence, with the settling time constrained by $T_{s-p\max} = T_{s\max}/\chi_s$, where $0 < \chi_s < 1$.

The residual set of tracking error $e$ is characterized by

$$e \in \left\{V_s \leq \min\left\{\left(\frac{L_{pra}}{k_{s\alpha}(1-\chi_s)}\right)^{\frac{1}{\lambda_f}}, \left(\frac{L_{pra}}{k_{s\beta}(1-\chi_s)}\right)^{\frac{1}{\mu_f}}\right\}\right\} \tag{3.35}$$

where $L_{pra} = \lambda_{\max}\left(H^{-1}K_L\right)\left(\frac{1}{2}\left\|L_\xi\right\|^2 + \frac{1}{2}\left\|L_{sd}\right\|^2\right)$, $k_{s\alpha} = k_{KH}(2\lambda_{\min}(H))^{\frac{\alpha+1}{2}}$, $k_{s\beta} = k_{KH}(3n)^{\frac{1-\beta}{2}}(2\lambda_{\min}(H))^{\frac{\beta+1}{2}}$, $\lambda_f = \frac{1+\alpha}{2}$, and $\mu_f = \frac{1+\beta}{2}$. Hence, the proof of Theorem 3.3 is finalized.

To guarantee that the sliding mode manifold is attained, the velocity command for each quadrotor UAV is formulated as

$$\begin{aligned}v_{di} = & \frac{1}{k_{Gi}}\left(\sum_{j=1}^{n} a_{ij}(v_j + \dot{\Delta}_{ij}) + b_i(v_0 + \dot{\Delta}_i)\right) + \frac{1}{k_{Gi}}\left(\left(I_{3\times 3} + \mathrm{diag}(\xi_i)^2\right)K_{arc}^{-1}\right) \\ & \left(-K_\lambda \mathrm{sig}^\lambda(s_i) - K_\mu \mathrm{sig}^\mu(s_i) - K_S s_i - k_{dot}\dot{s}_i - K_{\Gamma i}\left(\mathrm{sig}^\alpha(\xi_i) + \mathrm{sig}^\beta(\xi_i) + k_\xi \xi_i\right)\right)\end{aligned} \tag{3.36}$$

Here, $k_{Gi} = \sum_{j=1}^{n} a_{ij} + b_i$, $K_{arc} = \mathrm{diag}\{\mathrm{k_{arc,1}}, \mathrm{k_{arc,2}}, \mathrm{k_{arc,3}}\}$, $K_{\Gamma i} = \mathrm{diag}\{\Gamma_{\mathrm{i},1}, \Gamma_{\mathrm{i},2}, \Gamma_{\mathrm{i},3}\}$, $K_\lambda = \mathrm{diag}\{\mathrm{k}_{\lambda,1}, \mathrm{k}_{\lambda,2}, \mathrm{k}_{\lambda,3}\}$ with $K_\lambda > 0$, $K_s = \mathrm{diag}(\{\mathrm{k_{s,1}}, \mathrm{k_{s,2}}, \mathrm{k_{s,3}}\}$ where $K_s > 0$, $k_{dot} > 0$, $0 < \lambda, \alpha < 1$, $\mu, \beta > 1$ and $K_\mu = \mathrm{diag}\{\mathrm{k}_{\mu,1}, \mathrm{k}_{\mu,2}, \mathrm{k}_{\mu,3}\}$ with $K_\mu > 0$. $v_0$ represents the velocity of the virtual leader.

**Theorem 3.4** *Provided that the velocity command for each quadrotor UAV is formulated as Eq. (3.36), the sliding mode manifold S exhibits practical fixed-time convergence, and the formation error converges in the vicinity of the sliding manifold within a finite time, that is, S remains bounded.*

***Proof*** Choose Lyapunov candidate $V = \frac{1}{2}S^{\mathrm{T}}S$, then, we can obtain:

$$
\begin{aligned}
\dot{V} &= \sum_{i=1}^{n} s_i^{\mathrm{T}} \left(-K_\lambda \mathrm{sig}^\lambda(s_i) - K_\mu \mathrm{sig}^\mu(s_i) - K_s s_i\right) - k_{dot}\dot{V} \\
&\quad - \sum_{i=1}^{n} s_i^{\mathrm{T}} K_{arc}\left(I_3 + \mathrm{diag}(\xi_{\mathrm{i}})^2\right)^{-1}(k_{Gi}e_{vi}) \\
&\le -\frac{1}{1+k_{dot}}\sum_{i=1}^{n}\sum_{\kappa=1}^{3} k_{\lambda,\kappa}\left(\left|s_{i,\kappa}\right|^2\right)^{\frac{\lambda+1}{2}} - \frac{1}{1+k_{dot}}\sum_{i=1}^{n}\sum_{\kappa=1}^{3} k_{\mu,\kappa}\left(\left|s_{i,\kappa}\right|^2\right)^{\frac{\mu+1}{2}} \\
&\quad - \sum_{i=1}^{n}\sum_{\kappa=1}^{3}\frac{k_{s,\kappa}}{1+k_{dot}}\left|s_{i,\kappa}\right|^2 + \sum_{i=1}^{n}\sum_{\kappa=1}^{3}\frac{k_{arc,\kappa}k_{Gi}}{(1+k_{dot})}\left|s_{i,\kappa}\right|\left|e_{vi,\kappa}\right| \\
&\le -\frac{1}{1+k_{dot}}\sum_{i=1}^{n}\sum_{\kappa=1}^{3} k_{\lambda,\kappa}\left(\left|s_{i,\kappa}\right|^2\right)^{\frac{\lambda+1}{2}} - \frac{1}{1+k_{dot}}\sum_{i=1}^{n}\sum_{\kappa=1}^{3} k_{\mu,\kappa}\left(\left|s_{i,\kappa}\right|^2\right)^{\frac{\mu+1}{2}} \\
&\quad - \sum_{i=1}^{n}\sum_{\kappa=1}^{3}\frac{k_{s,\kappa}}{1+k_{dot}}\left|s_{i,\kappa}\right|^2 + \sum_{i=1}^{n}\sum_{\kappa=1}^{3}\frac{k_{Gi}}{(1+k_{dot})}\left(\frac{1}{2}\left|k_{arc,\kappa}s_{i,\kappa}\right|^2 + \frac{1}{2}\left|e_{vi,\kappa}\right|^2\right) \\
&= -\frac{1}{1+k_{dot}}\sum_{i=1}^{n}\sum_{\kappa=1}^{3} k_{\lambda,\kappa}\left(\left|s_{i,\kappa}\right|^2\right)^{\frac{\lambda+1}{2}} - \frac{1}{1+k_{dot}}\sum_{i=1}^{n}\sum_{\kappa=1}^{3} k_{\mu,\kappa}\left(\left|s_{i,\kappa}\right|^2\right)^{\frac{\mu+1}{2}} \\
&\quad - \sum_{i=1}^{n}\sum_{\kappa=1}^{3}\frac{k_{s,\kappa}}{1+k_{dot}}\left|s_{i,\kappa}\right|^2 + \sum_{i=1}^{n}\sum_{\kappa=1}^{3}\frac{k_{Gi}}{2(1+k_{dot})}\left(\left|k_{arc,\kappa}^2 s_{i,\kappa}\right|^2 + \left|e_{vi,\kappa}\right|^2\right)
\end{aligned}
\tag{3.37}
$$

Define $k_G = [k_{G1}, k_{G2}, \ldots, k_{Gn}]^{\mathrm{T}}$, and choose $k_{s,\kappa} \ge \frac{1}{2}k_{Gi}k_{arc,\kappa}^2$, Eq. (3.37) becomes

$$
\begin{aligned}
\dot{V} &\le -\frac{\lambda_{\min}(K_\lambda)}{1+k_{dot}}\sum_{i=1}^{n}\sum_{\kappa=1}^{3}\left(\left|s_{i,\kappa}\right|^2\right)^{\frac{\lambda+1}{2}} \\
&\quad -\frac{\lambda_{\min}(K_\mu)}{1+k_{dot}}\sum_{i=1}^{n}\sum_{\kappa=1}^{3}\left(\left|s_{i,\kappa}\right|^2\right)^{\frac{\mu+1}{2}} + \frac{\|k_G\|_\infty}{2(1+k_{dot})}\sum_{i=1}^{n}\|e_{vi}\|^2
\end{aligned}
\tag{3.38}
$$

Proofs in Theorem 3.2 have established that the velocity tracking error $e_{vi}$ is confined within $L_v$. Thus, by virtue of Lemma 2.2, the ensuing inequality holds

$$
\begin{aligned}
\dot{V} \leq & -\frac{\lambda_{\min}\left(K_{\lambda}\right)}{1+k_{dot}} \sum_{i=1}^{n}\left(\left(s_i^{\mathrm{T}} s_i\right)^{\frac{\lambda+1}{2}}\right) \\
& -\frac{\lambda_{\min}\left(K_{\mu}\right) 3^{\frac{1-\mu}{2}}}{1+k_{dot}} \sum_{i=1}^{n}\left(s_i^{\mathrm{T}} s_i\right)^{\frac{\mu+1}{2}}+\frac{\left\|k_G\right\|_{\infty}}{2\left(1+k_{dot}\right)} \sum_{i=1}^{n}\left\|e_{vi}\right\|^2 \\
\leq & -\frac{\lambda_{\min}\left(K_{\lambda}\right) 2^{\frac{\lambda+1}{2}}}{1+k_{dot}}\left(\frac{1}{2} S^{\mathrm{T}} S\right)^{\frac{\lambda+1}{2}}+\frac{\left\|k_G\right\|_{\infty}}{2\left(1+k_{dot}\right)} \sum_{i=1}^{n} L_v^2 \\
& -\frac{\lambda_{\min}\left(K_{\mu}\right) 3^{\frac{1-\mu}{2}} n^{\frac{1-\mu}{2}} 2^{\frac{\mu+1}{2}}}{1+k_{dot}}\left(\frac{1}{2} S^{\mathrm{T}} S\right)^{\frac{\mu+1}{2}} \\
\leq & -\gamma_{\lambda} V^{\frac{\lambda+1}{2}}-\gamma_{\mu} V^{\frac{\mu+1}{2}}+L_p
\end{aligned} \tag{3.39}
$$

where $\gamma_{\lambda}=\frac{\lambda_{\min}(K_{\lambda})2^{\frac{\lambda+1}{2}}}{1+k_{dot}}$, $\gamma_{\mu}=\frac{\lambda_{\min}(K_{\mu})(3n)^{\frac{1-\mu}{2}}2^{\frac{\mu+1}{2}}}{1+k_{dot}}$, and $L_p=\frac{\|k_G\|_{\infty}}{2(1+k_{dot})}\sum_{i=1}^{n}L_v^2$. In light of Lemma 3.2, the sliding mode manifold $S$ achieves practical fixed-time convergence, with the settling time bounded by $T_{p\max}=\frac{2}{\gamma_{\mu}\chi_p(\mu-1)}+\frac{2}{\gamma_{\lambda}\chi_p(1-\lambda)}$, where $0<\chi_p<1$.

The residual set of the sliding mode manifold $S$ is characterized by

$$
S \in \left\{V \leq \min\left\{\left(\frac{L_p}{\gamma_{\lambda}(1-\chi_f)}\right)^{\frac{1}{\lambda_p}}, \left(\frac{L_p}{\gamma_{\mu}(1-\chi_f)}\right)^{\frac{1}{\mu_p}}\right\}\right\} \tag{3.40}
$$

in this case, $\lambda_p=\frac{1+\lambda}{2}$ and $\mu_p=\frac{1+\mu}{2}$, finalizing the proof of Theorem 3.4.

It should be emphasized that as $S$ is bounded, the formation error $\xi$ is likewise bounded, fulfilling the condition stipulated in Theorem 3.3 where $|s_{i,\kappa}|\leq L_{si,\kappa}$. During the proof of Theorem 3.3, the assumption that $\left|\dot{s}_{i,\kappa}\right|\leq L_{sdi,\kappa}$ is justifiable since $\dot{s}_{i,\kappa}$ consists of formation error and UAVs' relative velocity, both of which are inherently constrained in engineering practice.

Proofs in Theorems 3.3 and 3.4 demonstrate that when the velocity command is implemented within a finite time, it can steer each quadrotor UAV to the sliding mode manifold and ensure formation convergence.

Express the sliding mode manifold in the form

$$
k_{arc,\kappa}\arctan(\xi_{i,\kappa}) = s_{i,\kappa} - \int_0^t \Gamma_{i,\kappa}\left(\operatorname{sig}^{\alpha}\left(\xi_{i,\kappa}\right)+k_{\xi}\xi_{i,\kappa}+\operatorname{sig}^{\beta}\left(\xi_{i,\kappa}\right)\right)\mathrm{d}t \tag{3.41}
$$

Previous analysis has established that $s_{i,\kappa}$ is bounded. Suppose the $i$-th quadrotor UAV were to collide with other UAVs, then $\Gamma_{i,\kappa}$ would approach infinity, and the term $s_{i,\kappa}-\int_0^t \Gamma_{i,\kappa}\left(\operatorname{sig}^{\alpha}\left(\xi_{i,\kappa}\right)+k_{\xi}\xi_{i,\kappa}+\operatorname{sig}^{\beta}\left(\xi_{i,\kappa}\right)\right)\mathrm{d}t$ would likewise approach infinity. (Notably, in the event of a collision, the formation error $\xi_{i,\kappa}$ cannot be zero, as the desired formation is designed to be collision-free.) This contradicts the fact that $k_{arc,\kappa}\arctan(\xi_{i,\kappa})$ is limited to $\frac{\pi}{2}k_{arc,\kappa}$. Hence, it is concluded that no collisions arise within the formation.

**Remark 3.6** An inspection of the sliding mode manifold structure reveals that the gradient function of the APF furnishes supplementary gain for the convergence of formation error. In the presence of collision tendencies between UAVs, the APF supplies extra control input to guide each quadrotor UAV toward the desired position while keeping them apart. By incorporating the fixed-time strategy, the agility of such response is ensured, enabling the UAV formation to be promptly restored.

**Remark 3.7** The rationale behind separating the velocity and formation controllers is to ensure that information from neighboring UAVs is exclusively utilized at the top level of the cascaded scheme. This approach mitigates the impact of potential communication errors on subsequent dynamics, provided that safety protocols are implemented to eradicate abnormal velocity commands.

## 3.4 Numerical Simulation

To demonstrate the fixed-time convergence characteristics of the formation, numerical simulations of the formation controller, command filter, and velocity controller were conducted prior to hardware implementation.

The command filter parameters were selected as follows: $k_c = 0.6, \alpha_1 = 0.9, \beta_1 = 1.1$. For the velocity controller, the parameters were chosen to be $k_{v1} = 1, k_{v2} = 0.1, \alpha_2 = 0.9, \beta_2 = 1.1$. The formation controller parameters were set as $k_{arc,1} = k_{arc,2} = k_{arc,3} = 0.01, \Gamma_{0,1} = \Gamma_{0,2} = \Gamma_{0,3} = 0.001, k_\xi = 0.1, \alpha = 0.9, \beta = 1.1, Q_1 = Q_2 = Q_3 = 0.072$, $R_{in} = 0.5$, $R_{out} = 1$, $k_{\lambda,1} = k_{\lambda,2} = k_{\mu,1} = k_{\mu,2} = 0.02$, $k_{\lambda,3} = k_{\mu,3} = 0.01$, $k_{s,1} = k_{s,2} = 0.02$, $k_{s,3} = 0.01$, $k_{dot} = 0.001$, $\lambda = \alpha = 0.9$, $\mu = \beta = 1.1$.

**Remark 3.8** During the selection of controller parameters, $k_{arc}$ needs to be initially adjusted to define the scale of the sliding manifold, while other parameters including $\Gamma_0$, $Q_{1,2,3}$, $k_\lambda$ and $k_\mu$ can be defined as products of $k_{arc}$ and relevant coefficients. Once these parameters are preliminarily defined in terms of order of magnitude, fine-tuning with minor modifications can be performed to achieve enhanced system performance.

The formation topology is described as follows. (Notably, the communication weight between UAV 2 and UAV 3 is set to a relatively low value, illustrating that neighbor information is exclusively utilized for collision avoidance.)

$$A = \begin{bmatrix} 0 & 1 & 1 \\ 1 & 0 & 0.1 \\ 1 & 0.1 & 0 \end{bmatrix}, B = \begin{bmatrix} 4 & 0 & 0 \\ 0 & 0 & 0 \\ 0 & 0 & 0 \end{bmatrix} \tag{3.42}$$

To verify the formation convergence under varied initial configurations, the Y-axis position was selected for testing. The initial geometric setup is specified as $\Delta_1 = [0, 0, 0]^T$, $\Delta_2 = [5, \Delta_y, 0]^T$, $\Delta_3 = [0, -\Delta_y, 0]^T$, with $\Delta_y$ taking values 5, 10, and 15. The target formation geometry is defined as $\Delta_1 = [0, 0, 0]^T$, $\Delta_2 = [1, 0, 0]^T$, $\Delta_3 = [-1, 0, 0]^T$, while the virtual leader's trajectory is given by $p_d = [0, 0, 0.5t]^T$. (Notably, the command filter was applied to the geometric transition of $\Delta_{ij}$ to ensure the existence of $\dot{\Delta}_{ij}$.)

The formation error along the Y-axis is depicted in Fig. 3.1, with corresponding modules designed using finite-time techniques simulated for comparative analysis. Results demonstrate that regardless of diverse initial configurations, the formation error converges within 12 s, exhibiting a faster convergence rate than that achieved under finite-time control methods.

**Remark 3.9** The practical fixed-time theory employed in the control framework does not yield an exact bound for the convergence time, but it ensures swifter convergence compared to other controller design techniques under identical parameters. Moreover, the mathematical characterization of the time bound qualitatively illustrates the relationship between parameter selection and convergence speed.

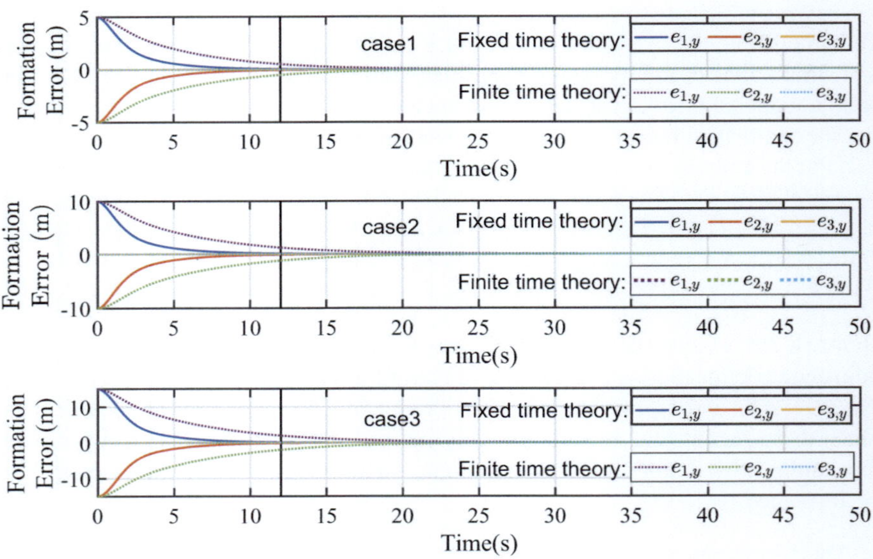


**Fig. 3.1** Numerical simulation of formation convergence

## 3.5 Hardware Implementation

To conform to the current trends in drone technology and create a high-fidelity simulation environment for the algorithm, the highly modular open-source Pixhawk platform was selected to execute the proposed control strategy. The fault-disturbance estimation observer, velocity controller, and formation controller have been integrated into the PX4 autopilot framework.

Normalization of control signals represents a standard engineering approach to alleviate negative impacts from singular data, establish motor duty cycles, and decouple the design of attitude/position controllers from motor dynamics–thereby enabling controller application to generalized cases. As implemented in the PX4 architecture, remote-control inputs, control torques, thrust values, and desired motor speeds are all normalized to simplify code development.

The normalization technique is not limited to the PX4 architecture; in any scenario where algorithm design and implementation efficiency are prioritized, the normalization process remains both appropriate and necessary.

The control thrust of each quadrotor UAV $T_{ui}$ is scaled to $T_{ui\text{-norm}} \in [0, 1]$, which requires normalizing both the desired acceleration and gravitational acceleration. The hovering throttle $T_{\text{hover}}$ configured in PX4 defines the transformation from gravitational acceleration to normalized control thrust $m_i g \rightarrow T_{\text{hover}}$, typically manifested as the integer output of the Z-axis PID velocity controller during hover (when direct access from flight control firmware is unavailable). Consequently, the desired acceleration is normalized as $\frac{m_i a_{ui}}{m_i g} = \frac{m_i a_{ui\text{-norm}}}{T_{\text{hover}}}$.

In the PX4 framework, the control torque output from the controller undergoes normalization to $\tau_{ui\text{-norm}} \in [-1, 1]$. To establish the correlation between this normalized control torque and the actual control torque $\tau_{ui}$ generated by the propellers, an analysis of the motor response model becomes essential. (In the subsequent analysis, it is assumed that actuator faults in the motors have already been compensated for, implying a fault-free motor model where $\tau_i = \tau_{ui}$ and $T_i = T_{ui}$.)

Using the $i$-th quadrotor UAV as an illustration, the normalized torque and thrust signals are transformed into motor duty cycles via the control allocation matrix $U$.

$$\begin{bmatrix} w_{1i\_c} & w_{2i\_c} & w_{3i\_c} & w_{4i\_c} \end{bmatrix}^{\mathrm{T}} = U\begin{bmatrix} \tau_{ui-norm} & T_{ui-norm} \end{bmatrix}^{\mathrm{T}} \tag{3.43}$$

$$U = \begin{bmatrix} -\frac{\sqrt{2}}{2} & \frac{\sqrt{2}}{2} & 1 & 1 \\ \frac{\sqrt{2}}{2} & -\frac{\sqrt{2}}{2} & 1 & 1 \\ \frac{\sqrt{2}}{2} & \frac{\sqrt{2}}{2} & -1 & 1 \\ -\frac{\sqrt{2}}{2} & -\frac{\sqrt{2}}{2} & -1 & 1 \end{bmatrix} \tag{3.44}$$

Given $w_{0i\_c} = T_{ui\text{-norm}}$ and $\delta_{jwi}(j = 1 \ldots 4) = w_{ji\_c} - w_{0i\_c}$, based on Eqs. (3.43) and (3.44), the relationship between the increment in the desired motor speed and the normalized control torque can be formulated as follows

$$\left[\delta_{1wi}\ \delta_{2wi}\ \delta_{3wi}\ \delta_{4wi}\right]^{\mathrm{T}} = U_\tau \tau_{ui\text{-norm}} \tag{3.45}$$

Here, $U_\tau$ is constituted by the initial three columns of matrix $U$

$$U_\tau = \begin{bmatrix} -\frac{\sqrt{2}}{2} & \frac{\sqrt{2}}{2} & 1 \\ \frac{\sqrt{2}}{2} & -\frac{\sqrt{2}}{2} & 1 \\ \frac{\sqrt{2}}{2} & \frac{\sqrt{2}}{2} & -1 \\ -\frac{\sqrt{2}}{2} & -\frac{\sqrt{2}}{2} & -1 \end{bmatrix} \tag{3.46}$$

Recalling the motor dynamics presented in Eq. 3.8, it can be re-expressed as $w_{ji\_d} = C_r\left(w_{0i\_c} + \delta_{jwi}\right) + W_b$. By defining $w_{0i} = C_r w_{0i\_c} + W_b$, the relationship between the actual control torque and the motor speed can be derived as per Eq. (3.6)

$$\begin{bmatrix} \tau_{ui,\phi} \\ \tau_{ui,\theta} \\ \tau_{ui,\psi} \end{bmatrix} = E_\tau \begin{bmatrix} (w_{0i} + C_r\delta_{1wi})^2 \\ (w_{0i} + C_r\delta_{2wi})^2 \\ (w_{0i} + C_r\delta_{3wi})^2 \\ (w_{0i} + C_r\delta_{4wi})^2 \end{bmatrix} \tag{3.47}$$

where $E_\tau$ is composed of the first three rows of matrix $E$.

$$E_\tau = \begin{bmatrix} -\frac{\sqrt{2}}{4}dC_t & \frac{\sqrt{2}}{4}dC_t & \frac{\sqrt{2}}{4}dC_t & -\frac{\sqrt{2}}{4}dC_t \\ \frac{\sqrt{2}}{4}dC_t & -\frac{\sqrt{2}}{4}dC_t & \frac{\sqrt{2}}{4}dC_t & -\frac{\sqrt{2}}{4}dC_t \\ C_m & C_m & -C_m & -C_m \end{bmatrix} \tag{3.48}$$

During normal flight operations, $C_r\delta_{jwi}$ remains negligible in comparison to $w_{0i}$. By ignoring the term $(C_r\delta_{jwi})^2$, Eq. (3.47) can be further simplified to

$$\begin{aligned} \begin{bmatrix} \tau_{ui,\phi} \\ \tau_{ui,\theta} \\ \tau_{ui,\psi} \end{bmatrix} &= E_\tau \left( \begin{bmatrix} w_{0i}^2 \\ w_{0i}^2 \\ w_{0i}^2 \\ w_{0i}^2 \end{bmatrix} + \begin{bmatrix} 2w_{0i}C_r\delta_{1wi} \\ 2w_{0i}C_r\delta_{2wi} \\ 2w_{0i}C_r\delta_{3wi} \\ 2w_{0i}C_r\delta_{4wi} \end{bmatrix} \right) \\ &= E_\tau \begin{bmatrix} 2w_{0i}C_r\delta_{1wi} \\ 2w_{0i}C_r\delta_{2wi} \\ 2w_{0i}C_r\delta_{3wi} \\ 2w_{0i}C_r\delta_{4wi} \end{bmatrix} = 2w_{0i}C_r E_\tau \begin{bmatrix} \delta_{1wi} \\ \delta_{2wi} \\ \delta_{3wi} \\ \delta_{4wi} \end{bmatrix} \end{aligned} \tag{3.49}$$

substituting $\delta_{jwi}(j = 1\ldots4)$ with Eq. (3.45)

$$\begin{aligned}\begin{bmatrix}\tau_{ui,\phi}\\ \tau_{ui,\theta}\\ \tau_{ui,\psi}\end{bmatrix} &= 2w_{0i}C_r E_\tau U_\tau \begin{bmatrix}\tau_{ui,\phi-norm}\\ \tau_{ui,\theta-norm}\\ \tau_{ui,\psi-norm}\end{bmatrix}\\ &= \begin{bmatrix}2dC_tC_rw_{0i}\tau_{ui,\phi-norm}\\ 2dC_tC_rw_{0i}\tau_{ui,\theta-norm}\\ 8C_mC_rw_{0i}\tau_{ui,\psi-norm}\end{bmatrix}\end{aligned} \tag{3.50}$$

Therefore, the relationship between the normalized control torque and the actual control torque has been established.

**Remark 3.10** During the normalization process, the nonlinearity and response time of motors are overlooked, which unavoidably leads to discrepancies in the normalization coefficient during rapid maneuvers. Such unmodeled motor dynamics can also be regarded as part of the aggregated disturbance, which can be estimated and mitigated by the fault-disturbance estimation observer.

**Remark 3.11** The implementation procedure is critical to the application of theoretical designs, in which the mathematical form of the control scheme needs to be encoded and compiled. Yet, such a procedure is rarely discussed in research studies on applications. The normalization of control output described above takes into account the process of control allocation and the control effectiveness of motors, which comprehensively addresses the entire dynamics of a quadrotor and satisfies the requirements of hardware-based flight control stacks.

The simulation platform, as presented in [6], consists of a real-time simulation computer and multiple Pixhawk 2.4.8 boards. The customized PX4 architecture (Firmware 1.7.3) operates on these Pixhawk boards, generating motor pulse width modulation signals for the high-fidelity simulated quadrotor UAVs in the simulation computer. The computer then feeds back the simulated sensor data to the Pixhawk boards.

To facilitate data recording and analysis and to ensure the stability of the attitude subsystem, the observer and controller within the attitude subsystems of three quadrotor UAVs are initially evaluated under hovering conditions to demonstrate their efficacy. The fault-disturbance estimation observer, baseline controller, and fault injection modules are integrated by substituting the original rate controller in the PX4 code, operating at a frequency of 250 Hz.

The simulated UAVs are configured identically with the following parameters: $m_i = 1.4\,\text{kg}$, $d = 0.45\,\text{m}$, $g = 9.8\,\text{N/kg}$, $J_i = \text{diag}\{0.0211, 0.0219, 0.0366\}$, $C_t = 1.105\times10^{-5}\,\text{kg}\cdot\text{m}^2$, $C_m = 1.779\times10^{-7}\,\text{kg}\cdot\text{m}^2$, $C_r = 1148\,\text{rad/s}$, $W_b = -141.4\,\text{rad/s}$, $T_{hover} = 0.608$, $\varepsilon = 0.05s$.

The parameters for the fault-disturbance estimation observer within the attitude subsystem are selected as follows $k_{m0} = k_{m1} = 20, k_{m2} = 15, \alpha_f = 0.95, \beta_f = 1.05$.

The PID angular velocity controller parameters are configured as Proportional gains: $P_{\text{roll}} = P_{\text{pitch}} = 0.15$, $P_{\text{yaw}} = 0.3$; Integral gains: $I_{\text{roll}} = I_{\text{pitch}} = 0.05$, $I_{\text{yaw}} = 0.1$; Derivative gains: $D_{\text{roll}} = D_{\text{pitch}} = 0.003$, $D_{\text{yaw}} = 0$.

The proportional attitude controller parameters are set to $P_{\text{roll\_angle}} = P_{\text{pitch\_angle}} = 6.5$. The yaw angle remains uncontrolled because the yaw angle rate is stabilized (this flight mode is provided by PX4).

Actuator faults are introduced by modifying the total thrust and control torque.

$$\left[\tau_{ui-norm}, T_{ui-norm}\right]^{\mathrm{T}} = U^{-1} P_i U \left[\overline{\tau}_{ui-norm}, \overline{T}_{ui-norm}\right]^{\mathrm{T}} \tag{3.51}$$

Here, $\left[\overline{\tau}_{ui\text{-norm}}, \overline{T}_{ui\text{-norm}}\right]^{\mathrm{T}}$ represents the initial normalized control torque and thrust.

The fault injection matrix $P_i$ for each quadrotor UAV is specified as $P_1 = \text{diag}\{1, 0.8, 0.9, 1\}$, $P_2 = \text{diag}\{0.7, 1, 1, 0.8\}$, $P_3 = \text{diag}\{0.9, 1, 0.6, 1\}$.

The roll-pitch response of the UAVs is depicted in Fig. 3.2. The estimation of fault torque is shown in Fig. 3.3 (Notably, the estimated fault torque follows the same normalization as the control torque). For comparative analysis, the roll-pitch response without the fault-disturbance estimation observer is presented in Fig. 3.4.

Results from the hovering test demonstrate that the fault-disturbance observer can swiftly identify and mitigate the additional torque induced by actuator faults, even under the influence of sensor noise. Comparative experiments indicate that the fixed-time fault-disturbance observer significantly enhances the baseline controller's

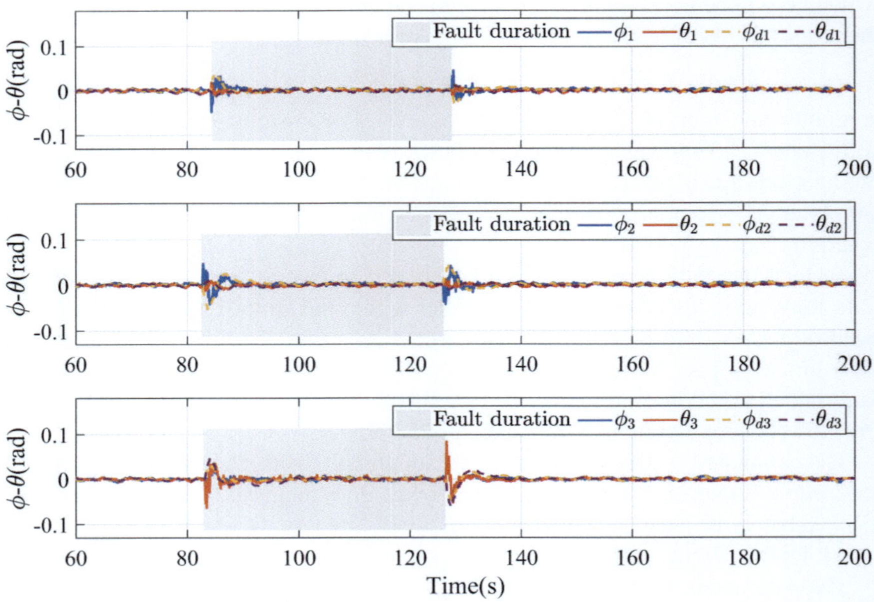


**Fig. 3.2** Roll-pitch response of quadrotor UAVs with fault-disturbance estimation observer

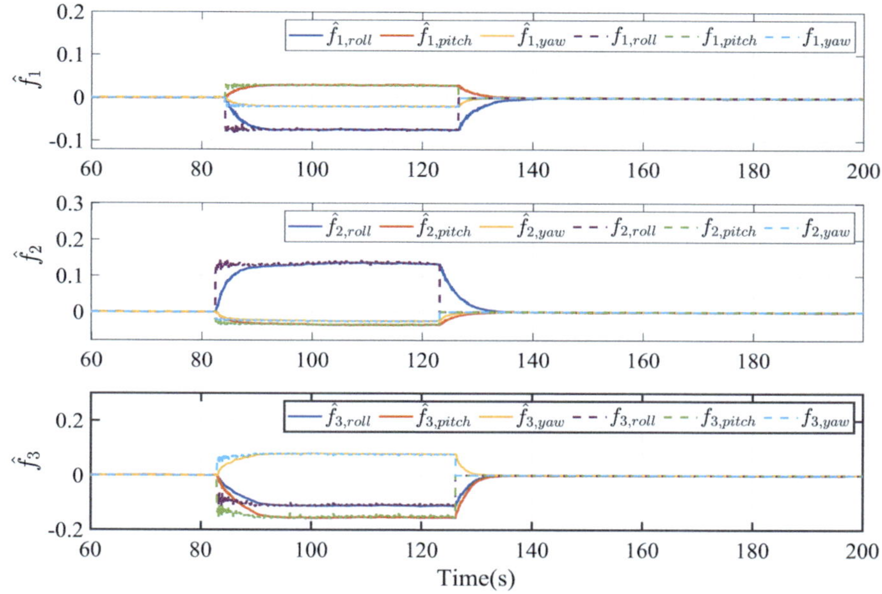


**Fig. 3.3** Estimation of faulty torque on quadrotor UAVs

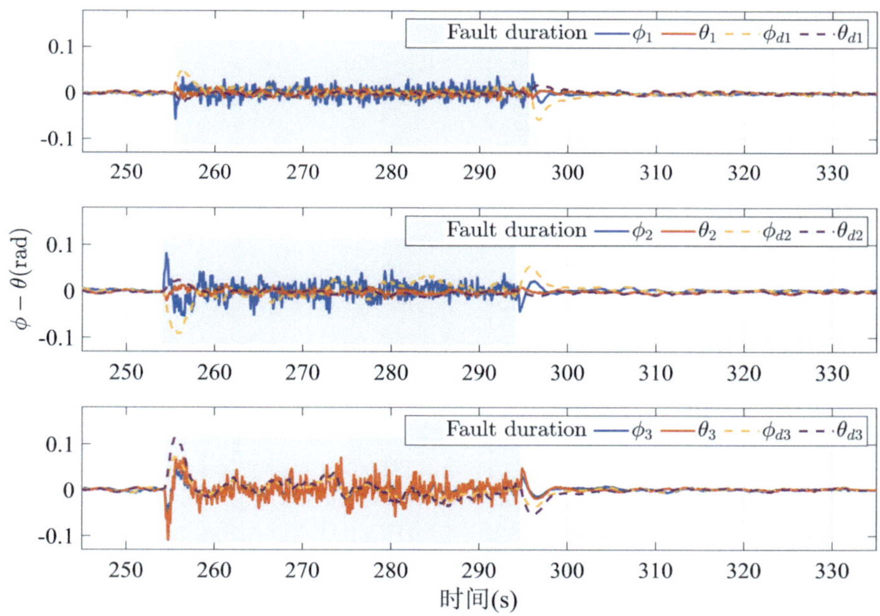


**Fig. 3.4** Roll-pitch response of quadrotor UAVs without fault-disturbance estimation observer

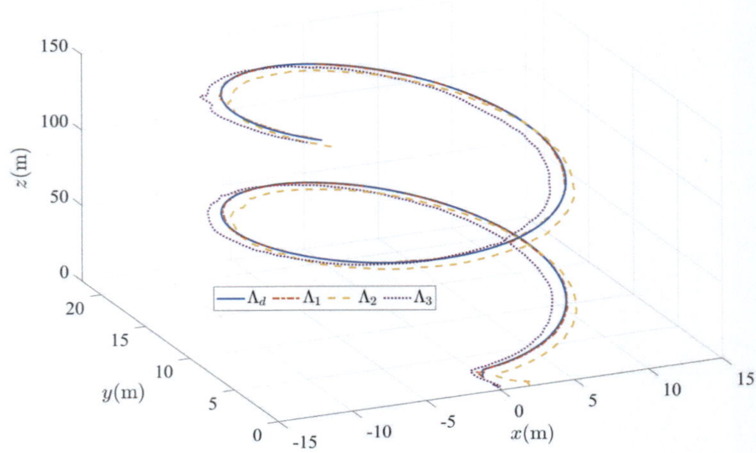


**Fig. 3.5** 3D trajectory of quadrotor UAVs

performance. Given that the observer design is independent of the controller architecture, the fixed-time observer-based FTC represents a practical augmentation to the original control scheme and a viable engineering solution for addressing actuator faults.

Building upon the foundation established by numerical simulations and hovering tests, the entire control scheme was implemented. The observer and controller for the position subsystem were integrated into the simulation companion computer at a 25 Hz frequency using the offboard protocol. (Formation controllers, requiring more computing resources but lower operational frequencies, are typically executed on companion computers during real flights.)

The formation controller parameters remained consistent with those in the numerical simulation, and the initial geometric configuration is described as $\Delta_1 = [0, 0, 0]^{\mathrm{T}}$, $\Delta_2 = [2, -2, 0]^{\mathrm{T}}$, $\Delta_3 = [0, -2, 0]^{\mathrm{T}}$. After 5 s, geometry changes into $\Delta_1 = [0, 0, 0]^{\mathrm{T}}$, $\Delta_2 = [0, -1, 0]^{\mathrm{T}}$, $\Delta_3 = [-1, 0, 0]^{\mathrm{T}}$. Trajectory of the virtual leader is designed as $p_d = [10\sin(0.04t), 10\sin(0.04t - \frac{\pi}{2}) + 12, 0.5t]^{\mathrm{T}}$.

At time instants $t = 100$ s, 170 s, and 240 s, the actuator faults characterized in the hover test are induced in UAV 2, UAV 1, and UAV 3 sequentially, with each quadrotor undergoing a 30 s fault duration. The 3D trajectories of the UAVs are depicted in Fig. 3.5. The detailed trajectory of $xyz$ axis is shown in Fig. 3.6.

The relative distance between UAVs is shown in Fig. 3.7. The estimation of the lumped disturbance in the position subsystem is shown in Fig. 3.8.

The experiments demonstrate that the distributed fixed-time sliding-mode control approach successfully achieves leader-follower formation control and collision avoidance. In the event of actuator faults, the fault-disturbance estimation observer ensures UAVs retain maneuverability: the collision tendency activates the repulsive function to keep UAVs apart, enabling rapid recovery of the desired formation.

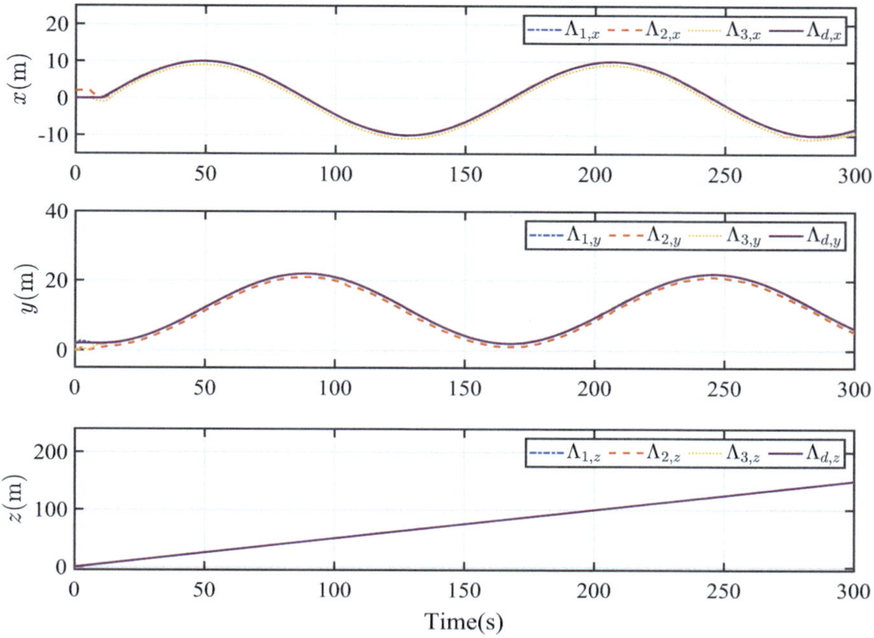


**Fig. 3.6** Detailed formation trajectory of $xyz$ axis

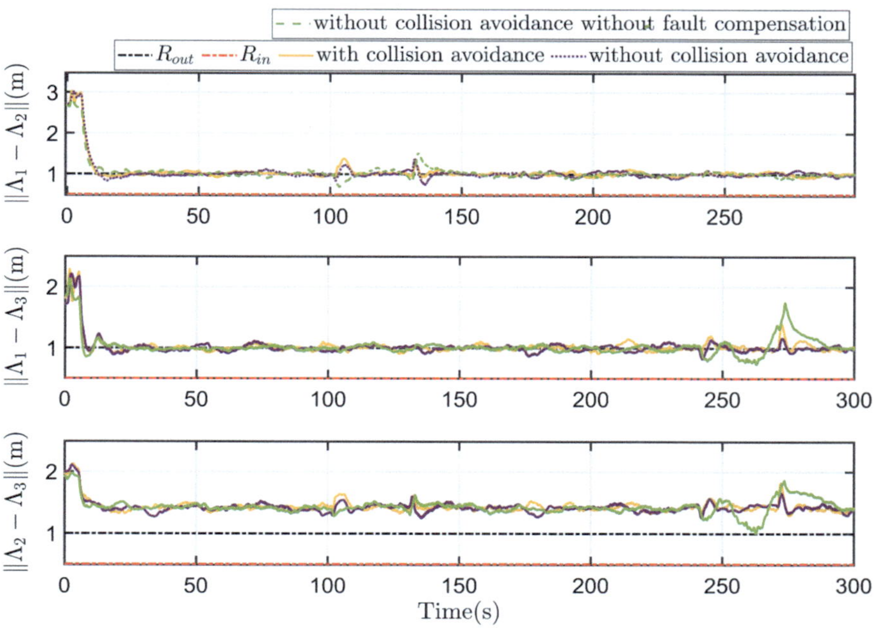


**Fig. 3.7** Relative distance between quadrotor UAVs

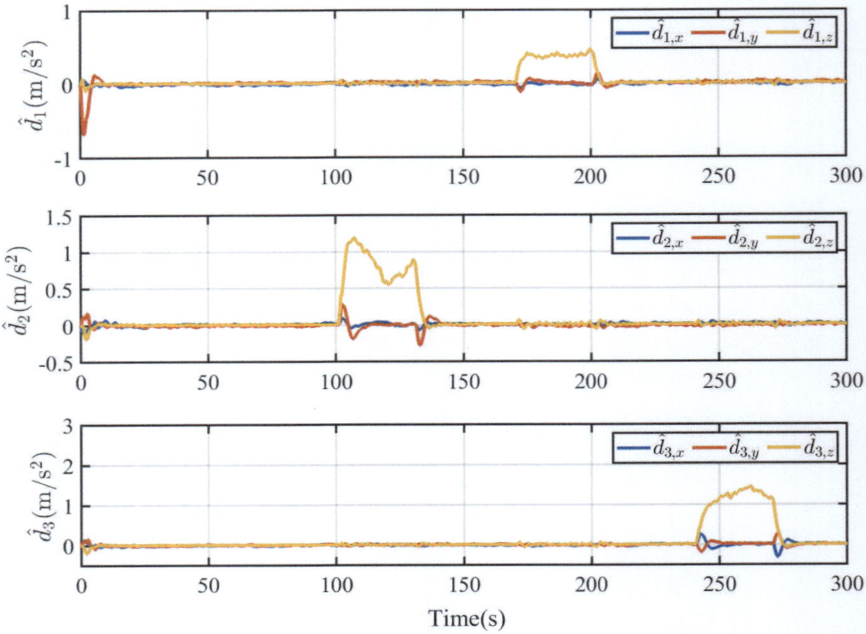


**Fig. 3.8** Estimation of lumped disturbance in position subsystem

Comparative experiments excluding the fault-disturbance observer and APF function show that without FTC and collision avoidance designs, inter-UAV distances dangerously decrease, making formation integrity hard to maintain. Statistically, with FTC and collision avoidance, quadrotors 1 & 2, 1 & 3, and 2 & 3 maintain minimum distances of 0.91 m, 0.90 m, and 1.30 m, respectively. Removing collision avoidance reduces these to 0.73, 0.83, and 1.25 m, while eliminating both FTC and collision avoidance further decreases them to 0.68, 0.74, and 1.00 m.

Additionally, the formation recovers swiftly from actuator faults because the fixed-time technique guarantees the scheme's response speed. The proposed strategy withstands noise and disturbances in the HITL-simulated environment, validating the collision-free FTC architecture for practical applications.

**Remark 3.12** The modules developed in this chapter enhance the base control framework of quadrotor UAVs, and can be independently integrated into practical platforms. For instance, considering PX4's support for offboard velocity set-point control, the formation controller alone can be deployed on an upper-layer companion computer to enable rapid formation control and collision avoidance–thereby expanding the application scope of the proposed strategy.

## 3.6 Conclusions

This chapter presents a fixed-time fault-tolerant formation control protocol integrated with collision avoidance. The fixed-time strategy ensures fast convergence for fault-disturbance estimation, velocity tracking, and formation realization. The coupled risks of actuator faults and collision tendencies are addressed via a distributed hierarchical control architecture. The protocol comprehensively considers the full dynamics of quadrotors, the practical viability of the algorithm, and the convenience for hardware implementation.

## References


1. Arul SH, Manocha D (2021) Swarmcco: probabilistic reactive collision avoidance for quadrotor swarms under uncertainty. IEEE Robot Autom Lett 6(2):2437–2444
2. Behjat A, Paul S, Chowdhury S (2019) Learning reciprocal actions for cooperative collision avoidance in quadrotor unmanned aerial vehicles. Robot Auton Syst 121:103270
3. Cheng W, Zhang K, Jiang B (2022) Distributed adaptive fixed-time fault-tolerant formation control for heterogeneous multiagent systems with a leader of unknown input. IEEE Trans Cybern 53(11):7285–7294
4. Cui G, Xu H, Chen X, Yu J (2023) Fixed-time distributed adaptive formation control for multiple QUAVs with full-state constraints. IEEE Trans Aerosp Electron Syst 59(4):4192–4206
5. Cui G, Yang W, Yu J, Li Z, Tao C (2022) Fixed-time prescribed performance adaptive trajectory tracking control for a QUAV. IEEE Trans Circuits Syst II Express Briefs 69(2):494–498
6. Dai X, Ke C, Quan Q, Cai K-Y (2021) Rflysim: Automatic test platform for UAV autopilot systems with fpga-based hardware-in-the-loop simulations. Aerosp Sci Technol 114:106727
7. Guo K, Zhang W, Zhu Y, Jia J, Yu X, Zhang Y (2022) Safety control for quadrotor UAV against ground effect and blade damage. IEEE Trans Industr Electron 69(12):13373–13383
8. Huang Y, Liu W, Li B, Yang Y, Xiao B (2020) Finite-time formation tracking control with collision avoidance for quadrotor UAVs. J Franklin Inst 357(7):4034–4058
9. Li B, Gong W, Yang Y, Xiao B (2023) Distributed fixed-time leader-following formation control for multiquadrotors with prescribed performance and collision avoidance. IEEE Trans Aerosp Electron Syst 59(5):7281–7294
10. Li Q, Wei J, Gou Q, Niu Z (2021) Distributed adaptive fixed-time formation control for second-order multi-agent systems with collision avoidance. Inf Sci 564:27–44
11. Liu Y, Dong X, Shi P, Ren Z, Liu J (2023) Distributed fault-tolerant formation tracking control for multiagent systems with multiple leaders and constrained actuators. IEEE Trans Cybern 53(6):3738–3747
12. Ma D, Xia Y, Shen G, Jiang H, Hao C (2020) Practical fixed-time disturbance rejection control for quadrotor attitude tracking. IEEE Trans Industr Electron 68(8):7274–7283
13. Mylvaganam T, Sassano M, Astolfi A (2017) A differential game approach to multi-agent collision avoidance. IEEE Trans Autom Control 62(8):4229–4235
14. Ni J, Shi P (2021) Adaptive neural network fixed-time leader–follower consensus for multiagent systems with constraints and disturbances. IEEE Trans Cybern 51(4):1835–1848
15. Omar HM (2022) Hardware-in-the-loop simulation of time-delayed anti-swing controller for quadrotor with suspended load. Appl Sci 12(3):1706
16. Polyakov A (2012) Nonlinear feedback design for fixed-time stabilization of linear control systems. IEEE Trans Autom Control 57(8):2106–2110
17. Sun J, Pu Z, Yi J, Liu Z (2020) Fixed-time control with uncertainty and measurement noise suppression for hypersonic vehicles via augmented sliding mode observers. IEEE Trans Industr Inf 16(2):1192–1203

18. Wu K, Hu J, Ding Z, Arvin F (2023) Finite-time fault-tolerant formation control for distributed multi-vehicle networks with bearing measurements. IEEE Trans Autom Sci Eng 21(2):1346–1357
19. Xu H, Cui G, Ma Q, Li Z, Hao W (2023) Fixed-time disturbance observer-based distributed formation control for multiple QUAVs. IEEE Trans Circuits Syst II Express Briefs 70(6):2181–2185
20. Xu L-X, Ma H-J, Guo D, Xie A-H, Song D-L (2020) Backstepping sliding-mode and cascade active disturbance rejection control for a quadrotor UAV. IEEE/ASME Trans Mechatron 25(6):2743–2753
21. Xu L-X, Wang Y-L, Wang X, Peng C (2024) Distributed active disturbance rejection formation tracking control for quadrotor uavs. IEEE Trans Cybern 54(8):4678–4689
22. Yang P, Zhang A, Bi W, Li M (2023) Cooperative group formation control for multiple quadrotors system with finite- and fixed-time convergence. ISA Trans 138:186–196
23. Yu Y, Chen C, Guo J, Chadli M, Xiang Z (2024) Adaptive formation control for unmanned aerial vehicles with collision avoidance and switching communication network. IEEE Trans Fuzzy Syst 32(3):1435–1445
24. Zhou D, Wang Z, Schwager M (2018) Agile coordination and assistive collision avoidance for quadrotor swarms using virtual structures. IEEE Trans Rob 34(4):916–923
25. Zhou S, Guo K, Yu X, Guo L, Xie L (2021) Fixed-time observer based safety control for a quadrotor UAV. IEEE Trans Aerosp Electron Syst 57(5):2815–2825
26. Zou A, Ruiter AH, Kumar KD (2016) Distributed finite-time velocity-free attitude coordination control for spacecraft formations. Automatica 67:46–53

# Chapter 4
# Fixed-Time FTC with Prescribed Performance for UAVs-UGVs

## 4.1 Introduction

Most existing formation control strategies for multiple unmanned systems guarantee only asymptotic convergence of the tracking error to a residual set, whose size is determined by unknown but bounded factors, including model uncertainties and external disturbances [7, 13, 22]. To guarantee prescribed transient and steady-state performance, the prescribed performance control (PPC) approach was first introduced in [1]. Subsequently, this method was extended to develop distributed controllers for multiple unmanned systems [2, 12, 18, 20]. Leveraging finite-time control theory, the work in [10] proposed a novel finite-time performance function (FTPF) for a class of nonstrict-feedback nonlinear systems, enabling finite-time tracking convergence. Building on FTPF and disturbance observer techniques, reference [15] achieved finite-time control for nonlinear systems. However, finite-time prescribed performance formation control remains unexplored for UAV-UGV systems.

During the practical operation of multiple unmanned systems, faults are likely to arise due to the intricate configuration of components such as actuators, sensors, and other system parts [8, 19, 21]. For instance, quadrotors in real-world operation inevitably encounter disturbances and actuator faults during flight, which can degrade system performance or even cause catastrophic faults [6, 17, 23, 24]. Furthermore, in UAV-UGV cooperative systems, actuator faults in one or more agents can propagate unevenly across the communication network, affecting the entire system's performance. Thus, developing fault-tolerant formation control methods for heterogeneous UAV-UGV systems is essential. Recent studies [4, 9, 11, 25] have proposed several effective FTC approaches for heterogeneous multiple unmanned systems. However, to the best of our knowledge, fixed-time fault-tolerant formation control with finite-time prescribed performance for UAV-UGV systems remains unexplored.

 
K. Zhang et al., *Fixed-Time/Prescribed-Time Fault-Tolerant Control for Multiple Unmanned Systems*, Unmanned System Technologies,
https://doi.org/10.1007/978-981-95-2913-1_4

Motivated by the above analysis, this chapter aims to address the fixed-time fault-tolerant formation control problem for UAV-UGV systems with finite-time prescribed performance. The main contributions of this work can be summarized as follows

(1) This chapter introduces, for the first time, the FTPF concept to address cooperative formation control in UAV-UGV systems. A novel fixed-time adaptive prescribed performance formation controller that integrates FTPF with fixed-time control theory is designed for UAVs-UGVs. The proposed approach simultaneously ensures that the completion of formation tasks within a predetermined time despite actuator faults and disturbances. Meanwhile, the proposed methods strict adherence to predefined performance metrics including convergence rate, overshoot, and steady-state error.

(2) From a safety perspective, timely estimation and accommodation of actuator faults coupled with effective disturbance suppression are critical requirements. This chapter addresses the problem that online update protocols for estimating unknown fault parameters and the upper bounds of lumped uncertainties. Then, a fixed-time compensation controller is designed for UAVs-UGVs. The integrated approach significantly enhances the robustness and fault tolerance of the UAV-UGV formation system.

(3) To minimize communication overhead, this chapter employs a directed graph topology and develop a fully distributed formation control protocol for the $xy$ axis that operates without global knowledge of either the network topologies or disturbance or fault characteristics. Furthermore, this chapter introduces a novel fixed-time fault-tolerant tracking controller in the $z$ axis for UAVs to enhance overall system maneuverability, which has not been considered in the existing literature.

## 4.2 Problem Formulation and Preliminaries

### *4.2.1 Problem Formulation*

Consider a heterogeneous multiagent formation system consisting of one leader labeled as 0 and $N$ ($N = N_1 + N_2$) follower agents labeled as $i = 1, 2, ..., N$ on a directed graph. Specifically, it is assumed that the number of UAVs is $N_1$ and the number of UGVs is $N_2$. The quadrotor dynamics is modeled as the UAV and a wheeled mobile robot is used as the UGV. For convenience, let $F_1 = \{1, 2, ..., N_1\}$, $F_2 = \{N_1 + 1, N_1 + 2, ..., N_1 + N_2\}$ and $F = \{1, 2, ..., N\}$. The UGV is designated as the leader agent due to: (1) its more challenging operational environment, (2) its ability to represent multiple practical scenarios, and (3) its versatile dynamics model, which can be formulated as follows [16]

$$\begin{cases} \dot{x}_0 = v_0 \\ \dot{v}_0 = u_0 \end{cases} \tag{4.1}$$

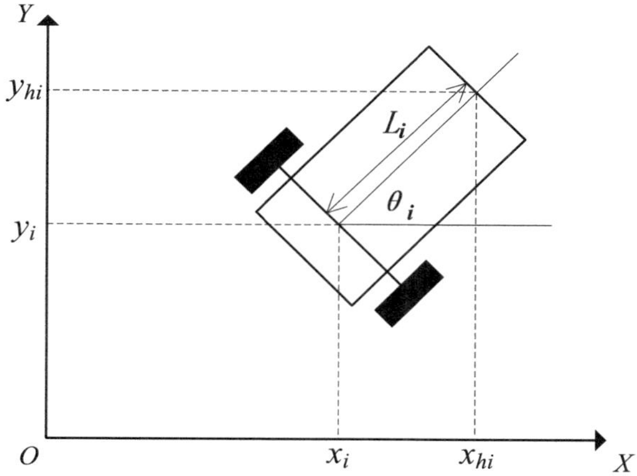


**Fig. 4.1** Schematic diagram of the $i$th UGV

where $x_0 \in R^2$,$v_0 \in R^2$, and $u_0 \in R^2$ are the position, velocity, and control input of the leader, respectively. Once the leader is selected, it can dynamically adjust the desired formation configuration to accomplish time-varying formation tasks.

Consider a differentially driven two-wheeled mobile robot as the UGV, whose simplified mechanical structure is illustrated in Fig. 4.1. The dynamic model of the $i$th UGV can be formulated as follows

$$\begin{cases} \dot{x}_i = v_i \cos\theta_i \\ \dot{y}_i = v_i \sin\theta_i \\ \dot{\theta}_i = \omega_i \end{cases} \tag{4.2}$$

where $x_i$ and $y_i$ denote the position of the $i$th UGV, $v_i$ is the linear velocity, $\omega_i$ is the angular velocity, and $\theta_i$ is the angular orientation of the $i$th robot.

As illustrated in Fig. 4.1, we designate $(x_{hi}, y_{hi})$ as reference points to account for the non-holonomic constraints, while $(x_i, y_i)$ denote the inertial position of the $i$th UGV. Since the reference point is kinematically consistent, we focus on controlling the position of $(x_{hi}, y_{hi})$. The hand point (virtual reference point) can be mathematically expressed as follows

$$\begin{cases} x_{hi} = x_i + L_i \cos\theta_i \\ y_{hi} = y_i + L_i \sin\theta_i \end{cases} \tag{4.3}$$

where $L_i$ is the distance between the reference point and inertial point of the $i$th UGV.

Then, by computing the second-order time derivative of Eq. (4.2), we obtain the following dynamic relationship

$$\begin{bmatrix} \ddot{x}_{hi} \\ \ddot{y}_{hi} \end{bmatrix} = \begin{bmatrix} \cos\theta_i & -L_i \sin\theta_i \\ \sin\theta_i & L_i \cos\theta_i \end{bmatrix} \begin{bmatrix} \dot{v}_i \\ \dot{\omega}_i \end{bmatrix} + \begin{bmatrix} -v_i\omega_i \sin\theta_i & -L_i\omega_i^2 \sin\theta_i \\ v_i\omega_i \sin\theta_i & -L_i\omega_i^2 \cos\theta_i \end{bmatrix} \tag{4.4}$$

Then, if

$$\begin{bmatrix} \dot{v}_i \\ \dot{\omega}_i \end{bmatrix} = \begin{bmatrix} \cos\theta_i & \sin\theta_i \\ -\frac{1}{L_i}\sin\theta_i & \frac{1}{L_i}\cos\theta_i \end{bmatrix} \begin{bmatrix} u_{xi} \\ u_{yi} \end{bmatrix} \tag{4.5}$$

$$\begin{bmatrix} f_{xi} \\ f_{yi} \end{bmatrix} = \begin{bmatrix} -v_i\omega_i \sin\theta_i - L_i\omega_i^2 \cos\theta_i \\ v_i\omega_i \cos\theta_i - L_i\omega_i^2 \sin\theta \end{bmatrix} \tag{4.6}$$

Hence, the $i$th UGV dynamic model can be represented as

$$\begin{bmatrix} \ddot{x}_{hi} \\ \ddot{y}_{hi} \end{bmatrix} = \begin{bmatrix} u_{xi} \\ u_{y_i} \end{bmatrix} + \begin{bmatrix} f_{xi} \\ f_{yi} \end{bmatrix} \tag{4.7}$$

where $u_{xi}$ and $u_{yi}$ represent the transformed control inputs for the reference point, while $f_{xi}$ and $f_{yi}$ denote lumped nonlinear functions treated as bounded disturbances.

For each UAV system ($i \in F_1$), the complete nonlinear dynamics relative to inertial frame $E$ are formulated as [17]

$$\begin{cases} \ddot{x}_i = (\cos\phi_i \sin\theta_i \cos\psi_i + \sin\phi_i \sin\psi_i)u_{i1}/m_i - \xi_{xi}\dot{x}_i/m_i \\ \ddot{y}_i = (\cos\phi_i \sin\theta_i \sin\psi_i - \sin\phi_i \sin\psi_i)u_{i1}/m_i - \xi_{yi}\dot{y}_i/m_i \\ \ddot{z}_i = -g + (\cos\theta_i \cos\phi_i)u_{i1}/m_i - \xi_{zi}\dot{z}_i/m_i \\ \ddot{\phi}_i = \dot{\theta}_i\dot{\psi}_i\frac{I_{yi}-I_{zi}}{I_{xi}} - \frac{I_{ri}}{I_{xi}}\dot{\theta}_i\bar{w}_i + \frac{1}{I_{xi}}u_{i2} - \frac{\xi_{\phi i}}{I_{xi}}\dot{\phi}_i \\ \ddot{\theta}_i = \dot{\phi}_i\dot{\psi}_i\frac{I_{zi}-I_{xi}}{I_{yi}} - \frac{I_{ri}}{I_{yi}}\dot{\phi}_i\bar{w}_i + \frac{1}{I_{yi}}u_{i3} - \frac{\xi_{\theta i}}{I_{yi}}\dot{\theta}_i \\ \ddot{\psi}_i = \dot{\phi}_i\dot{\theta}_i\frac{I_{xi}-I_{yi}}{I_{zi}} + \frac{1}{I_{zi}}u_{i4} - \frac{\xi_{\psi i}}{I_{zi}}\dot{\psi}_i \end{cases} \tag{4.8}$$

where $x_i$, $y_i$ and $z_i$ denote the Euclidean positions, $\phi_i$, $\theta_i$ and $\psi_i$ are the Euler angles of the $i$th UAV, $m_i$ is the mass, $g$ is the acceleration due to gravity, $I_{ri}$ represents the inertia of rotor, $\xi_{xi}, \xi_{yi}, \xi_{zi}$ and $\xi_{\phi i}, \xi_{\theta i}$ and $\xi_{\psi i}$ denote the aerodynamic damping coefficient, $\bar{w}_i = w_{i4} + w_{i3} - w_{i2} - w_{i1}$ is the overall residual angular of all propellers being considered as a bounded disturbance, $u_{i1}, u_{i2}, u_{i3}$ and $u_{i4}$ are the control inputs of the $i$th quadrotor, which are correlated with

$$\begin{cases} u_{i1} = a_i(w_{i1}^2 + w_{i2}^2 + w_{i3}^2 + w_{i4}^2) \\ u_{i2} = a_i l_i(w_{i4}^2 - w_{i2}^2) \\ u_{i3} = a_i l_i(w_{i3}^2 - w_{i1}^2) \\ u_{i4} = b_i(w_{i2}^2 + w_{i4}^2 - w_{i1}^2 - w_{i3}^2) \end{cases} \tag{4.9}$$

where $l_i$ is the interval between the rotor shaft and the centroid, $a_i$ and $b_i$ are the reverse moment coefficient and drag force coefficient of the $i$th UAV, respectively.

The formation objective for UAVs consists of: regulating the centroid position relative to the leader, and independently stabilizing the attitude subsystem. Building on Eq. (4.9), we derive the following formation sub-dynamics model

$$\begin{cases} \ddot{x}_i = u_{xi}/m_i + f_{xi} \\ \ddot{y}_i = u_{yi}/m_i + f_{yi} \\ \ddot{z}_i = u_{zi}/m_i + f_{zi} - g \end{cases} \tag{4.10}$$

where $u_{xi}$, $u_{yi}$, and $u_{zi}$, as defined in Eq. (4.11), are the virtual control inputs in three-dimensional space.

$$\begin{cases} u_{xi} = (\cos\phi_i \sin\theta_i \cos\psi_i + \sin\phi_i \sin\psi_i)u_{i1} \\ u_{yi} = (\cos\phi_i \sin\theta_i \sin\psi_i - \sin\phi_i \sin\psi_i)u_{i1} \\ u_{zi} = (\cos\theta_i \cos\phi_i)u_{i1} \end{cases} \tag{4.11}$$

and $f_{xi}$, $f_{yi}$, and $f_{zi}$ are given in Eq. (4.12).

$$\begin{cases} f_{xi} = -\xi_{xi}\dot{x}_i/m_i \\ f_{yi} = -\xi_{yi}\dot{y}_i/m_i \\ f_{zi} = -\xi_{zi}\dot{z}_i/m_i. \end{cases} \tag{4.12}$$

### 4.2.2 Preliminaries

To formalize the control algorithm design, we adopt the following notation. Let $R$ and $R^n$ stand for the set of real numbers and $n$-dimensional real vectors, respectively. During the fixed-time controller design, denote $\mathrm{sig}(x)^\alpha = |x|^\alpha \mathrm{sign}(x)$, where $x \in R$, $\alpha > 0$, and $\mathrm{sign}(x)$ is the standard signum function. For a vector $x = [x_1, x_2, ...x_n] \in R^n$, define $\mathrm{sig}(x)^\alpha = \left[|x_1|^\alpha \mathrm{sign}(x_1), |x_2|^\alpha \mathrm{sign}(x_2), ..., |x_n|^\alpha \mathrm{sign}(x_n)\right]$.

Let $G = (V, E, A)$ describe a direct graph with $N$ nodes of MASs, where $V = \{v_1, ..., v_N\}$ and $E \in \{e_{ij} \in (v_i, v_j) : v_i, v_j \in V; i \notin j\}$ denote the set of nodes and edges, respectively. And $A = \left[a_{ij}\right] \in R^{N\times N}$ represents the adjacency matrix of the graph with nonnegative weights $a_{ij}$. The weight $a_{ij} > 0$ denotes that the agent $j$ can obtain information from agent $i$, and $a_{ij} = 0$ otherwise. It is assumed that $a_{ii} = 0$. The in-degree matrix of graph $G$ is described by $D = \mathrm{diag}\{\mathrm{deg_{in}}(v_1), ..., \mathrm{deg_{in}}(v_N)\}$, where $\mathrm{deg_{in}}(v_i) = \sum_{j=1}^{N} a_{ij}$. The Laplacian matrix $L$ of graph $G$ is defined as $L = D - A$. A directed graph $G$ contains a directed spanning tree if and only if there exists a root node possessing directed paths to all other nodes in the graph. The leader-follower interconnection matrix $B = \mathrm{diag}(b_1, ..., b_N)$ encodes the connectivity between leader and followers, where $b_i > 0$ if the $i$th follower receives information directly from the leader.

**Lemma 4.1** ([5]) *Consider the nth-order integrator system*

$$\begin{cases} \dot{x}_1(t) = x_2(t) \\ \dot{x}_2(t) = x_3(t) \\ \quad \cdot \\ \quad \cdot \\ \quad \cdot \\ \dot{x}_n(t) = u(t),\ x(0) = x_0 \end{cases} \tag{4.13}$$

*where $x = [x_1, ..., x_n]^{\mathrm{T}}$ is the state vector and $u \in R$ is the control input. Let the constants $k_i > 0\,(i = 1, 2, ..., n)$ such that both the polynomial $s^n + k_n s^{n-1} + ... + k_2 s + k_1$ and $s^n + 3k_n s^{n-1} + ... + 3k_2 s + 3k_1$ are Hurwitz with respect to the Laplace operator $s$. There exists a constant $\sigma \in \left(\frac{n-2}{n-1}, 1\right)$ such that, for every $\tau \in (\sigma, 1)$, the integrator system of Eq. (4.13) can be stabilized at the zero within a fixed time by using the following feedback control law*

$$u(x) = -\sum_{i=1}^{n} k_i \left[\mathrm{sig}(x_i)^{\tau_i} + \mathrm{sig}\,(x_i) + \mathrm{sig}(x_i)^{\delta_i}\right] \tag{4.14}$$

*with parameters $\tau_i$ and $\delta_i$ satisfying, for $j = 0, 1, ..., n-1$,*

$$\tau_{n-j} = \frac{\tau}{(j+1) - j\tau},\ \delta_{n-j} = \frac{2-\tau}{j\tau - (j-1)}. \tag{4.15}$$

## 4.3 Main Results

### *4.3.1 Fixed-Time Sliding Mode Surface Design*

Due to the heterogeneous degrees of freedom between UAVs and UGVs, the formation control problem is decomposed into distributed cooperative control in the $xy$ axis for all agents and decentralized altitude tracking control for UAVs in the $z$ axis. To address practical challenges, the system dynamics in both axes are formulated to explicitly account for external disturbances and actuator faults, yielding the following compact models

$$\ddot{x}_i = A_i\,(I - \rho_i)\,u_i + \Delta_i \quad i \in F \tag{4.16}$$

and

$$\ddot{z}_i = (1 - \rho_{zi})\,u_{zi}/m_i + \Delta_{zi} - g \quad i \in F_1 \tag{4.17}$$

where $x_i \in R^2$ and $u_i \in R^2$ denote the position and control input of the $i$th follower, respectively; $\rho_i \in R^2$ is the loss efficiency factor; $\Delta_i = f_i + d_i$ is the lumped uncertainty term, where $f_i \in R^2$ is the unknown nonlinear vector function and $d_i \in R^2$ represents the external disturbance. The diagonal matrix $A_i = \text{diag}\{1/m_i, 1/m_i\}$ if and only if $i \in F_1$, otherwise $A_i = I_2$.

**Control Objective**: The objective of fixed-time cooperative formation control is to ensure all followers converge to their desired positions relative to the leader within a predetermined time interval, regardless of initial conditions. Mathematically, this requires satisfying the following relationships for any bounded initial states of the followers

$$\begin{cases} \lim\limits_{t \to T_{\max}} (x_i(t) - (x_0(t) - h_i(t))) = 0 \\ \lim\limits_{t \to T_{\max}} (z_i(t) - p_i(t)) = 0 \end{cases} \tag{4.18}$$

where $T_{\max} > 0$ denotes a predefined convergence time (independent of initial conditions). The time-varying formation in the $xy$ axis is characterized by a piecewise continuously differentiable vector $h_i(t) \in R^2$. Then the expected time-varying formation of all followers in the $xy$ axis can be described by a vector $h = [h_1, ..., h_N]^{\mathrm{T}}$. Meanwhile, $p_i(t) \in R$ represents the expected height generated by the leader trajectory.

**Remark 4.1** Due to the presence of disturbances and actuator faults, exact convergence of formation errors to zero is generally unattainable. In practice, formation tasks are considered achieved when the errors converge to a sufficiently small neighborhood of the origin, i.e., become uniformly ultimately bounded (UUB). This relaxed stability criterion is widely acceptable for most real-world applications. Motivated by this, we employ the practical fixed-time stability theory to develop a FTC scheme that guarantees UUB convergence of formation errors within a fixed settling time.

To guarantee prescribed performance tracking, we introduce the following formal definition of FTPF.

**Definition 4.1** A continuously differentiable function $\varpi(t)$ is termed a FTPF if it satisfies the following conditions

(1) $\varpi(t) > 0$;
(2) $\dot{\varpi}(t) \leq 0$;
(3) $\lim\limits_{t \to t_f} \varpi(t) = \varpi_{t_f}$, which is a predesigned positive constant;
(4) when $t \geq t_f$, $\varpi(t) = \varpi_{t_f}$ with $t_f$ being the settling time.

In accordance with Definition 4.1, the FTPF is formally characterized as follows

$$\varpi(t) = \begin{cases} \left(\varpi_0^{\kappa} - \kappa\eta t\right)^{\frac{1}{\kappa}} + \varpi_{t_f}, 0 \leq t < t_f \\ \varpi_{t_f}, t \geq t_f \end{cases} \tag{4.19}$$

where $\varpi_0$, $\varpi_{t_f}$, $\kappa$, and $\eta$ are designed positive constants and $\kappa = \frac{c}{v} \in (0, 1]$ with $c$ and $v$ are positive odd and even integer, respectively. Furthermore, $\varpi_0 + \varpi_{t_f} = \varpi(0)$ is the initial value, and $t_f = \frac{\varpi_0^{\kappa}}{\kappa\eta}$ is the settling time.

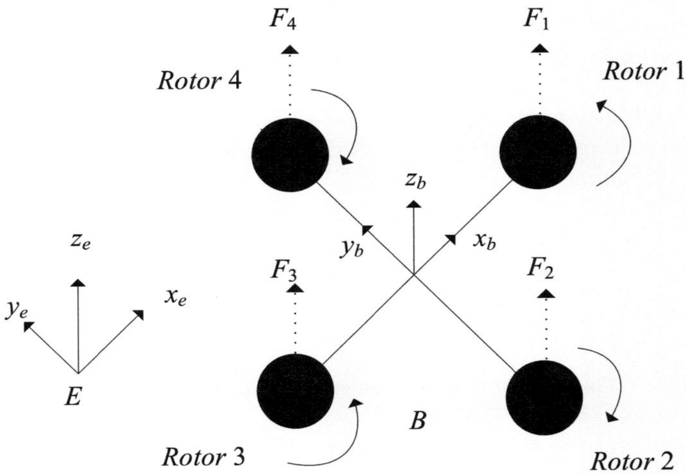


**Fig. 4.2** Schematic diagram of the $i$th UAV

**Remark 4.2** Although the existing exponential prescribed performance function $\eta(t) = (\eta_0 - \eta_\infty) e^{-a_0 t} + \eta_\infty$ (see [2, 18]) can regulate tracking errors, they lack finite-time convergence guarantees. In contrast, as demonstrated in Figs. 4.1 and 4.2, the proposed FTPF ensures that tracking errors converge to predefined bounds within a fixed time interval, making it particularly suitable for time-critical applications (Figs. 4.3 and 4.4).

To guarantee achievement of the prescribed performance objectives, the following constraint must be satisfied

$$-\varpi_i(t) < e_i(t) < \varpi_i(t) \tag{4.20}$$

where $e_i(t)$ denotes the tracking error. Let $h_i \in R^2$ $(i \in F)$ denote the expected formation in the X-Y axis, then the distributed formation error of the $i$th follower can be designed as

$$e_i(t) = \sum_{j=1}^{N} a_{ij} \left((x_i + h_i) - \left(x_j + h_j\right)\right) + b_i \left(x_i + h_i - x_0\right). \tag{4.21}$$

**Remark 4.3** Let $\xi_i(t)$ represents the formation error of the $i$th agent, and $\xi_i(t) = x_i(t) + h_i(t) - x_0(t)$. Moreover, from the definitions of $\xi_i(t)$ and $e_i(t)$, one can get that $e(t) = ((L + B) \otimes I_2)\, \xi(t)$, where $\xi(t) = \left[\xi_1^{\mathrm{T}}(t), ..., \xi_N^{\mathrm{T}}(t)\right]^{\mathrm{T}}$ and $e(t) = \left[e_1^{\mathrm{T}}(t), ..., e_N^{\mathrm{T}}(t)\right]^{\mathrm{T}}$. Assuming the formation topologies contain a directed spanning tree, Lemma 1 implies that the fixed-time formation objective is achievable when $\lim_{t \to T_{\max}} e(t) \to 0$.

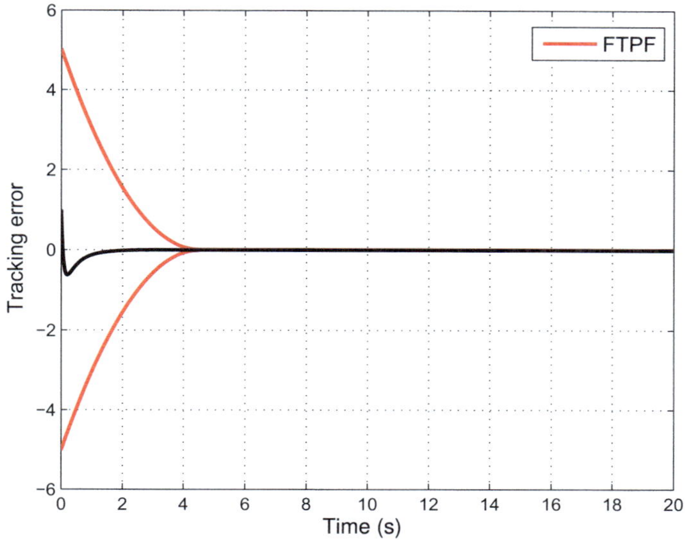


**Fig. 4.3** Tracking error with FTPF

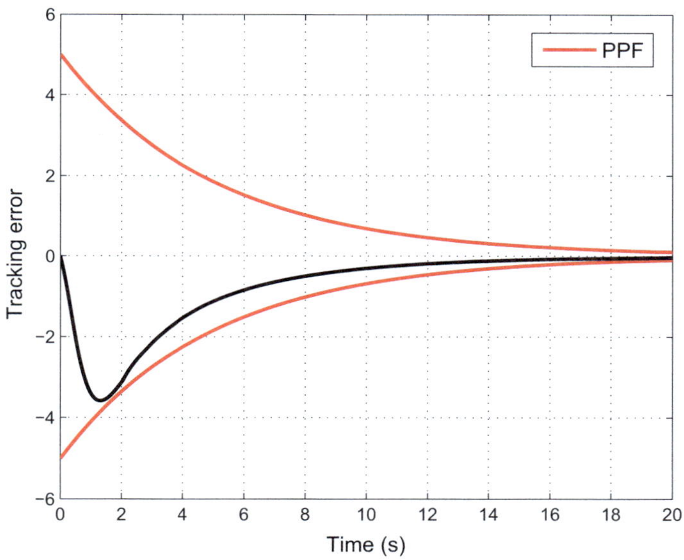


**Fig. 4.4** Tracking error with prescribed performance function (PPF) proposed in [18]

To ensure the FTPF properties are maintained, we implement the following state transformation

$$e_i(t) = \Omega\left(\nu_i(t)\right)\varpi_i(t) \tag{4.22}$$

in which $\nu_i(t)$ is the new transformed error and $\Omega\left(\nu_i(t)\right)$ is a smooth and strictly increasing function defined as below

$$\Omega\left(\nu_i(t)\right) = \frac{2}{\pi}\arctan\left(\nu_i(t)\right). \tag{4.23}$$

Then, the time derivative of $e_i(t)$ is obtained as follows

$$\dot{e}_i(t) = \frac{2}{\pi}\frac{\dot{\nu}_i(t)}{1+\nu_i^2(t)}\varpi_i(t) + \frac{2}{\pi}\arctan(\nu_i(t))\dot{\varpi}_i(t) \tag{4.24}$$

which yields

$$\dot{\nu}_i(t) = \frac{\pi\left(1+\nu_i^2(t)\right)}{2\varpi_i(t)}\left(\dot{e}_i(t) - \frac{2}{\pi}\arctan(\nu_i(t))\dot{\varpi}_i(t)\right). \tag{4.25}$$

Let

$$r_i(t) = \frac{\pi\left(1+\nu_i^2(t)\right)}{2\varpi_i(t)} \tag{4.26}$$

and

$$\gamma_i(t) = \frac{2}{\pi}\arctan(\nu_i(t))\dot{\varpi}_i(t). \tag{4.27}$$

Consequently, the formation error dynamics under transformation can be expressed as

$$\dot{\nu}_i(t) = r_i(t)\dot{e}_i(t) - r_i(t)\gamma_i(t). \tag{4.28}$$

In this section, we analyze the formation system in the absence of lumped uncertainties $\Delta_i$. Building on Lemma 4.1, we propose an integral-type fixed-time sliding mode surface to ensure stability for the heterogeneous UAV-UGV system. The switching manifold for the $i$th follower is designed as follows

$$s_i = \dot{\nu}_i - \int_{t_0}^{t} u_{ie}\mathrm{d}r \tag{4.29}$$

where

$$u_{ie} = -\sum_{j=1}^{n} k_{i,j}\left[\left(\mathrm{sig}(\sigma_{i,j})^{\tau_{i,j}} + \mathrm{sig}\left(\sigma_{i,j}\right) + \mathrm{sig}(\sigma_{i,j})^{\delta_{i,j}}\right)\right] \tag{4.30}$$

with $t_0$ is the initial time, and $\sigma_{i,1} = \nu_i$, $\sigma_{i,2} = \dot{\nu}_i$, and $n = 2$. The parameters $k_{i,j}$, $\tau_{i,j}$, $\delta_{i,j}$ are selected as given in Lemma 4.1.

**Theorem 4.1** *If the formation states of the UAGs-UGVs of Eqs. (4.16) and (4.17) can arrive the constructed sliding mode surface of Eqs. (4.29) and (4.30), then the expected formation tracking can be realized within a fixed time.*

***Proof*** Upon reaching the sliding surface, i.e., $s_i = 0$ and $\dot{s}_i = 0$, the switching manifold dynamics can be equivalently expressed as the following error system.

$$\begin{cases} \dot{\sigma}_{i,1} = \sigma_{i,2} \\ \dot{\sigma}_{i,2} = -\sum\limits_{j=1}^{n} k_{i,j}\left[\left(\mathrm{sig}(\sigma_{i,j})^{\tau_{i,j}} + \mathrm{sig}\left(\sigma_{i,j}\right) + \mathrm{sig}(\sigma_{i,j})^{\delta_{i,j}}\right)\right]. \end{cases} \tag{4.31}$$

Based on Lemma 4.1, it can be readily deduced that the formation error system described by Eq. (4.31) is globally fixed-time stable at the origin. Specifically, there exists a constant settling time $T_s$ such that the transformed formation tracking error $\nu_i$ converges to zero for all $t \geq T_s$. Furthermore, as analyzed in Remark 4.4, the formation errors $e_i(t)$ also exhibit fixed-time convergence with guaranteed finite-time prescribed performance characteristics.

**Remark 4.4** Since the inequality $-\varpi_i(t) < \Omega(\nu_i(t))\varpi_i(t) < \varpi_i(t)$ holds, the formation error $e_i(t)$ always satisfies $-\varpi_i(t) < e_i(t) < \varpi_i(t)$. This implies that if $\nu_i(t)$ remains bounded for all $t \geq 0$, the prescribed transient and steady-state performance can be guaranteed. Consequently, the finite-time prescribed performance formation control problem reduces to stabilizing the transformed error $\nu_i(t)$.

**Remark 4.5** Similar to the transformations applied in the $xy$-plane, the error transformation and fixed-time sliding mode surface can also be constructed for the $z$-axis. However, it should be noted that the tracking error in this case is redefined as $e_i(t) = z_i(t) - h_{z_i}(t)$, where $h_{z_i}(t)$ where $h_{z_i}(t)$ represents the desired trajectory of the $i$th agent ($i \in F_1$) along the $z$-axis.

### *4.3.2 Fixed-Time Formation Tracking Controller Design*

This section presents a fixed-time formation controller based on the simplified dynamical model introduced in [16]. This simplified model serves as a foundation for understanding the controller design methodology that will be developed in the following section. The UAV-UGV system dynamics can be simplified as follows

$$\ddot{x}_i = A_i u_i, \quad i \in F \tag{4.32}$$

and

$$\ddot{z}_i = u_{zi}/m_i + g, \quad i \in F_1 \tag{4.33}$$

To address the fixed-time fault-tolerant formation control problem under consideration, we introduce the following key assumption.

**Assumption 4.1** The directed graph $G$ contains a spanning tree with the leader agent as its root node.

**Theorem 4.2** *For the UAV-UGV system described by Eqs. (4.32) and (4.33) under Assumption 4.1, the proposed control scheme comprising: (i) fixed-time sliding mode manifolds Eqs. (4.29) and (4.30), (ii) distributed formation controllers Eqs. (4.34)–(4.35), and (iii) decentralized tracking controllers Eqs. (4.36)–(4.37) guarantees fixed-time convergence to the desired 3D formation configuration, with all formation errors globally asymptotically stabilizing to zero within a predetermined settling time.*

$$u_i = \left(r_i \sum_{j=1}^{N} (a_{ij} + b_i) A_i\right)^{-1} \left[-\dot{r}_i \dot{e}_i - r_i \sum_{j=1}^{N} (a_{ij} + b_i) \ddot{h}_i + r_i \sum_{j=1}^{N} a_{ij} A_j u_j + r_i \sum_{j=1}^{N} a_{ij} \ddot{h}_j + r_i b_i u_0 + \dot{r}_i \gamma_i + r_i \dot{\gamma}_i + u_{ie}\right] + u_{im} \tag{4.34}$$

*and*

$$u_{im} = \left(r_i \sum_{j=1}^{N} (a_{ij} + b_i) A_i\right)^{-1} \left[-\chi_i \mathrm{sig}(s_i)^{\alpha} - \lambda_i \mathrm{sig}(s_i)^{\beta} - \varepsilon_i \mathrm{sign}(s_i)\right] \tag{4.35}$$

*where* $\alpha$, $\beta$, $\chi_i$, $\lambda_i$, $k_i$ *are positive controller parameters, and* $0 < \alpha < 1$, $\beta > 1$, *and the decentralized control scheme in the Z axis is designed as follows*

$$u_{zi} = m_i \left(\ddot{h}_{zi} - g\right) + \frac{m_i}{r_{zi}} \left(\dot{r}_{zi} \gamma_{zi} + r_{zi} \dot{\gamma}_{zi} + u_{ie} - \dot{r}_{zi} \dot{e}_{zi}\right) + u_{im} \tag{4.36}$$

*and*

$$u_{im} = m_i * \left[-\phi_i \mathrm{sig}(s_{zi})^{p} - c_i \mathrm{sig}(s_{zi})^{q} - d_i \mathrm{sign}(s_{zi})\right] / r_{zi} \tag{4.37}$$

*where* $\phi_i > 0$, $c_i > 0$, $d_i > 0$ *are designed controller parameters, and* $0 < p < 1$, $q > 1$.

***Proof*** For the dynamical system described by Eq. (4.32), consider the candidate Lyapunov function $V_1 = \frac{1}{2} \sum_{i=1}^{N} {s_i}^T s_i$. The time derivative of $V_1$ along the system trajectories can be obtained by differentiating the switching manifold expression in Eq. (4.29) yielding

$$
\begin{aligned}
\dot{s}_i &= \dot{r}_i \dot{e}_i + r_i \ddot{e}_i - \dot{r}_i \gamma_i - r_i \dot{\gamma}_i - u_{ie} \\
&= \dot{r}_i \dot{e}_i + r_i \sum_{j=1}^{N} \left(a_{ij} + b_i\right) A_i u_i + r_i \sum_{j=1}^{N} \left(a_{ij} + b_i\right) \ddot{h}_i \\
&\quad - r_i \sum_{j=1}^{N} a_{ij} A_j u_j - r_i \sum_{j=1}^{N} a_{ij} \ddot{h}_j - r_i b_i u_0 - \dot{r}_i \gamma_i - r_i \dot{\gamma}_i - u_{ie}
\end{aligned}
\tag{4.38}
$$

Then, taking time derivative of $V_1$ along Eq. (4.38) with Eqs. (4.34) and (4.35) gives

$$
\begin{aligned}
\dot{V}_1 &= \sum_{i=1}^{N} {s_i}^{\mathrm{T}} \dot{s}_i \\
&= \sum_{i=1}^{N} {s_i}^{\mathrm{T}} \left[-\chi_i \mathrm{sig}(s_i)^{\alpha} - \lambda_i \mathrm{sig}(s_i)^{\beta} - k_i \mathrm{sign}(s_i)\right] \\
&\le -\chi_{\min} \sum_{i=1}^{N} \|s_i\|^{\alpha+1} - \lambda_{\min} \sum_{i=1}^{N} \|s_i\|^{\beta+1} \\
&\le -\chi_{\min} (2V_1)^{\frac{\alpha+1}{2}} - \lambda_{\min} (2N)^{\frac{1-\beta}{2}} (2V_1)^{\frac{\beta+1}{2}}
\end{aligned}
\tag{4.39}
$$

where $\chi_{\min} = \min\{\chi_1, \ldots, \chi_N\}$, and $\lambda_{\min} = \min\{\lambda_1, \ldots, \lambda_N\}$. Lemma 2.2 establishes the second inequality, which, through application of fixed-time stability theory, proves the global fixed-time stability of the equilibrium $V_1 = 0$. Consequently, all transformed formation errors are guaranteed to reach the sliding manifold within a bounded convergence time which is independent of initial conditions.

$$
T_{c1} \le T_{1\max} = \frac{\pi \mu_1}{2\sqrt{2^{\frac{\alpha+2}{2}} N^{\frac{1-\beta}{2}} \chi_{\min} \lambda_{\min}}}. \tag{4.40}
$$

For the vertical formation tracking control problem, considering the altitude error $e_i = z_i - h_{zi}$ for each agent, the time derivative of the sliding surface variable $s_i$ can be derived as follows

$$
\dot{s}_i = \dot{r}_i \dot{e}_i + r_i \left(u_{zi}/m_i + g - \ddot{h}_{zi}\right) - \dot{r}_i \gamma_i - r_i \dot{\gamma}_i - u_{ie}. \tag{4.41}
$$

Using the controller law of Eqs. (4.36)–(4.37) and following the same procedures as Eqs. (4.39)–(4.40), it can be proved that there exists a fixed time $T_{2\max} = \frac{\pi \mu_2}{2\sqrt{2^{\frac{p+q+2}{2}} \phi_{\min} c_{\min}}}$, where $\phi_{\min} = \min\left\{\phi_1, \ldots, \phi_{N_1}\right\}$, and $c_{\min} = \min\left\{c_1, \ldots, c_{N_1}\right\}$ such that the transformed tracking error in $z$ axis converges to zero within fixed time $T_{2\max}$. For the sake of space, the detailed proof is omitted herein.

### 4.3.3 Distributed Adaptive Fixed-Time FTC Design

Practical unmanned vehicle systems are inherently susceptible to both external disturbances and actuator faults. This necessitates the development of a robust FTC scheme for the UAV-UGV system described by Eqs. (4.16) and (4.17). However, the control design faces two key challenges:

(1) The upper bound of the lumped uncertainties $\varDelta_i$ is typically unknown.

(2) The effectiveness factor $\rho_i$ for partial actuator faults is difficult to quantify precisely.

To address these challenges, we employ an adaptive control architecture that automatically estimates these unknown parameters in real-time.

**Assumption 4.2** The efficiency factor $\rho_i$ is unknown, possibly time-varying, but bounded, and there exists an unknown constant $\rho_{if}$ such that $0 \leq \rho_{if} \leq \rho_i < 1$.

**Assumption 4.3** The lumped uncertainty $\varDelta_i$ is bounded such that $\|\varDelta_i\| < \varDelta_m$, where $\varDelta_m$ is an unknown positive scalar.

**Theorem 4.3** *The UAV-UGV system dynamics governed by Eqs. (4.16) and (4.17) incorporate fixed-time sliding surfaces Eqs. (4.29) and (4.30) under actuator fault and uncertainty conditions specified in Assumptions 4.2 and 4.3. Through a hierarchical control architecture combining: (i) distributed adaptive fault-compensation formation protocols Eqs. (4.42)–(4.44) with online parameter estimation Eqs. (4.45)–(4.46) for planar coordination, and (ii) decentralized adaptive fault-tolerant tracking controllers Eqs. (4.47)–(4.49) featuring real-time adaptation mechanisms Eqs. (4.50)–(4.51) for altitude regulation, all transformed formation errors demonstrate uniform ultimate boundedness convergence to a compact set within a predetermined settling time in three-dimensional space.*

$$
\begin{aligned}
u_i = &\left(r_i \sum_{j=1}^{N} \left(a_{ij} + b_i\right) A_i \left(I - \hat{\rho}_i\right)\right)^{-1} \left[-r_i \sum_{j=1}^{N} \left(a_{ij} + b_i\right) \ddot{h}_i \right. \\
&\left. - \dot{r}_i \dot{e} + r_i \sum_{j=1}^{N} a_{ij} A_j u_j + r_i \sum_{j=1}^{N} a_{ij} \ddot{h}_j + r_i b_i u_0 + \dot{r}_i \gamma_i + r_i \dot{\gamma}_i + u_{ie}\right] + u_{ip} + u_{iq}
\end{aligned} \tag{4.42}
$$

$$
u_{ip} = \left(r_i \sum_{j=1}^{N} \left(a_{ij} + b_i\right) A_i \left(I - \hat{\rho}_i\right)\right)^{-1} \left[-t_i \operatorname{sig}(s_i)^{\upsilon} - l_i \operatorname{sig}(s_i)^{\kappa} - r_i \hat{\zeta}_i \frac{s_i}{\|s_i\|}\right] \tag{4.43}
$$

$$
\begin{aligned}
u_{iq} = &\left(r_i \sum_{j=1}^{N} \left(a_{ij} + b_i\right) A_i \left(I - \hat{\rho}_i\right)\right)^{-1} \left(-\frac{4^{\kappa} s_i \varGamma_i}{32\|s_i\|^2 \varLambda_i} \left\|\hat{\theta}_i\right\|^{2\kappa} - \frac{2^{\upsilon} s_i \varGamma_i}{16\|s_i\|^2 \varLambda_i} \left\|\hat{\theta}_i\right\|^{2\upsilon} \right. \\
&\left. - \frac{4^{\kappa} s_i \mathrm{M}_i}{32\|s_i\|^2 \mathrm{N}_i} \left\|\hat{\zeta}_i\right\|^{2\kappa} - \frac{2^{\upsilon} s_i \mathrm{M}_i}{16\|s_i\|^2 \mathrm{N}_i} \left\|\hat{\zeta}_i\right\|^{2\upsilon}\right)
\end{aligned} \tag{4.44}
$$

*with updated by*

$$\dot{\hat{\theta}}_i = -\Gamma_i\hat{\theta}_i - \Lambda_i \mathrm{diag}(u_i)^{\mathrm{T}} A_i{}^{\mathrm{T}} \left( \sum_{j=1}^{N} (a_{ij} + b_i) \right)^{\mathrm{T}} r_i{}^{\mathrm{T}} s_i \tag{4.45}$$

$$\dot{\hat{\zeta}}_i = -\mathrm{M}_i\hat{\zeta}_i + \mathrm{N}_i r_i{}^{\mathrm{T}} \|s_i\| \tag{4.46}$$

*where* $\kappa$, $\upsilon$, $t_i$, $l_i$, $\Gamma_i$, $\Lambda_i$, $\mathrm{M}_i$, *and* $\mathrm{N}_i$ *are designed positive parameters, and* $0 < \upsilon < 1$, $\kappa > 1$. *The adaptive items* $\hat{\zeta}_i$ *and* $\hat{\theta}_i$ *are the estimated value of the parameters* $\zeta_i$ *and* $\theta_i$. *Moreover,* $\hat{\theta}_i$ *fits for* $\hat{\rho}_i u_i = \mathrm{diag}(u_i)\hat{\theta}_i$ *where* $\hat{\rho}_i$ *is the estimation of the fault factor* $\rho_i$, *and* $\zeta_i$ *will be defined later. The decentralized tracking controller in the z axis is designed as follows*

$$u_{zi} = \left( \frac{r_{zi}}{m_i} (1 - \hat{\rho}_{zi}) \right)^{-1} \left( \dot{r}_{zi}\gamma_{zi} + r_{zi}\dot{\gamma}_{zi} + u_{ie} - \dot{r}_{zi}\dot{e}_{zi} + r_{zi}\ddot{h}_{zi} - r_{zi}g \right) + u_{is} + u_{it} \tag{4.47}$$

$$u_{is} = \left( \frac{r_{zi}}{m_i} (1 - \hat{\rho}_{zi}) \right)^{-1} \left[ -t_{zi}\mathrm{sig}(s_{zi})^{\kappa_z} - l_{zi}\mathrm{sig}(s_{zi})^{\upsilon_z} - r_{zi}\hat{\zeta}_{zi} \frac{s_{zi}}{\|s_{zi}\|} \right] \tag{4.48}$$

$$\begin{aligned} u_{it} = {} & \left( \tfrac{r_{zi}}{m_i} (1 - \hat{\rho}_{zi}) \right)^{-1} * \left( -\frac{4^{\kappa_z} s_{zi} \Gamma_{zi}}{32\|s_{zi}\|^2 \Lambda_{zi}} \left\| \hat{\theta}_{zi} \right\|^{2\kappa_z} - \frac{2^{\upsilon_z} s_{zi} \Gamma_{zi}}{16\|s_{zi}\|^2 \Lambda_{zi}} \left\| \hat{\theta}_{zi} \right\|^{2\upsilon_z} \right. \\ & \left. - \frac{4^{\kappa_z} s_{zi} \mathrm{M}_{zi}}{32\|s_{zi}\|^2 \mathrm{N}_{zi}} \left\| \hat{\zeta}_{zi} \right\|^{2\kappa_z} - \frac{2^{\upsilon_z} s_{zi} \mathrm{M}_{zi}}{16\|s_{zi}\|^2 \mathrm{N}_{zi}} \left\| \hat{\zeta}_{zi} \right\|^{2\upsilon_z} \right) \end{aligned} \tag{4.49}$$

*with updated by*

$$\dot{\hat{\theta}}_{zi} = -\Gamma_{zi}\hat{\theta}_{zi} - \Lambda_{zi}\mathrm{diag}(u_{zi})^{\mathrm{T}} r_{zi}{}^{\mathrm{T}} s_{zi}/m_i \tag{4.50}$$

$$\dot{\hat{\zeta}}_{zi} = -\mathrm{M}_{zi}\hat{\zeta}_{zi} + \mathrm{N}_{zi} r_{zi}{}^{\mathrm{T}} \|s_{zi}\| \tag{4.51}$$

*where* $\kappa_z$, $\upsilon_z$, $\Gamma_{zi}$, $\Lambda_{zi}$, $\mathrm{M}_{zi}$, *and* $\mathrm{N}_{zi}$ *have the same definitions as* $\kappa$, $\upsilon$, $\Gamma_i$, $\Lambda_i$, $\mathrm{M}_i$, *and* $\mathrm{N}_i$, *respectively.*

***Proof*** For system of Eq. (4.16), the Lyapunov candidate function is defined as $V_2 = \sum_{i=1}^{N} V_{2i}$, where

$$V_{2i} = \frac{1}{2} s_i{}^{\mathrm{T}} s_i + \frac{1}{2\Lambda_i} \tilde{\theta}_i^{\mathrm{T}} \tilde{\theta}_i + \frac{1}{2\mathrm{N}_i} \tilde{\zeta}_i^{\mathrm{T}} \tilde{\zeta}_i \tag{4.52}$$

where $\tilde{\theta}_i = \theta_i - \hat{\theta}_i$, and $\tilde{\zeta}_i = \zeta_i - \hat{\zeta}_i$.

Firstly, the derivative of the sliding surface can be obtained

$$\dot{s}_i = \dot{r}_i\dot{e}_i + r_i\sum_{j=1}^{N}\left(a_{ij}+b_i\right)A_i\left(I-\rho_i\right)u_i + r_i\sum_{j=1}^{N}a_{ij}\left(\Delta_i-\Delta_j\right) - r_i\sum_{j=1}^{N}a_{ij}A_ju_j + r_i\sum_{j=1}^{N}\left(a_{ij}+b_i\right)\ddot{h}_i - r_i\sum_{j=1}^{N}a_{ij}\ddot{h}_j + r_ib_i\Delta_i - r_ib_iu_0 - \dot{r}_i\gamma_i - r_i\dot{\gamma}_i - u_{ie} \tag{4.53}$$

From Assumption 4.3, the following inequality is obtained

$$\sum_{j=1}^{N}a_{ij}\left(\Delta_i-\Delta_j\right) + b_i\Delta_i \le \sum_{j=1}^{N}a_{ij}\left(\Delta_m+\Delta_m\right) + b_i\Delta_m = \left(2d_i+b_i\right)\Delta_m \tag{4.54}$$

Then, define the upper bound of the lumped uncertainty as $\zeta_i = \left(2d_i+b_i\right)\Delta_m \ge 0$. Taking the derivative of Eq. (4.52) has

$$\begin{aligned}
\dot{V}_{2i} = {} & {s_i}^{\mathrm{T}}\dot{s}_i - \frac{1}{\Lambda_i}\tilde{\theta}_i^{\mathrm{T}}\dot{\hat{\theta}}_i - \frac{1}{\mathrm{N}_i}\tilde{\zeta}_i^{\mathrm{T}}\dot{\hat{\zeta}}_i \\
& - r_i\sum_{j=1}^{N}\left(a_{ij}+b_i\right)A_i\tilde{\rho}_iu_i + r_i\zeta_i + r_i\sum_{j=1}^{N}\left(a_{ij}+b_i\right)\ddot{h}_i \\
\le {} & {s_i}^{\mathrm{T}}\Bigg(\dot{r}_i\dot{e}_i + r_i\sum_{j=1}^{N}\left(a_{ij}+b_i\right)A_i\left(I-\hat{\rho}_i\right)u_i \\
& - r_i\sum_{j=1}^{N}a_{ij}A_ju_j - r_i\sum_{j=1}^{N}a_{ij}\ddot{h}_j - r_ib_iu_0 - \dot{r}_i\gamma_i - r_i\dot{\gamma}_i \\
& - u_{ie} - \frac{1}{\Lambda_i}\tilde{\theta}_i^{\mathrm{T}}\dot{\hat{\theta}}_i - \frac{1}{\mathrm{N}_i}\tilde{\zeta}_i^{\mathrm{T}}\dot{\hat{\zeta}}_i\Bigg)
\end{aligned} \tag{4.55}$$

By implementing the adaptive controller from Eqs. (4.42)–(4.44) in Eq. (4.55) and applying the adaptive laws specified in Eqs. (4.45) and (4.46), we obtain

$$
\begin{aligned}
\dot{V}_{2i} \le\ & {s_i}^{\mathrm{T}}\left(-r_i\sum_{j=1}^{N}\left(a_{ij}+b_i\right)A_i\tilde{\rho}_i u_i + r_i\zeta_i - t_i\operatorname{sig}(s_i)^{\kappa} - l_i\operatorname{sig}(s_i)^{\upsilon} - r_i\hat{\zeta}_i\frac{s_i}{\|s_i\|}\right.\\
& \left. -\frac{4^{\kappa}s_i\Gamma_i}{32\|s_i\|^2\Lambda_i}\left\|\hat{\theta}_i\right\|^{2\kappa} - \frac{2^{\upsilon}s_i\Gamma_i}{16\|s_i\|^2\Lambda_i}\left\|\hat{\theta}_i\right\|^{2\upsilon} - \frac{4^{\kappa}s_i\mathrm{M}_i}{32\|s_i\|^2\mathrm{N}_i}\left\|\hat{\zeta}_i\right\|^{2\kappa} - \frac{2^{\upsilon}s_i\mathrm{M}_i}{16\|s_i\|^2\mathrm{N}_i}\left\|\hat{\zeta}_i\right\|^{2\upsilon}\right)\\
& -\frac{1}{\Lambda_i}\tilde{\theta}_i^{\mathrm{T}}\dot{\hat{\theta}}_i - \frac{1}{\mathrm{N}_i}\tilde{\zeta}_i^{\mathrm{T}}\dot{\hat{\zeta}}_i\\
\le\ & -t_i{s_i}^{\mathrm{T}}\operatorname{sig}(s_i)^{\kappa} - l_i{s_i}^{\mathrm{T}}\operatorname{sig}(s_i)^{\upsilon} + r_i\tilde{\zeta}_i\|s_i\| - {s_i}^{\mathrm{T}}r_i\sum_{j=1}^{N}\left(a_{ij}+b_i\right)A_i\tilde{\rho}_i u_i\\
& -\frac{4^{\kappa}\Gamma_i}{32\Lambda_i}\left\|\hat{\theta}_i\right\|^{2\kappa} - \frac{2^{\upsilon}\Gamma_i}{16\Lambda_i}\left\|\hat{\theta}_i\right\|^{2\upsilon} - \frac{4^{\kappa}\mathrm{M}_i}{32\mathrm{N}_i}\left\|\hat{\zeta}_i\right\|^{2\kappa} - \frac{2^{\upsilon}\mathrm{M}_i}{16\mathrm{N}_i}\left\|\hat{\zeta}_i\right\|^{2\upsilon}\\
& -\frac{1}{\Lambda_i}\tilde{\theta}_i^{\mathrm{T}}\dot{\hat{\theta}}_i - \frac{1}{\mathrm{N}_i}\tilde{\zeta}_i^{\mathrm{T}}\dot{\hat{\zeta}}_i\\
\le\ & -t_i{s_i}^{\mathrm{T}}\operatorname{sig}(s_i)^{\kappa} - l_i{s_i}^{\mathrm{T}}\operatorname{sig}(s_i)^{\upsilon} - \frac{4^{\kappa}\Gamma_i}{32\Lambda_i}\left\|\hat{\theta}_i\right\|^{2\kappa} - \frac{2^{\upsilon}\Gamma_i}{16\Lambda_i}\left\|\hat{\theta}_i\right\|^{2\upsilon}\\
& -\frac{4^{\kappa}\mathrm{M}_i}{32\mathrm{N}_i}\left\|\hat{\zeta}_i\right\|^{2\kappa} - \frac{2^{\upsilon}\mathrm{M}_i}{16\mathrm{N}_i}\left\|\hat{\zeta}_i\right\|^{2\upsilon} + \frac{\Gamma_i}{\Lambda_i}\tilde{\theta}_i^{\mathrm{T}}\hat{\theta}_i + \frac{\mathrm{M}_i}{\mathrm{N}_i}\tilde{\zeta}_i^{\mathrm{T}}\hat{\zeta}_i\\
\le\ & -2^{\frac{1-\kappa}{2}}t_i\|s_i\|^{\kappa+1} - l_i\|s_i\|^{\upsilon+1} + \frac{\Gamma_i}{\Lambda_i}\left(\tilde{\theta}_i^{T}\theta_i - \left\|\tilde{\theta}_i\right\|^2\right) + \frac{\mathrm{M}_i}{\mathrm{N}_i}*\left(\tilde{\zeta}_i^{T}\zeta_i - \left\|\tilde{\zeta}_i\right\|^2\right)\\
& -\frac{4^{\kappa}\Gamma_i}{32\Lambda_i}\left\|\hat{\theta}_i\right\|^{2\kappa} - \frac{2^{\upsilon}\Gamma_i}{16\Lambda_i}\left\|\hat{\theta}_i\right\|^{2\upsilon} - \frac{4^{\kappa}\mathrm{M}_i}{32\mathrm{N}_i}\left\|\hat{\zeta}_i\right\|^{2\kappa} - \frac{2^{\upsilon}\mathrm{M}_i}{16\mathrm{N}_i}\left\|\hat{\zeta}_i\right\|^{2\upsilon}\\
\le\ & -2^{\frac{1-\kappa}{2}}t_i\|s_i\|^{\kappa+1} - l_i\|s_i\|^{\upsilon+1} - \frac{4^{\kappa}s_i\Gamma_i}{32\|s_i\|^2\Lambda_i}\left\|\hat{\theta}_i\right\|^{2\kappa} - \frac{2^{\upsilon}\Gamma_i}{16\Lambda_i}\left\|\hat{\theta}_i\right\|^{2\upsilon}\\
& -\frac{4^{\kappa}\mathrm{M}_i}{32\mathrm{N}_i}\left\|\hat{\zeta}_i\right\|^{2\kappa} - \frac{2^{\upsilon}\mathrm{M}_i}{16\mathrm{N}_i}\left\|\hat{\zeta}_i\right\|^{2\upsilon} + \frac{\Gamma_i}{2\Lambda_i}\left(\|\theta_i\|^2 - \left\|\tilde{\theta}_i\right\|^2\right) + \frac{\mathrm{M}_i}{2\mathrm{N}_i}\left(\|\zeta_i\|^2 - \left\|\tilde{\zeta}_i\right\|^2\right)\\
\le\ & -2^{\frac{1-\kappa}{2}}t_i\|s_i\|^{\kappa+1} - l_i\|s_i\|^{\upsilon+1} - \frac{4^{\kappa}\Gamma_i}{32\Lambda_i}\left\|\hat{\theta}_i\right\|^{2\kappa} - \frac{2^{\upsilon}\Gamma_i}{16\Lambda_i}\left\|\hat{\theta}_i\right\|^{2\upsilon} - \frac{4^{\kappa}\mathrm{M}_i}{32\mathrm{N}_i}\left\|\hat{\zeta}_i\right\|^{2\kappa}\\
& -\frac{2^{\upsilon}\mathrm{M}_i}{16\mathrm{N}_i}\left\|\hat{\zeta}_i\right\|^{2\upsilon} - \frac{\Gamma_i}{4\Lambda_i}\left(\left\|\tilde{\theta}_i\right\| - \frac{1}{2}\left\|\tilde{\theta}_i\right\|^{\kappa}\right)^2 - \frac{\Gamma_i}{4\Lambda_i}\left(\left\|\tilde{\theta}_i\right\| - \frac{1}{2}\left\|\tilde{\theta}_i\right\|^{\upsilon}\right)^2\\
& -\frac{\mathrm{M}_i}{4\mathrm{N}_i}\left(\left\|\tilde{\zeta}_i\right\| - \frac{1}{2}\left\|\tilde{\zeta}_i\right\|^{\kappa}\right)^2 - \frac{\mathrm{M}_i}{4\mathrm{N}_i}\left(\left\|\tilde{\zeta}_i\right\| - \frac{1}{2}\left\|\tilde{\zeta}_i\right\|^{\upsilon}\right)^2\\
& -\frac{\Gamma_i}{4\Lambda_i}\left\|\tilde{\theta}_i\right\|^{\kappa+1} - \frac{\Gamma_i}{4\Lambda_i}\left\|\tilde{\theta}_i\right\|^{\upsilon+1} - \frac{\mathrm{M}_i}{4\mathrm{N}_i}\left\|\tilde{\zeta}_i\right\|^{\kappa+1} - \frac{\mathrm{M}_i}{4\mathrm{N}_i}\left\|\tilde{\zeta}_i\right\|^{\upsilon+1}\\
& +\frac{\Gamma_i}{16\Lambda_i}\left\|\tilde{\theta}_i\right\|^{2\kappa} + \frac{\Gamma_i}{16\Lambda_i}\left\|\tilde{\theta}_i\right\|^{2\upsilon} + \frac{\mathrm{M}_i}{16\mathrm{N}_i}\left\|\tilde{\zeta}_i\right\|^{2\kappa} + \frac{\mathrm{M}_i}{16\mathrm{N}_i}\left\|\tilde{\zeta}_i\right\|^{2\upsilon}\\
& +\frac{\Gamma_i}{2\Lambda_i}\|\theta_i\|^2 + \frac{\mathrm{M}_i}{2\mathrm{N}_i}\|\zeta_i\|^2
\end{aligned}
\tag{4.56}
$$

Then, one has

$$
\begin{aligned}
\dot{V}_{2i} \leq & -2^{\frac{1-\kappa}{2}} t_i \|s_i\|^{\kappa+1} - l_i \|s_i\|^{\upsilon+1} - \frac{4^{\kappa} \Gamma_i}{32 \Lambda_i} \left\|\hat{\theta}_i\right\|^{2\kappa} \\
& - \frac{2^{\upsilon} \Gamma_i}{16 \Lambda_i} \left\|\hat{\theta}_i\right\|^{2\upsilon} - \frac{4^{\kappa} \mathrm{M}_i}{32 \mathrm{N}_i} \left\|\hat{\zeta}_i\right\|^{2\kappa} - \frac{2^{\upsilon} \mathrm{M}_i}{16 \mathrm{N}_i} \left\|\hat{\zeta}_i\right\|^{2\upsilon} \\
& - \frac{\Gamma_i}{4 \Lambda_i} \left( \left\|\tilde{\theta}_i\right\| - \frac{1}{2} \left\|\tilde{\theta}_i\right\|^{\kappa} \right)^2 - \frac{\Gamma_i}{4 \Lambda_i} \left( \left\|\tilde{\theta}_i\right\| - \frac{1}{2} \left\|\tilde{\theta}_i\right\|^{\upsilon} \right)^2 \\
& - \frac{\mathrm{M}_i}{4 \mathrm{N}_i} \left( \left\|\tilde{\zeta}_i\right\| - \frac{1}{2} \left\|\tilde{\zeta}_i\right\|^{\kappa} \right)^2 - \frac{\mathrm{M}_i}{4 \mathrm{N}_i} \left( \left\|\tilde{\zeta}_i\right\| - \frac{1}{2} \left\|\tilde{\zeta}_i\right\|^{\upsilon} \right)^2 \\
& - \frac{\Gamma_i}{4 \Lambda_i} \left\|\tilde{\theta}_i\right\|^{\kappa+1} - \frac{\Gamma_i}{4 \Lambda_i} \left\|\tilde{\theta}_i\right\|^{\upsilon+1} - \frac{\mathrm{M}_i}{4 \mathrm{N}_i} \left\|\tilde{\zeta}_i\right\|^{\kappa+1} - \frac{\mathrm{M}_i}{4 \mathrm{N}_i} \left\|\tilde{\zeta}_i\right\|^{\upsilon+1} \\
& + \frac{4^{\kappa} \Gamma_i}{32 \Lambda_i} \left( \|\theta_i\|^{2\kappa} + \left\|\hat{\theta}_i\right\|^{2\kappa} \right) + \frac{2^{\upsilon} \Gamma_i}{16 \Lambda_i} \left( \|\theta_i\|^{2\upsilon} + \left\|\hat{\theta}_i\right\|^{2\upsilon} \right) \\
& + \frac{4^{\kappa} \mathrm{M}_i}{32 \mathrm{N}_i} \left( \|\zeta_i\|^{2\kappa} + \left\|\hat{\zeta}_i\right\|^{2\kappa} \right) + \frac{2^{\upsilon} \mathrm{M}_i}{16 \mathrm{N}_i} \left( \|\zeta_i\|^{2\upsilon} + \left\|\hat{\zeta}_i\right\|^{2\upsilon} \right) \\
& + \frac{\Gamma_i}{2 \Lambda_i} \|\theta_i\|^2 + \frac{\mathrm{M}_i}{2 \mathrm{N}_i} \|\zeta_i\|^2 \\
\leq & -2^{\frac{1-\kappa}{2}} t_i \|s_i\|^{\kappa+1} - l_i \|s_i\|^{\upsilon+1} - \frac{\Gamma_i}{4 \Lambda_i} \left( \left\|\tilde{\theta}_i\right\| - \frac{1}{2} \left\|\tilde{\theta}_i\right\|^{\kappa} \right)^2 \\
& - \frac{\Gamma_i}{4 \Lambda_i} \left( \left\|\tilde{\theta}_i\right\| - \frac{1}{2} \left\|\tilde{\theta}_i\right\|^{\upsilon} \right)^2 - \frac{\mathrm{M}_i}{4 \mathrm{N}_i} \left( \left\|\tilde{\zeta}_i\right\| - \frac{1}{2} \left\|\tilde{\zeta}_i\right\|^{\kappa} \right)^2 \\
& - \frac{\mathrm{M}_i}{4 \mathrm{N}_i} \left( \left\|\tilde{\zeta}_i\right\| - \frac{1}{2} \left\|\tilde{\zeta}_i\right\|^{\upsilon} \right)^2 - \frac{\Gamma_i}{4 \Lambda_i} \left\|\tilde{\theta}_i\right\|^{\kappa+1} - \frac{\Gamma_i}{4 \Lambda_i} \left\|\tilde{\theta}_i\right\|^{\upsilon+1} \\
& - \frac{\mathrm{M}_i}{4 \mathrm{N}_i} \left\|\tilde{\zeta}_i\right\|^{\kappa+1} - \frac{\mathrm{M}_i}{4 \mathrm{N}_i} \left\|\tilde{\zeta}_i\right\|^{\upsilon+1} + \frac{4^{\kappa} \Gamma_i}{32 \Lambda_i} \|\theta_i\|^{2\kappa} + \frac{2^{\upsilon} \Gamma_i}{16 \Lambda_i} \|\theta_i\|^{2\upsilon} \\
& + \frac{4^{\kappa} \mathrm{M}_i}{32 \mathrm{N}_i} \|\zeta_i\|^{2\kappa} + \frac{2^{\upsilon} \mathrm{M}_i}{16 \mathrm{N}_i} \|\zeta_i\|^{2\upsilon} + \frac{\Gamma_i}{2 \Lambda_i} \|\theta_i\|^2 + \frac{\mathrm{M}_i}{2 \mathrm{N}_i} \|\zeta_i\|^2 \\
\leq & -2^{\frac{1-\kappa}{2}} t_i \left( \|s_i\|^2 \right)^{\frac{\kappa+1}{2}} - \frac{\Gamma_i}{4 \Lambda_i} \left( \left\|\tilde{\theta}_i\right\|^2 \right)^{\frac{\kappa+1}{2}} - \frac{\mathrm{M}_i}{4 \mathrm{N}_i} \left( \left\|\tilde{\zeta}_i\right\|^2 \right)^{\frac{\kappa+1}{2}} \\
& - l_i \left( \|s_i\|^2 \right)^{\frac{\upsilon+1}{2}} - \frac{\Gamma_i}{4 \Lambda_i} \left( \left\|\tilde{\theta}_i\right\|^2 \right)^{\frac{\upsilon+1}{2}} - \frac{\mathrm{M}_i}{4 \mathrm{N}_i} \left( \left\|\tilde{\zeta}_i\right\|^2 \right)^{\frac{\upsilon+1}{2}} \\
& + \frac{4^{\kappa} \Gamma_i}{32 \Lambda_i} \|\theta_i\|^{2\kappa} + \frac{2^{\upsilon} \Gamma_i}{16 \Lambda_i} \|\theta_i\|^{2\upsilon} + \frac{\Gamma_i}{2 \Lambda_i} \|\theta_i\|^2 \\
& + \frac{4^{\kappa} \mathrm{M}_i}{32 \mathrm{N}_i} \|\zeta_i\|^{2\kappa} + \frac{2^{\upsilon} \mathrm{M}_i}{16 \mathrm{N}_i} \|\zeta_i\|^{2\upsilon} + \frac{\mathrm{M}_i}{2 \mathrm{N}_i} \|\zeta_i\|^2 \\
\leq & -3^{\frac{1-\kappa}{2}} \tau_{1i} V_{2i}^{\frac{\kappa+1}{2}} - \tau_{2i} V_{2i}^{\frac{\upsilon+1}{2}} + \vartheta_i
\end{aligned}
\tag{4.57}
$$

where

$$\tau_{1i} = \min\{2t_i, \frac{\Gamma_i}{4\Lambda_i} * (2\Lambda_i)^{\frac{\kappa+1}{2}}, \frac{\mathrm{M}_i}{4\mathrm{N}_i} * (2\mathrm{N}_i)^{\frac{\kappa+1}{2}}\},$$

$$\tau_{2i} = \min\{l_i * 2^{\frac{\upsilon+1}{2}}, \frac{\Gamma_i}{4\Lambda_i} * (2\Lambda_i)^{\frac{\upsilon+1}{2}}, \frac{\mathrm{M}_i}{4\mathrm{N}_i} * (2\mathrm{N}_i)^{\frac{\upsilon+1}{2}}\},$$

$$\vartheta_i = \frac{4^{\kappa}\Gamma_i}{32\Lambda_i}\|\theta_i\|^{2\kappa} + \frac{2^{\upsilon}\Gamma_i}{16\Lambda_i}\|\theta_i\|^{2\upsilon} + \frac{\Gamma_i}{2\Lambda_i}\|\theta_i\|^2 + \frac{4^{\kappa}\mathrm{M}_i}{32\mathrm{N}_i}\|\zeta_i\|^{2\kappa} + \frac{2^{\upsilon}\mathrm{M}_i}{16\mathrm{N}_i}\|\zeta_i\|^{2\upsilon} + \frac{\mathrm{M}_i}{2\mathrm{N}_i}\|\zeta_i\|^2.$$

Then, the following conclusion can be derived that

$$\begin{aligned} \dot{V}_2 &\le -\sum_{i=1}^{N} 3^{\frac{1-\kappa}{2}} \tau_{1i} V_{2i}^{\frac{\kappa+1}{2}} - \sum_{i=1}^{N} \tau_{2i} V_{2i}^{\frac{\upsilon+1}{2}} + \sum_{i=1}^{N} \vartheta_i \\ &\le -3^{\frac{1-\kappa}{2}} \tau_1 N^{\frac{1-\kappa}{2}} V_2^{\frac{\kappa+1}{2}} - \tau_2 V_2^{\frac{\kappa+1}{2}} + \vartheta \end{aligned} \tag{4.58}$$

where $\tau_1 = \min\{\tau_{1i}\}$, $\tau_2 = \min\{\tau_{2i}\}$, and $\vartheta = \sum_{i=1}^{N} \vartheta_i$. Applying Lemma 3.2 establishes that the UAV-UGV formation errors achieve practical fixed-time stability within settling time $T_3$. Through analogous analysis, the tracking control system Eq. (4.52) is similarly shown to attain practical fixed-time convergence within time $T_4$. As practical fixed-time stability constitutes an extension of conventional fixed-time theory, and since all equilibrium points within the residual set satisfy global stability conditions, the proposed controller guarantees globally stable operation. This completes the proof.

**Remark 4.6** Theorem 4.3 demonstrates that the transformed errors achieve practical fixed-time stability despite actuator faults and lumped uncertainties. Analysis reveals that the bounded set $\Omega$ of $V_2$ is determined by: (i) fault characteristics, (ii) uncertainty bounds, and (iii) control parameters $\kappa$, $\upsilon$, $t_i$, $l_i$, $\Gamma_i$, $\Lambda_i$, $\mathrm{M}_i$, and $\mathrm{N}_i$ $(i \in F)$. Through proper parameter selection based on formation tracking requirements, these errors can be regulated to converge within an arbitrarily small neighborhood of the origin.

**Remark 4.7** This work presents a novel decentralized adaptive framework that fundamentally differs from traditional observer-based fault-tolerant methods. The proposed solution exploits the structural property $\hat{\rho}_i u_i = \mathrm{diag}(u_i)\,\hat{\theta}_i$ to derive distributed estimation mechanisms, where: (i) Eqs. (4.45) and (4.46) provide real-time identification of both planar $\rho_i$ and vertical $\rho_{zi}$ actuator efficiency factors, while (ii) Eqs. (4.50) and (4.51) enable autonomous bounding of aggregated disturbances $\zeta_i$ and $\zeta_{zi}$ in respective spatial dimensions. Crucially, the architecture eliminates dependencies on prior knowledge of disturbance characteristics, fault parameters, or global network topologies, thereby achieving enhanced robustness and fault tolerance through completely self-adjusting control loops for heterogeneous UAV-UGV teams.

**Remark 4.8** Simulation results demonstrate that the proposed adaptive algorithm achieves a minimum execution period of 0.0006 s, corresponding to a maximum supported harmonic signal frequency of 830 Hz. Given the typical operational bandwidth requirements of UAV-UGV formations are significantly below this threshold, the algorithm proves fully implementable on the target platform. The computational complexity exhibits substantial reduction once the system reaches the sliding surface. While the convergence time is tunable, engineering practice requires careful consideration of the trade-off between rapid convergence and control input magnitude. The optimal compromise between time complexity and control performance should be determined based on the specific computational capabilities of the deployed industrial computer.

## 4.4 Simulation Experiments

The effectiveness of the proposed control algorithms is verified through simulation studies of a heterogeneous multi-agent system comprising five nodes: a single leader (agent 0) coordinating with two UGVs (agents 1, 3) and two UAVs (agents 2, 4). System parameters for the follower agents follow the experimental configurations documented in [3]. As illustrated in Fig. 4.5, the inter-agent communication network adopts a directed graph structure with the following connectivity weights

$$A=\begin{bmatrix} 0 & 0 & 0 & 0.1 \\ 1 & 0 & 0 & 0 \\ 0 & 1 & 0 & 0 \\ 0 & 0 & 0.7 & 0 \end{bmatrix},\ B=\begin{bmatrix} 1 & 0 & 0 & 0 \\ 0 & 0 & 0 & 0 \\ 0 & 0 & 0 & 0 \\ 0 & 0 & 0 & 0 \end{bmatrix}.$$

In addition, the initial states of followers have been set to $x_0(0) = [0, 0]^{\mathrm{T}}, x_1(0) = [0.5, 0.4]^{\mathrm{T}}$, $x_2(0) = [0.1, -0.4, 0]^{\mathrm{T}}$, $x_3(0) = [0.35, 0.2]^{\mathrm{T}}$, $x_4(0) = [-0.3, 0.6, 0]^{\mathrm{T}}$. The positions of the HMAS need to maintain a predesigned formation configuration, which is chosen as $h_1 = [0, -0.5]^{\mathrm{T}}$, $h_2 = [-0.5, -0.25, 2t]^{\mathrm{T}}$, $h_3 = [0.5, 0]^{\mathrm{T}}$, and $h_4 = [1, 1, 3t]^{\mathrm{T}}$. The control input of the leader is given as $u_0 = [0.5\sin(0.01\mathrm{t}), 0.01]^{\mathrm{T}}$. Moreover, the chosen parameters of FTPF are as follows: $\varpi_0 = 5, \kappa = 0.5$,

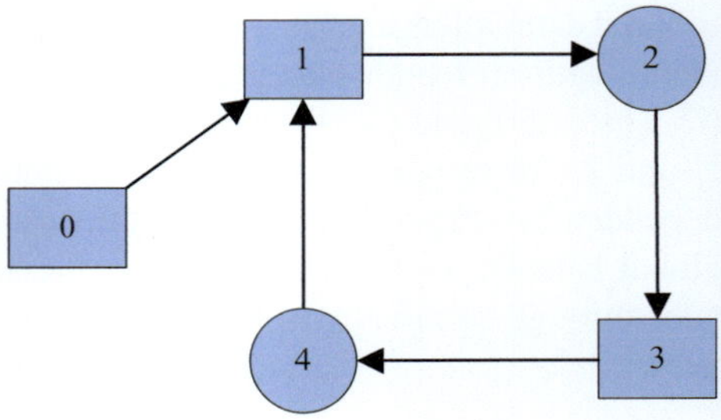


**Fig. 4.5** Directed communication topology graph

$\eta = 1, \varpi_{t_f} = 0.008$. It should be pointed that the settling time of the prescribed performance can be obtained by $t_f = \frac{\varpi_0^\kappa}{\kappa\eta}$, and then we can obtain $t_f = 4.4721s$.

### 4.4.1 Simulation for Distributed Fixed-Time Formation Tracking Controller

We first implement the nominal control scheme from Theorem 4.2, which does not account for actuator faults or system uncertainties. The controller parameters are configured as: $\alpha = p = 0.4$, $\beta = q = 1.5$, $\chi_i = \phi_i = 10$, $\lambda_i = c_i = 5$, $\varepsilon_i = d_i = 0.1$, $k_{i,1} = 9$, and $k_{i,2} = 5$. These values were systematically determined through iterative tuning guided by the theoretical framework established in the supporting lemma and theorem, ensuring satisfactory closed-loop performance while maintaining stability margins.

Figures 4.6, 4.7 and 4.8 demonstrate the formation tracking errors for the heterogeneous UAV-UGV system. The results confirm that: (1) all follower errors $e_i$ remain strictly within their prescribed finite-time performance boundaries, validating the theoretical predictions; and (2) the fixed-time controllers successfully drive all tracking errors to zero within a guaranteed convergence time that is shorter than the FTPF settling time. These outcomes collectively verify the efficacy of the proposed fixed-time control architecture in achieving prescribed performance specifications under nominal operating conditions.

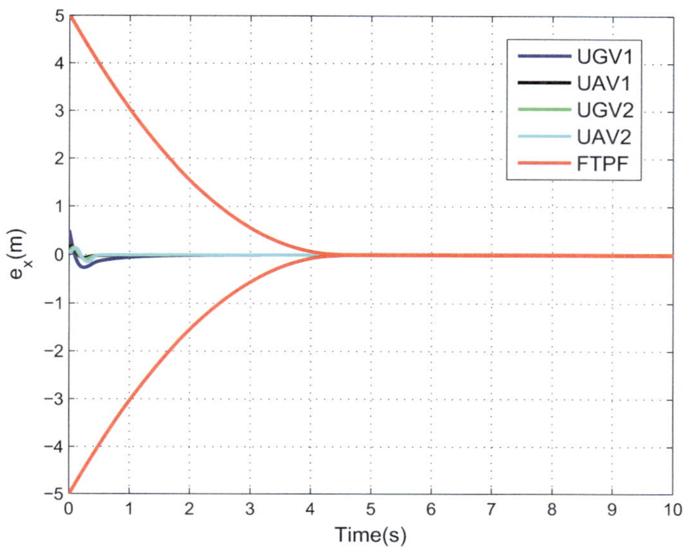


**Fig. 4.6** Formation tracking error in the $x$ axis

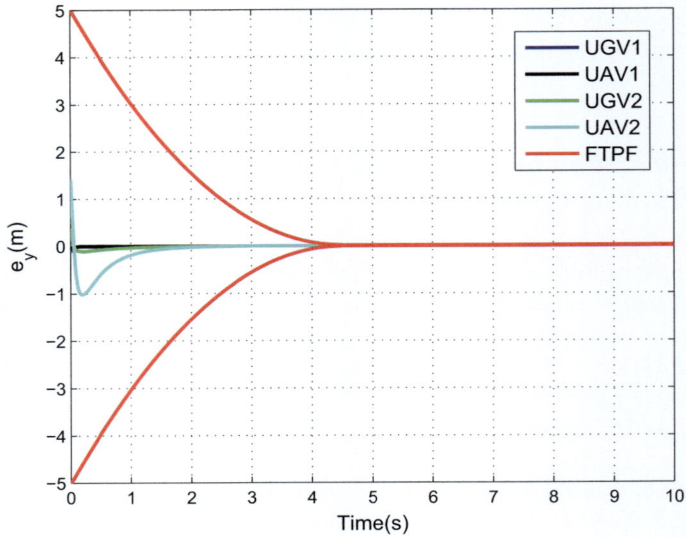


**Fig. 4.7** Formation tracking error in the $y$ axis

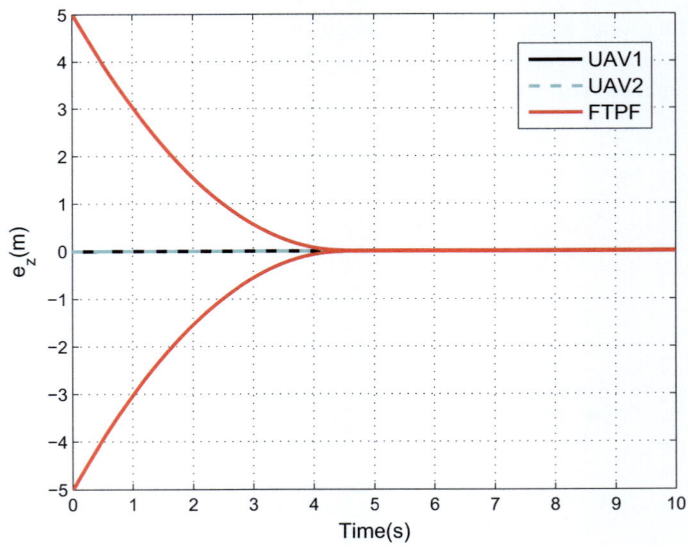


**Fig. 4.8** Formation tracking error in the $z$ axis

### 4.4.2 *Simulation for Distributed Adaptive Fixed-Time Fault-Tolerant Controller*

To evaluate the proposed adaptive algorithm performance under adverse conditions, we introduce partial actuator faults in agents 1 and 2 while maintaining normal operation in the remaining agents. The specific fault characteristics are modeled as follows

$$\rho_{1x} = \begin{cases} 0, t \leq 4.5s \\ 0.2{+}0.015 * (t - 4.5), 4.5s < t \leq 14.5s \\ 0.35, t > 14.5s, \end{cases}$$

$$\rho_{1y} = \begin{cases} 0, t \leq 4.5s \\ 0.4, 4.5s < t \leq 12.5s \\ 0.5, t > 12.5s, \end{cases} \rho_{2x} = \begin{cases} 0, t \leq 7s \\ 0.25, t > 7s, \end{cases}$$

$$\rho_{2y} = \begin{cases} 0, t \leq 7s \\ 0.35, t > 7s, \end{cases} \rho_{2z} = \begin{cases} 0, t \leq 12s \\ 0.2, t > 12s. \end{cases}$$

The adaptive control parameters are configured as: $\kappa = 1.5$, $\upsilon = 0.4$, $t_i = 30$, $l_i = 20$, $k_{i,1} = 9$, and $k_{i,2} = 5$. Initial conditions for all adaptive estimation laws are set to zero, with individual follower gains specified as: $\Gamma_1 = 0.46$, $\Lambda_1 = 0.003$, $\Gamma_2 = 1.16$, $\Lambda_2 = 0.007$, $\Gamma_3 = 0.5$, $\Lambda_3 = 0.001$, $\Gamma_4 = 1.2$, $\Lambda_4 = 0.001$, and $\Gamma_{z2} = \Gamma_{z4} = 0.85$, $\Lambda_{z2} = \Lambda_{z4} = 0.0007$.

As demonstrated in Figs. 4.9, 4.10 and 4.11, the UGVs-UAVs system maintains accurate formation trajectory tracking with predefined performance characteristics despite actuator faults and lumped uncertainty. This performance is achieved through the effective online estimation of these disturbances via adaptive laws, where the transient behavior of tracking errors is governed by the FTPF. These results confirm the system practical capability to execute formations under adverse conditions. Figure 4.12 illustrates the convergence of fault factors, demonstrating the effectiveness of the adaptive laws in estimating actuator efficiency loss parameters. The evolution of the adaptive parameter $\hat{\zeta}_i$ is further detailed in Fig. 4.13. Finally, Fig. 4.14 presents the 3-D formation trajectories of all five agents, visually validating the system collaborative performance.

## 4.5 Conclusions

This chapter addresses the fixed-time fault-tolerant formation control problem for heterogeneous UAVs-UGVs systems with finite-time prescribed performance. To ensure transient performance, a FTPF is introduced. Leveraging bi-limit homogeneity and fixed-time stability theory, a fixed-time sliding mode surface is designed, followed by the development of a fixed-time formation controller under normal

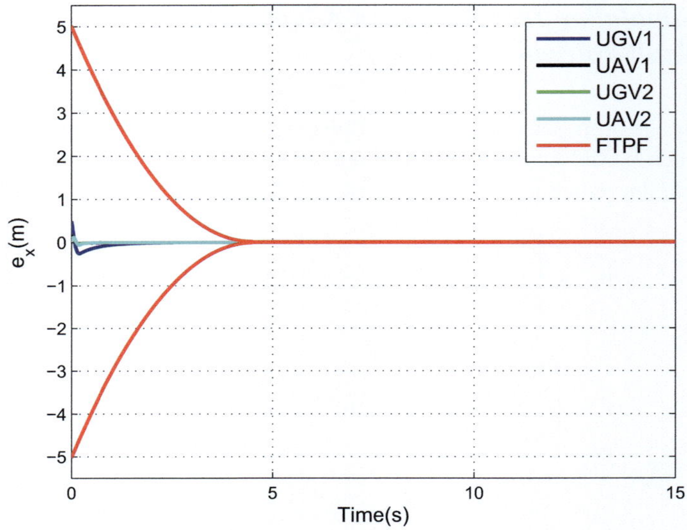


**Fig. 4.9** Formation tracking error in the $x$ axis

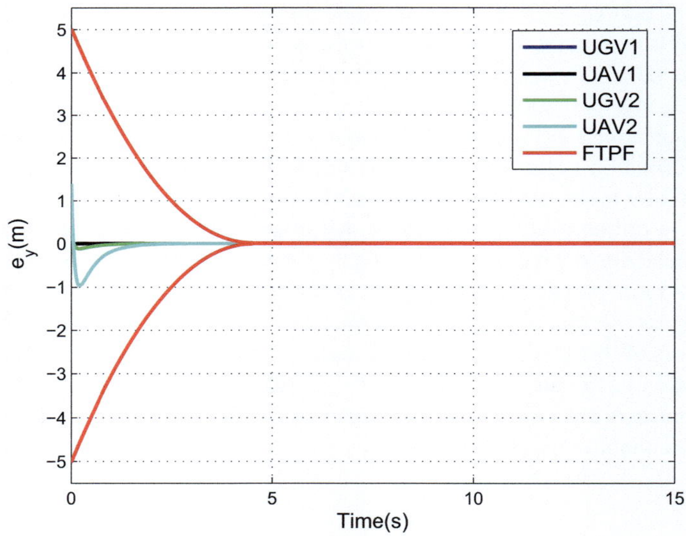


**Fig. 4.10** Formation tracking error in the $y$ axis

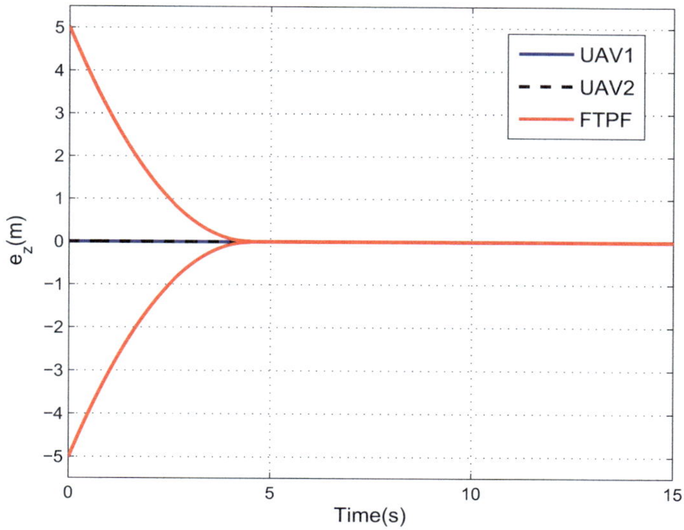


**Fig. 4.11** Formation tracking error in the $z$ axis

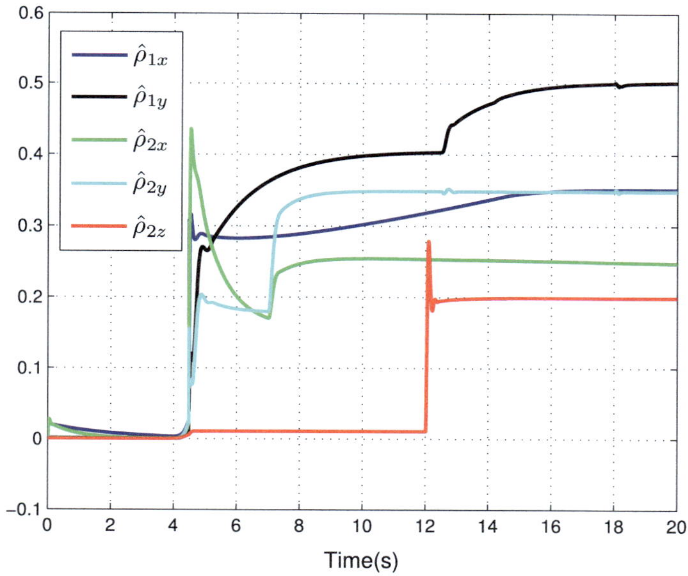


**Fig. 4.12** Fault estimation results

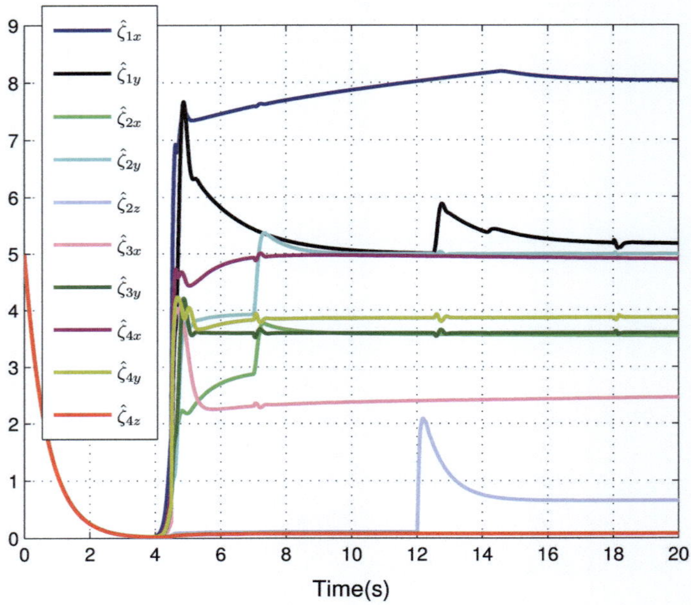


**Fig. 4.13** Estimate parameter $\hat{\zeta}_i$ updating process

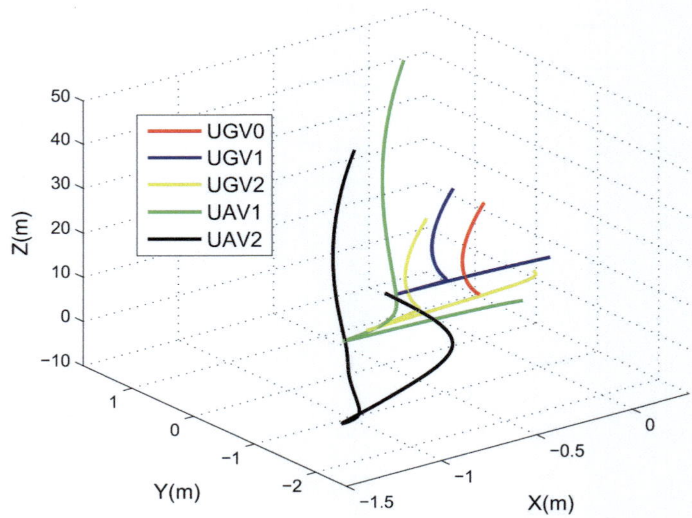


**Fig. 4.14** 3-D position trajectory of the formation system

operating conditions. Furthermore, an adaptive fixed-time formation algorithm is proposed to achieve fault-tolerant tracking in the presence of actuator faults. Simulation results validate the effectiveness of the proposed control schemes. The adaptive fixed-time formation controller demonstrates not only fast convergence and fault tolerance with guaranteed finite-time prescribed performance but also accurate estimation of actuator fault parameters.

## References


1. Bechlioulis CP, Rovithakis GA (2008) Robust adaptive control of feedback linearizable mimo nonlinear systems with prescribed performance. IEEE Trans Autom Control 53(9):2090–2099
2. Chen F, Dimarogonas DV (2021) Leader–follower formation control with prescribed performance guarantees. IEEE Trans Control Netw Syst 8(1):450–461
3. Cheng W, Zhang K, Jiang B, Ding SX (2021) Fixed-time fault-tolerant formation control for heterogeneous multi-agent systems with parameter uncertainties and disturbances. IEEE Trans Circuits Syst I Regul Pap 68(5):2121–2133
4. Deng C, Wen C (2020) Distributed resilient observer-based fault-tolerant control for heterogeneous multiagent systems under actuator faults and dos attacks. IEEE Trans Control Netw Syst 7(3):1308–1318
5. Gao Z, Guo G (2020) Fixed-time sliding mode formation control of auvs based on a disturbance observer. IEEE/CAA J Automatica Sinica 7(2):539–545
6. Guo L, Xiang YU, Zhang X, Zhang Y (2020) Safety control system technologies for uavs: review and prospect. Scientia Sinica Informationis 50(2):184–194
7. Huang S, Zong G, Niu B, Xu N, Zhao X (2024) Dynamic self-triggered fuzzy bipartite time-varying formation tracking for nonlinear multiagent systems with deferred asymmetric output constraints. IEEE Trans Fuzzy Syst 32(5):2700–2712
8. Li T, Bai W, Liu Q, Long Y, Chen CLP (2023) Distributed fault-tolerant containment control protocols for the discrete-time multiagent systems via reinforcement learning method. IEEE Trans Neural Netw Learn Syst 34(8):3979–3991
9. Liu J, Dong J (2024) Distributed fault-tolerant control for nonlinear heterogeneous multiagent system with fault indication under safety constraint. IEEE Trans Syst Man Cybern Syst 54(6):3843–3853
10. Liu Y, Liu X, Jing Y, Zhang Z (2019) A novel finite-time adaptive fuzzy tracking control scheme for nonstrict feedback systems. IEEE Trans Fuzzy Syst 27(4):646–658
11. Liu Y, Wang Z (2024) Data-based output synchronization of discrete-time heterogeneous multiagent systems with sensor faults. IEEE Trans Cybern 54(1):265–272
12. Liu Y, Yang G-H (2020) Prescribed performance-based consensus of nonlinear multiagent systems with unknown control directions and switching networks. IEEE Trans Syst Man Cybern Syst 50(2):609–616
13. Peng X-J, He Y, Shen J (2024) Time-varying formation tracking control of multi-leader multiagent systems with sampled-data. IEEE Trans Autom Sci Eng 21(3):3182–3192
14. Polyakov A (2012) Nonlinear feedback design for fixed-time stabilization of linear control systems. IEEE Trans Autom Control 57(8):2106–2110
15. Qiu J, Wang T, Sun K, Rudas IJ, Gao H (2022) Disturbance observer-based adaptive fuzzy control for strict-feedback nonlinear systems with finite-time prescribed performance. IEEE Trans Fuzzy Syst 30(4):1175–1184
16. Rahimi R, Abdollahi F, Naqshi K (2014) Time-varying formation control of a collaborative heterogeneous mult-agent system. Robot Auton Syst 62(12):1799–1805
17. Song Y, He L, Zhang D, Qian J, Fu J (2019) Neuroadaptive fault-tolerant control of quadrotor uavs: a more affordable solution. IEEE Trans Neural Netw Learn Syst 30(7):1975–1983

18. Song Y, Wang Y, Wen C (2017) Adaptive fault-tolerant pi tracking control with guaranteed transient and steady-state performance. IEEE Trans Autom Control 62(1):481–487
19. Tong S, Zhou H, Li Y (2023) Neural network event-triggered formation fault-tolerant control for nonlinear multiagent systems with actuator faults. IEEE Trans Syst Man Cybern Syst 53(12):7571–7582
20. Zhang Y, Chadli M, Xiang Z (2023) Prescribed-time formation control for a class of multiagent systems via fuzzy reinforcement learning. IEEE Trans Fuzzy Syst 31(12):4195–4204
21. Zhang Y, Zhao B, Liu D, Zhang S (2024) Distributed fault tolerant consensus control of nonlinear multiagent systems via adaptive dynamic programming. IEEE Trans Neural Netw Learn Syst 35(7):9041–9053
22. Zhao J, Zhu K, Hu H, Yu X, Li X, Wang H (2023) Formation control of networked mobile robots with unknown reference orientation. IEEE/ASME Trans Mechatron 28(4):2200–2212
23. Zhou S, Guo K, Yu X, Guo L, Xie L (2021) Fixed-time observer based safety control for a quadrotor uav. IEEE Trans Aerosp Electron Syst 57(5):2815–2825
24. Zhou X, Yu X, Guo K, Zhou S, Guo L, Zhang Y, Peng X (2023) Safety flight control design of a quadrotor UAV with capability analysis. IEEE Trans Cybern 53(3):1738–1751
25. Zou W, Ahn CK, Xiang Z (2021) Fuzzy-approximation-based distributed fault-tolerant consensus for heterogeneous switched nonlinear multiagent systems. IEEE Trans Fuzzy Syst 29(10):2916–2925

# Part II
# Fixed-Time FTC for Multiple Unmanned Systems with Switching Topologies

# Chapter 5
# Hierarchical-Based Fixed-Time FTC for Fixed-Wing UAVs with Switching Topologies

## 5.1 Introduction

The communication topology is mainly categorized into fixed topology and switching topology. Fixed topology means that whether there is communication exchange between agents in the formation system is fixed [1, 2], while the switching topology means that whether there is communication exchange between agents in the formation system is changing with time, and its communication graph is changing with time [3–5]. The main kind of switching topology is expressed by a topology working for a certain time and then switching to another topology working for a certain time, and so on [6]. There can be more than one communication topology, but each communication topology must work for a certain period of time [7, 8]. In practice, due to problems such as complex environments or formation spacing beyond the reach of sensors the communication transmission between agents in fixed-wing UAVs formation may fail or reconnect the communication exchange. Therefore, the study of fixed-wing UAVs formation control under switching topology is of great practical importance [9].

At the same time, while actuator faults were considered, only partial faults were investigated, so this chapter investigates the problem of FTC of actuator faults where exists a more generalized problem, i.e., both partial faults and bias faults are considered. In practical engineering systems, the problem of input saturation constraints is unavoidable due to safety and physical limitations [10]. The input saturation problem of the actuator makes the actuator saturate rapidly after the actuator reaches the constraint value, which reduces the operating performance of the system and even makes the system unstable [11, 12]. Therefore, this chapter considers the input saturation constraint problem and designs a corresponding control scheme for it. And considering the convergence time problem, this chapter is based on the fixed time consistency principle, which makes convergence time bounded and convergence speed faster.

 
K. Zhang et al., *Fixed-Time/Prescribed-Time Fault-Tolerant Control for Multiple Unmanned Systems*, Unmanned System Technologies,
https://doi.org/10.1007/978-981-95-2913-1_5

Based on the above analysis, this chapter focuses on the formation control problem of fixed-wing UAVs in the longitudinal formation flight scenario. The main contributions are as follows

(1) Consider the challenges brought by switching communication topologies, actuator faults, input saturation, model uncertainties, and external disturbances. To address these issues, a formation FTC scheme based on integral terminal sliding mode (ITSM) is developed. To systematically mitigate the impact of fault propagation and complex system interactions, a hierarchical control structure is adopted. Specifically, the control system is divided into upper and lower layers, allowing for modular and task-specific controller design [13]. This architecture enhances the scalability and robustness of the control framework under dynamic formation reconfigurations and uncertain operating environments.
(2) To handle actuator faults, model uncertainties, and external disturbances in a unified manner, a nonlinear fixed-time disturbance observer is designed to estimate the combined uncertainty term, which encapsulates the effects of faults, model mismatch, and unknown disturbances. In addition, to compensate for input saturation, an auxiliary system is constructed to generate appropriate corrective signals that ensure admissible control actions. The stability and convergence of the closed-loop system are then rigorously analyzed using fixed-time consistency theory, which guarantees convergence of the system states within a predefined time bound, irrespective of initial conditions. Finally, the effectiveness and robustness of the proposed method are validated through comprehensive simulations, demonstrating strong tracking performance and fault-tolerant capability under various adverse conditions.

## 5.2 Problem Formulation and Preliminaries

### *5.2.1 Problem Formulation*

The dynamic model of fixed-wing UAVs is

$$\begin{cases} \dot{V}_i = \dfrac{T_i \cos\alpha_i - D_i}{m_i} - g \sin\gamma_i \\ \dot{h}_i = V_i \sin\gamma_i \\ \dot{\gamma}_i = \dfrac{L_i + T_i \sin\alpha_i}{m_i V_i} - \dfrac{g \cos\gamma_i}{V_i} \\ \dot{\alpha}_i = q_i - \dot{\gamma}_i \\ \dot{q}_i = \dfrac{\bar{M}_i}{I_y} \end{cases} \tag{5.1}$$

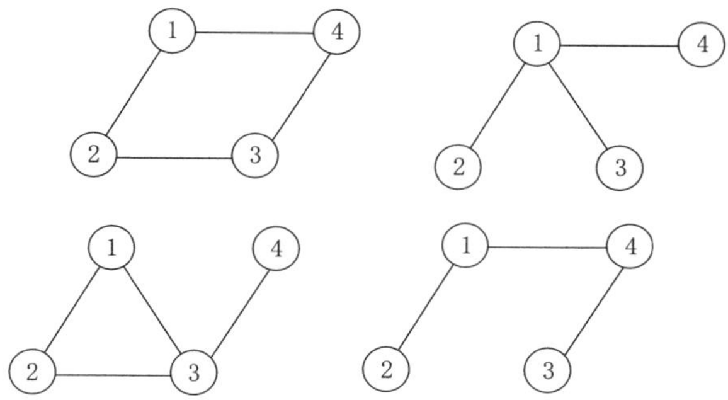


**Fig. 5.1** Switching communication topology

where $V_i$, $h_i$, $\gamma_i$, $\alpha_i$ and $q_i$ are the flight velocity, the flight altitude, the path angle, the angle of attack and pitch rate, respectively. $m_i$ is the mass of the $i$th fixed-wing UAVs. $T_i$, $D_i$, $L_i$ and $\bar{M}_i$ denote the thrust generated by the throttle input, the drag, the lift and the torque of pitch. $I_y$ represents the moment of inertia with respect to the pitch.

The object of this chapter is a formation system consisting of four fixed-wing UAVs as well as a virtual leader. The communication topology of the system is the switching topology, which is shown in Fig. 5.1, and as time changes, the communication graph is either of the topologies in Fig. 5.1. The two main control objectives are as the following

(1) The fixed-wing UAVs formation has the same speed as the leader in a fixed time, i.e., it satisfies

$$\lim_{t \to T_f} (V_i - V_0) = 0 \tag{5.2}$$

where $V_i$ and $V_0$ represent respectively the speed of the $i$th follower fixed-wing UAVs and the flight speed of the virtual leader.

(2) The fixed-wing UAVs formation reaches the desired formation altitude formation in a fixed time, i.e., it satisfies

$$\lim_{t \to T_f} ((h_i - \chi_i) - h_0) = 0, \quad \lim_{t \to T_f} ((h_i - \chi_i) - (h_j - \chi_j)) = 0 \tag{5.3}$$

where $h_i$, $\chi_i$ and $h_0$ denote respectively the height of the $i$th follower fixed-wing UAVs, the desired relative height distance and the height of the virtual leader height.

**Assumption 5.1** ([14]) The flight velocity of the virtual leader $V_0$ and its first-order derivative $\dot{V}_0$ are continuous and bounded, the height $h_0$ and its first-order derivative $\dot{h}_0$ are continuous and bounded with respect to the second-order derivative $\ddot{h}_0$.

**Assumption 5.2** ([14]) The virtual leader interacts with each follower drone with information. The Laplace matrix $L_{\varsigma(t)} + B_{\varsigma(t)}$ is symmetric positive definite invertible.

The formation control structure consists of a virtual leader and $N$ followers. The longitudinal dynamics model of the $i$th follower fixed-wing UAVs is established.

The velocity subsystem of the fixed-wing UAVs is

$$\dot{V}_i = f_{V_i} + \Delta f_{V_i} + d_{V_i} + g_{V_i} T_i^s = F_{V_i} + g_{V_i} T_i^s \tag{5.4}$$

where $f_{v_i} = -D_i/m_i - g \sin \gamma_i$, $g_{v_i} = \cos(\alpha_i)/m_i$. $\Delta f_{V_i}$ and $d_{V_i}$ donate respectively the model uncertainty due to uncertainty in aerodynamic parameters and bounded external disturbances. $F_{vi} = f_{vi} + \Delta f_{vi} + d_{vi}$. $T_i^s = \mathrm{sat}(T_i)$ is the constrained control input, and its expression is

$$\mathrm{sat}(T_i) = \begin{cases} T_{\mathrm{m}}, \ T_i \geq T_{\mathrm{m}} \\ T_i, \ -T_{\mathrm{m}} < T_i < T_{\mathrm{m}} \\ -T_{\mathrm{m}}, \ T_i \leq -T_{\mathrm{m}} \end{cases} \tag{5.5}$$

where $T_m$ is an upper bound on the control inputs defining $\Delta T_i = T_i^s - T_i$, $|\Delta T_i| \leq u_0$.

The mathematical model of the height subsystem subject to model uncertainty as well as actuator fault is

$$\begin{cases} \dot{h}_i = g_{hi}\gamma_i \\ \dot{\gamma}_i = f_{\gamma i} + \Delta f_{\gamma i} + d_{\gamma i} + g_{\gamma i}\alpha_i = F_{\gamma i} + g_{\gamma i}\alpha_i \\ \dot{\alpha}_i = f_{\alpha i} + \Delta f_{\alpha i} + d_{\alpha i} + g_{\alpha i} q_i = F_{\alpha i} + g_{\alpha i} q_i \\ \dot{q}_i = f_{qi} + \Delta f_{qi} + d_{qi} + g_{qi}\delta_{qi}^f = F_{qi} + g_{qi}\delta_{qi}^f \end{cases} \tag{5.6}$$

where $g_{hi} = V_i$, $f_{\gamma i} = -g \cos \gamma_i / V_i + \overline{q}_i S_{ref} C_{L0}/(m_i V_i)$, $g_{\gamma i} = \overline{q}_i S_{ref} C_{L\alpha}/(m_i V_i)$, $f_{\alpha i} = -(f_{\gamma i} + g_{\gamma i}\alpha i)$, $g_{ai} = 1$, $f_{qi} = \overline{q}_i S_{rej} \overline{c}[C_{m0} + C_{m\alpha}\alpha_i + C_{mq}\overline{c} q_i / 2V_i]/I_y$, $g_{qi} = \overline{q}_i S_{ref} \overline{c} C_{m\delta_e}/I_y$. $\Delta f_{\gamma i}$, $\Delta f_{\alpha i}$ and $\Delta f_{qi}$ donate the uncertainties of $f_{\gamma i}$, $f_{\alpha i}$ and $f_{qi}$ caused by the uncertainties in the aerodynamic parameters. $d_{\gamma i}$, $d_{\alpha i}$ and $d_{qi}$ are bounded external disturbance. $\delta_{ei}^f$ is an actuator with faults, and its fault model is $\delta_{ei}^f = \rho_i \delta_{ei} + \overline{\delta}_i$, $0 < \rho_i < 1$.

**Assumption 5.3** ([15]) The derivative of nonlinear function, uncertainty, and aggregate information of disturbances of the longitudinal model of fixed-wing UAV $F_{ji}$, $j = \gamma, \alpha, q$ is bounded, i.e., $\dot{F}_{ji} \leq \vartheta_{ji}$, $j = \gamma, \alpha, q$.

### 5.2.2 Preliminaries

**Lemma 5.1** ([16, 17]) *The disturbance observers are constructed as*

$$\begin{cases} \dot{\hat{x}} = \hat{F}_{xi} + g_{xi}u + \theta_{x1}\,\mathrm{sig}(e_{x0})^a + \theta_{x2}\,\mathrm{sig}(e_{x0})^b \\ \dot{\hat{F}}_{xi} = \theta_{x3}\,\mathrm{sig}(e_{x0})^a + \theta_{x4}\,\mathrm{sig}(e_{x0})^b \end{cases} \tag{5.7}$$

*where $e_{x0} = x - \hat{x}$, $0 < a < 1$, $b > 1$, $\hat{F}_{xi}$ is estimation of $F_{xi}$, and estimation error is defined as $e_{x1} = F_{xi} - \hat{F}_{xi}$. Parameters $\theta_{xi}(i = 1, 2, 3, 4)$ are positive parameters and their upper bounds are bounded by the boundary values of the first-order derivatives of $F_{xi}$ and the following matrices are Hurwitz matrices*

$$A_1 = \begin{bmatrix} -\theta_1 & 1 \\ -\theta_3 & 0 \end{bmatrix}, A_2 = \begin{bmatrix} -\theta_2 & 1 \\ -\theta_4 & 0 \end{bmatrix} \tag{5.8}$$

*Combining Eqs. (5.4), (5.6) and (5.7), the error dynamic system is drawn as*

$$\begin{cases} \dot{e}_{x0} = e_{x1} - \theta_{x1}\,\mathrm{sig}(e_{x0})^a - \theta_{x2}\,\mathrm{sig}(e_{x0})^b \\ \dot{e}_{x1} = \dot{F}_{xi} - \theta_{x3}\,\mathrm{sig}(e_{x0})^a - \theta_{x4}\,\mathrm{sig}(e_{x0})^b \end{cases} \tag{5.9}$$

*It can be obtained that the error $[e_{x0}, e_{x1}]^{\mathrm{T}}$ converges to zero at a fixed time $t_{x0}$ and the convergence time satisfies*

$$t_{x0} \leq \frac{\lambda_{\max}^{1-\alpha}(P_1)}{k_1(1-a)} + \frac{1}{k_2(b-1)\lambda_{\min}^{b-1}(P_2)} \tag{5.10}$$

*where $\lambda_{\max}(\cdot)$ and $\lambda_{\min}(\cdot)$ denote the maximum and minimum eigenvalues of the matrix, respectively. $k_1 = \lambda_{\min}(Q_1)/\lambda_{\max}(P_1)$, $Q_1$ and $P_1$ are symmetric positive definite matrices and satisfy $P_1A_1 + A_1^{\mathrm{T}}P_1 = -Q_1$. $k_2 = \lambda_{\min}(Q_2)/\lambda_{\max}(P_2)$, $Q_2$ and $P_2$ are symmetric positive definite matrices and satisfy $P_2A_2 + A_2^{\mathrm{T}}P_2 = -Q_2$.*

**Lemma 5.2** ([18]) *For any real number $x$ and $\varepsilon > 0$, one has*

$$0 < |x| - x\tanh\left(\frac{x}{\varepsilon}\right) \leq 0.2785\varepsilon \tag{5.11}$$

## 5.3 Main Results

In this section, the FTC design of the formation system for the switching topology is based on a hierarchical control structure. The upper layer control system designs fixed-time converged controllers based on ITSM for greater robustness. The lower level control system uses a fixed-time observer to estimate the model uncertainty and actuator fault information collectively, and introduces a fixed-time auxiliary system to solve the input saturation problem.

## 5.3.1 Fixed-Time FTC Design for Virtual Layer

The virtual system of the upper layer is divided into virtual speed formation system and virtual height formation system, which is designed by adopting the high-order sliding mode controller, which not only has stronger robustness but also avoids the chattering problem of the sliding mode, and has higher control accuracy [19].

### 5.3.1.1 Design of Integral Sliding Mode Formation Controller for Virtual Velocity Systems

The virtual speed formation system has $N$ agents, where the $i$th first-order agent is modeled as

$$\dot{\overline{x}}_{vi} = \overline{u}_{vi} \tag{5.12}$$

where $\overline{u}_{vi}$ is the control input to the system.

The tracking error of the $i$th first-order agent is defined as

$$\overline{e}_{iv1} = \overline{x}_{vi} - V_0 \tag{5.13}$$

where $V_0$ represents the leader's speed signal.

Integral terminal sliding mold surface is designed as [20]

$$S_{iv1} = \int_0^t \left(\dot{\overline{e}}_{iv1} + p_{11}\,\mathrm{sig}\,(\overline{e}_{iv1})^a + p_{12}\,\mathrm{sig}\,(\overline{e}_{iv1})^b\right) \mathrm{d}\tau \tag{5.14}$$

where $p_{11}$ and $p_{12}$ are adjustable positive parameters, $0 < a < 1$ and $b > 1$.

The sliding mold surface is differentiated and defined as

$$\begin{cases} \dot{S}_{iv1} = S_{iv2} \\ \dot{S}_{iv2} = \dfrac{\mathrm{d}}{\mathrm{d}t}\left(\overline{u}_{vi} - \dot{V}_0 + p_{11}\,\mathrm{sig}\,(\overline{e}_{iv1})^a + p_{12}\,\mathrm{sig}\,(\overline{e}_{iv1})^b\right) \end{cases} \tag{5.15}$$

According to the definition given in Eq. (5.15), a coordinate transformation yields

$$\begin{cases} \omega_{iv1} = S_{iv1} \\ \omega_{iv2} = S_{iv2} - \zeta_{iv1} \end{cases} \tag{5.16}$$

where $\zeta_{iv1}$ is the virtual control law to be designed.

The Lyapunov function is considered

$$\bar{W}_{V1} = \sum_{i=1}^{N} \frac{1}{2} \omega_{iv2}^2 \tag{5.17}$$

Derivation of $\bar{W}_{V1}$ is

$$\dot{\bar{W}}_{V1} = \sum_{i=1}^{N} \omega_{iv2}(\frac{\mathrm{d}}{\mathrm{d}t} \left(\overline{u}_{vi} - \dot{V}_0 + p_{11} \operatorname{sig}(\overline{e}_{iv1})^a + p_{12} \operatorname{sig}(\overline{e}_{iv1})^b\right) - \dot{\zeta}_{iv1}) \tag{5.18}$$

Control inputs are designed as

$$\begin{cases} \overline{u}_{vi} = \dot{V}_0 - p_{11} \operatorname{sig}(\overline{e}_{iv1})^a + p_{12} \operatorname{sig}(\overline{e}_{iv1})^b + \overline{u}_{vi}^* \\ \overline{u}_{vi}^* = \int_0^t \left(\dot{\zeta}_{iv1} - p_{21} \operatorname{sig}(\omega_{iv2})^{l_1} - p_{22} \operatorname{sig}(\omega_{iv2})^{l_2}\right) \mathrm{d}\tau \end{cases} \tag{5.19}$$

where $0 < l_1 < 1$, $l_2 > 1$, $p_{21}$ and $p_{22}$ are positive parameters.
Substituting into Eq. (5.19), the result is

$$\begin{aligned} \dot{\bar{W}}_{V1} &= \sum_{i=1}^{N} \omega_{iv2}(-p_{21} \operatorname{sig}(\omega_{iv2})^{l_1} - p_{22} \operatorname{sig}(\omega_{iv2})^{l_2}) \\ &= \sum_{i=1}^{N} (-p_{21}(\omega_{iv2}^2)^{\frac{l_1+1}{2}} - p_{22}(\omega_{iv2}^2)^{\frac{l_2+1}{2}}) \end{aligned} \tag{5.20}$$

The Lyapunov function is considered

$$\bar{W}_{V2} = \frac{1}{2} \omega_{v1}^{\mathrm{T}} \left(L_{\varsigma(t)} + B_{\varsigma(t)}\right) \omega_{v1} \tag{5.21}$$

where $\omega_{v1} = [\omega_{1v1}, \omega_{2v1}, ..., \omega_{Nv1}]^{\mathrm{T}}$, and $N$ is the number of agents. The matrix $(L_{\varsigma(t)} + B_{\varsigma(t)})$ is symmetric positive definite.
Derivation of $\bar{W}_{V2}$ is

$$\begin{aligned} \dot{\bar{W}}_{V2} &= \omega_{v1}^{\mathrm{T}} \left(L_{\varsigma(t)} + B_{\varsigma(t)}\right) \dot{\omega}_{v1} \\ &= \omega_{v1}^{\mathrm{T}} \left(L_{\varsigma(t)} + B_{\varsigma(t)}\right) (\omega_{v2} + \zeta_{v1}) \end{aligned} \tag{5.22}$$

where $\omega_{v2} = [\omega_{1v2}, \omega_{2v2}, ..., \omega_{Nv2}]^{\mathrm{T}}$, $\zeta_{v1} = [\zeta_{1v1}, \zeta_{2v1}, ..., \zeta_{Nv1}]^{\mathrm{T}}$.

Virtual control law $\zeta_{v1}$ is designed as

$$\zeta_{v1} = -p_{31}\,\mathrm{sig}\left(\left(L_{\varsigma(t)} + B_{\varsigma(t)}\right)\omega_{v1}\right)^{l_1} - p_{32}\,\mathrm{sig}\left(\left(L_{\varsigma(t)} + B_{\varsigma(t)}\right)\omega_{v1}\right)^{l_2} \tag{5.23}$$

where $p_{31}$ and $p_{32}$ are positive parameters.

#### 5.3.1.2 Design of Integral Sliding Mode Formation Controller for Virtual Altitude Systems

The $i$th second-order agent model in the virtual altitude formation system is

$$\begin{cases} \dot{\overline{x}}_{hi} = \overline{v}_{hi} \\ \dot{\overline{v}}_{hi} = \overline{u}_{hi} \end{cases} \tag{5.24}$$

where $\overline{u}_{hi}$ represents the control inputs to the system.

The tracking error of the $i$th second-order agents is defined as

$$\overline{e}_{ih1} = \overline{x}_{hi} - h_0 - \chi_i \tag{5.25}$$

where $h_0$ represents the leader's velocity signal, $\chi_i$ represents the desired relative height distance between the intelligence and the virtual leader.

Integral terminal sliding mold surface is designed as

$$S_{ih1} = \int_0^t \left(\dot{\overline{e}}_{ih1} + k_{11}\,\mathrm{sig}\left(\overline{e}_{ih1}\right)^a + k_{12}\,\mathrm{sig}\left(\overline{e}_{ih1}\right)^b\right)\mathrm{d}\tau \tag{5.26}$$

where $k_{11}$ and $k_{12}$ are adjustable positive parameters.

The sliding mold surface $S_{ih1}$ is differentiated and defined as

$$\begin{cases} \dot{S}_{ih1} = S_{ih2} \\ \dot{S}_{ih2} = S_{ih3} \\ \dot{S}_{ih3} = \dfrac{\mathrm{d}}{\mathrm{d}t}\overline{u}_{hi} - \ddot{h}_0 + ak_{11}|\overline{e}_{ih1}|^{a-1}\dot{\overline{e}}_{i1} + bk_{12}|\overline{e}_{ih1}|^{b-1}\dot{\overline{e}}_{ih1} \end{cases} \tag{5.27}$$

**Remark 5.1** It can be found through Eq. (5.27) that the designed sliding mode surface of the integration terminal has the form of a higher order sliding mode. Therefore, it is possible to design a controller similar to the higher-order sliding mode control structure to solve the vibration problem.

According to the definition given in Eq. (5.27), a coordinate transformation yields

$$\begin{cases} \omega_{ih1} = S_{ih1} \\ \omega_{ih2} = S_{ih2} - \zeta_{ih1} \\ \omega_{ih3} = S_{ih3} - \zeta_{ih2} \end{cases} \tag{5.28}$$

where $S_{ih1}$ and $S_{ih2}$ are the virtual control laws to be designed.

The following Lyapunov function is considered

$$\bar{W}_{h1} = \sum_{i=1}^{N} \frac{1}{2}\omega_{ih3}^{2} \tag{5.29}$$

The derivation of $\bar{W}_{h1}$ yields

$$\dot{\bar{W}}_{h1} = \sum_{i=1}^{N} \omega_{ih3}\left(\frac{\mathrm{d}}{\mathrm{d}t}\left(\overline{u}_{hi} - \ddot{h}_0 + ak_{11}|\overline{e}_{ih1}|^{a-1}\dot{\overline{e}}_{i1} + bk_{12}|\overline{e}_{ih1}|^{b-1}\dot{\overline{e}}_{ih1}\right) - \dot{\zeta}_{ih2}\right) \tag{5.30}$$

Control input is designed as

$$\begin{cases} \overline{u}_{hi} = \ddot{h}_0 - ak_{11}\left|\overline{e}_{ih1}\right|^{a-1}\dot{\overline{e}}_{ih1} - bk_{12}\left|\overline{e}_{ih1}\right|^{b-1}\dot{\overline{e}}_{ih1} + \overline{u}_{hi}^{*} \\ \overline{u}_{hi}^{*} = \displaystyle\int_0^t \left(\dot{\zeta}_{ih2} - k_{21}\,\mathrm{sig}\,(\omega_{ih3})^{l_1} - k_{22}\,\mathrm{sig}\,(\omega_{ih3})^{l_2}\right)\mathrm{d}\tau \end{cases} \tag{5.31}$$

where $k_{21}$ and $k_{22}$ are positive parameters.

Substituting into Eq. (5.31), the result is

$$\begin{aligned} \dot{\bar{W}}_{h1} &= \sum_{i=1}^{N} \omega_{ih3}(-k_{21}\,\mathrm{sig}(\omega_{ih3})^{l_1} - k_{22}\,\mathrm{sig}(\omega_{ih3})^{l_2}) \\ &= \sum_{i=1}^{N}(-k_{21}(\omega_{ih3}^{2})^{\frac{l_1+1}{2}} - k_{22}(\omega_{ih3}^{2})^{\frac{l_2+1}{2}}) \end{aligned} \tag{5.32}$$

The Lyapunov function is chosen as

$$\bar{W}_{h2} = \sum_{i=1}^{N} \frac{1}{2}\omega_{ih2}^{2} \tag{5.33}$$

The derivation of $\bar{W}_{h2}$ yields

$$\dot{\bar{W}}_{h2} = \sum_{i=1}^{N} \omega_{ih2}(\omega_{ih3} + \zeta_{ih2} - \dot{\zeta}_{ih1}) \tag{5.34}$$

Control law $\zeta_{ih2}$ is designed as

$$\zeta_{ih2} = \dot{\zeta}_{ih1} - k_{31}\,\mathrm{sig}(\omega_{ih2})^{l_1} - k_{32}\,\mathrm{sig}(\omega_{ih2})^{l_2} \tag{5.35}$$

where $k_{31}$ and $k_{32}$ are positive parameters. The Lyapunov function is chosen as

$$\bar{W}_{h3} = \frac{1}{2}\omega_{h1}^{\mathrm{T}}\left(L_{\varsigma(t)} + B_{\varsigma(t)}\right)\omega_{h1} \tag{5.36}$$

where $\omega_{h1} = [\omega_{1h1}, \omega_{2h1}, ..., \omega_{Nh1}]^{\mathrm{T}}$, $N$ is the number of agents and the matrix $(L_{\varsigma(t)} + B_{\varsigma(t)})$ is symmetric positive definite.

Derivation of $\bar{W}_{h2}$ is

$$\begin{aligned}\dot{\bar{W}}_{h3} &= \omega_{h1}^{\mathrm{T}}\left(L_{\varsigma(t)} + B_{\varsigma(t)}\right)\dot{\omega}_{h1} \\ &= \omega_{h1}^{\mathrm{T}}\left(L_{\varsigma(t)} + B_{\varsigma(t)}\right)(\omega_{h2} + \zeta_{h1})\end{aligned} \tag{5.37}$$

where $\omega_{h2} = [\omega_{1h2}, \omega_{2h2}, ..., \omega_{Nh2}]^{\mathrm{T}}$, $\zeta_{h1} = [\zeta_{1h1}, \zeta_{2h1}, ..., \zeta_{Nh1}]^{\mathrm{T}}$.

Virtual control law $\zeta_{h1}$ is designed as

$$\zeta_{h1} = -k_{41}\,\mathrm{sig}\left(\left(L_{\varsigma(t)} + B_{\varsigma(t)}\right)\omega_{h1}\right)^{l_1} - k_{42}\,\mathrm{sig}\left(\left(L_{\varsigma(t)} + B_{\varsigma(t)}\right)\omega_{h1}\right)^{l_2} \tag{5.38}$$

where $k_{41}$ and $k_{42}$ are positive parameters.

#### 5.3.1.3 Stability Analysis of Upper Level Systems

**Theorem 5.1** Consider the upper level virtual formation system based on switching topology with one leader and $N$ followers structure, its formation tracking target is consistent with the actual fixed-wing UAVs formation. If the designs of sliding mode surface are Eqs. (5.14) and (5.26), their controllers are designed as Eqs. (5.19), (5.23), (5.24), (5.25), and (5.38), then the upper level closed-loop system is fixed-time stable and the tracking error is able to converge to zero in a fixed time.

***Proof*** According to Lemma 2.2 and Eq. (5.20), the result is

$$\begin{aligned}\dot{\bar{W}}_{V1} &= \sum_{i=1}^{N}(-p_{21}(\omega_{iv2}^{2})^{\frac{l_1+1}{2}} - p_{22}(\omega_{iv2}^{2})^{\frac{l_2+1}{2}}) \\ &\le -p_{21}2^{\frac{l_1+1}{2}}\overline{W}_{V1}^{\frac{l_1+1}{2}} - p_{22}2^{\frac{l_2+1}{2}}N^{\frac{1-l_2}{2}}\overline{W}_{V1}^{\frac{l_2+1}{2}}\end{aligned} \tag{5.39}$$

By Lemma 2.1, it can be proven that $\bar{W}_{V1} = 0$ is fixed-time stable at the equilibrium point and converges in time $T_{s1} = 1/(p_{21}2^{(l_1-1)/2}(1-l_1)) + 1/(p_{22}2^{(l_2-1)/2} N^{(1-l_2)/2}(l_2-1))$.

When $t > T_{s1}$, $\omega_{iv2} = 0$. Combining Eqs. (5.22) and (5.23), it follows that

$$\begin{aligned}\dot{\bar{W}}_{V2} =& - p_{31}\left(\left(L_{\varsigma(t)} + B_{\varsigma(t)}\right)\omega_{v1}\right)^{\mathrm{T}} \operatorname{sgn}\left(\left(L_{\varsigma(t)} + B_{\varsigma(t)}\right)\omega_{v1}\right)^{l_1} \\ &- p_{32}\left(\left(L_{\varsigma(t)} + B_{\varsigma(t)}\right)\omega_{v1}\right)^{\mathrm{T}} \operatorname{sig}\left(\left(L_{\varsigma(t)} + B_{\varsigma(t)}\right)\omega_{v1}\right)^{l_2} \\ \leq& - p_{31}\left(\frac{2\lambda_{\min}^2(\boldsymbol{L}_{\varsigma(t)} + \boldsymbol{B}_{\varsigma(t)})}{\lambda_{\max}(\boldsymbol{L}_{\varsigma(t)} + \boldsymbol{B}_{\varsigma(t)})}\right)^{\frac{l_1+1}{2}} \bar{W}_{V2}^{\frac{l_1+1}{2}} \\ &- p_{32}\left(\frac{2\lambda_{\min}^2(\boldsymbol{L}_{\varsigma(t)} + \boldsymbol{B}_{\varsigma(t)})}{\lambda_{\max}(\boldsymbol{L}_{\varsigma(t)} + \boldsymbol{B}_{\varsigma(t)})}\right)^{\frac{l_2+1}{2}} \bar{W}_{V2}^{\frac{l_2+1}{2}}\end{aligned} \tag{5.40}$$

where $\lambda_{\max}(L_{\varsigma(t)} + B_{\varsigma(t)})$ and $\lambda_{\min}(L_{\varsigma(t)} + B_{\varsigma(t)})$ are the maximum and minimum eigenvalues of $(L_{\varsigma(t)} + B_{\varsigma(t)})$, respectively.

By Lemma 2.2, it can be shown that there exists a fixed time $T_{s2}$ as

$$\begin{aligned}T_{s1} =& 2/\left(p_{31}\left(\frac{2\lambda_{\min}^2(L_{\varsigma(t)} + B_{\varsigma(t)})}{\lambda_{\max}(L_{\varsigma(t)} + B_{\varsigma(t)})}\right)^{\frac{l_1+1}{2}}(1-l_1)\right) \\ &+ 2/\left(p_{32}\left(\frac{2\lambda_{\min}^2(L_{\varsigma(t)} + B_{\varsigma(t)})}{\lambda_{\max}(L_{\varsigma(t)} + B_{\varsigma(t)})}\right)^{\frac{l_2+1}{2}}(l_2-1)\right)\end{aligned} \tag{5.41}$$

such that $\omega_{v1}$ converges to 0 in $T_{s1} + T_{s2}$.

From Lemma 2.2 and Eq. (5.32), it follows that

$$\dot{W}_{h1} \leq -k_{21}2^{\frac{l_1+1}{2}} \bar{W}_{h1}^{\frac{l_1+1}{2}} - k_{22}2^{\frac{l_2+1}{2}} N^{\frac{1-l_2}{2}} \bar{W}_{h1}^{\frac{l_2+1}{2}} \tag{5.42}$$

By Lemma 2.2, it can be proven that $\bar{W}_{h1} = 0$ is fixed-time stable at the equilibrium point and converges in time $T_{s3} = 1/(k_{21}2^{(l_1-1)/2}(1-l_1)) + 1/(k_{22}2^{(l_2-1)/2} N^{(1-l_2)/2}(l_2-1))$.

Therefore, when $t > T_{s3}$, it has $\omega_{ih3} = 0$. Combining Lemma 2.2, Eqs. (5.34) and (5.35), it follows that

$$\begin{aligned}\dot{\bar{W}}_{h2} &= \sum_{i=1}^{N}(-k_{31}(\omega_{ih2}^2)^{\frac{l_1+1}{2}} - k_{32}(\omega_{ih2}^2)^{\frac{l_2+1}{2}}) \\ &\leq -k_{31}2^{\frac{l_1+1}{2}} \bar{W}_{h2}^{\frac{l_1+1}{2}} - k_{32}2^{\frac{l_2+1}{2}} N^{\frac{1-l_2}{2}} \bar{W}_{h2}^{\frac{l_2+1}{2}}\end{aligned} \tag{5.43}$$

By Lemma 2.2, it can be proven that $\bar{W}_{h2} = 0$ is fixed-time stable at the equilibrium point and there exists a fixed time $T_{s4} = 1/(k_{31}2^{(l_1-1)/2}(1-l_1)) + 1/(k_{32}2^{(l_2-1)/2}N^{(1-l_2)/2}(l_2-1))$ such that $\omega_{ih2}$ converges in $T_{s3} + T_{s4}$.

Therefore, when $t > T_{s3} + T_{s4}$, it has $\omega_{ih2} = 0$. Combining Eqs. (5.37) and (5.38), it follows that

$$\begin{aligned}\dot{\bar{W}}_{h3} =& -k_{41}\left(\left(L_{\varsigma(t)} + B_{\varsigma(t)}\right)\omega_{h1}\right)^{\mathrm{T}} \operatorname{sgn}\left(\left(L_{\varsigma(t)} + B_{\varsigma(t)}\right)\omega_{h1}\right)^{l_1} \\ & - k_{42}\left(\left(L_{\zeta(t)} + B_{\zeta(t)}\right)\omega_{h1}\right)^{\mathrm{T}} \operatorname{sig}\left(\left(L_{\zeta(t)} + B_{\zeta(t)}\right)\omega_{h1}\right)^{l_2} \\ \leq & -k_{41}\left(\frac{2\lambda_{\min}^2(\boldsymbol{L}_{\varsigma(t)} + \boldsymbol{B}_{\varsigma(t)})}{\lambda_{\max}(\boldsymbol{L}_{\varsigma(t)} + \boldsymbol{B}_{\varsigma(t)})}\right)^{\frac{l_1+1}{2}} \bar{W}_{h3}^{\frac{l_1+1}{2}} \\ & - k_{42}\left(\frac{2\lambda_{\min}^2(\boldsymbol{L}_{\varsigma(t)} + \boldsymbol{B}_{\varsigma(t)})}{\lambda_{\max}(\boldsymbol{L}_{\varsigma(t)} + \boldsymbol{B}_{\varsigma(t)})}\right)^{\frac{l_2+1}{2}} \bar{W}_{h3}^{\frac{l_2+1}{2}}\end{aligned} \tag{5.44}$$

By Lemma 2.2, it can be shown that there exists a fixed time $T_{s5}$ as

$$\begin{aligned}T_{s5} =& 2/(k_{41}(\frac{2\lambda_{\min}^2(\boldsymbol{L}_{\varsigma(t)} + \boldsymbol{B}_{\varsigma(t)})}{\lambda_{\max}(\boldsymbol{L}_{\varsigma(t)} + \boldsymbol{B}_{\varsigma(t)})})^{\frac{l_1+1}{2}}(1 - l_1)) \\ & + 2/(k_{42}(\frac{2\lambda_{\min}^2(\boldsymbol{L}_{\varsigma(t)} + \boldsymbol{B}_{\varsigma(t)})}{\lambda_{\max}(\boldsymbol{L}_{\varsigma(t)} + \boldsymbol{B}_{\varsigma(t)})})^{\frac{l_2+1}{2}}(l_2 - 1))\end{aligned} \tag{5.45}$$

such that $\omega_{h1}$ converges to 0 in $T_{s3} + T_{s4} + T_{s5}$.

According to the above analysis, the sliding mode surfaces $S_{iv1}$ and $S_{iH1}$ reach zero at a fixed time. When $S_{iv1} = S_{ih1} = 0$, it can be obtained that

$$\begin{cases}\dot{\bar{e}}_{iv1} = -p_{11}\operatorname{sig}(\bar{e}_{iv1})^a - p_{12}\operatorname{sig}(\bar{e}_{iv1})^b \\ \dot{\bar{e}}_{ih1} = -k_{11}\operatorname{sig}(\bar{e}_{ih1})^a - k_{12}\operatorname{sig}(\bar{e}_{ih1})^b\end{cases} \tag{5.46}$$

The Lyapunov function is chosen as

$$\bar{W}_0 = \frac{1}{2}\bar{e}_{iv1}^2 + \frac{1}{2}\bar{e}_{ih1}^2 \tag{5.47}$$

Combined with Eq. (5.46), the derivative of $W$ can be calculated as

$$\begin{aligned}\dot{\bar{W}}_0 =& -p_{11}\left(\bar{e}_{iv1}^2\right)^{(a+1)/2} - p_{12}\left(\bar{e}_{iv1}^2\right)^{(b+1)/2} \\ & - k_{11}\left(\bar{e}_{ih1}^2\right)^{(a+1)/2} - k_{12}\left(\bar{e}_{ih1}^2\right)^{(b+1)/2} \\ \leq & -\lambda_1 2^{(a+1)/2}\bar{W}_0^{(a+1)/2} - 2\lambda_2\bar{W}_0^{(b+1)/2}\end{aligned} \tag{5.48}$$

where $\lambda_1 = \min\{p_{11}, k_{11}\} > 0$, $\lambda_2 = \min\{p_{12}, k_{12}\} > 0$.

By Lemma 2.2, it can be shown that the tracking errors $\bar{e}_{iv1}$ and $\bar{e}_{ih1}$ converge to zero in fixed time.

Taken together, the above analysis proves that the upper-layer virtual formation system is capable of accomplishing the formation goal in fixed time under the switching communication topology. The closed-loop system of upper-layer virtual formation is fixed-time stable.

### 5.3.2 *Fixed-Time FTC Design for Actual Layer*

The lower system utilizes the output of the upper system to design an anti-saturation fault-tolerant tracking controller for the longitudinal model of a fixed-wing UAV. An integral sliding mode controller against input saturation and an anti-saturation fault-tolerant controller based on the inverse step structure are designed for the velocity subsystem and the altitude subsystem of the fixed-wing UAVs, respectively. Meanwhile, for actuator faults, a fixed-time disturbance observer is designed to estimate the actuator faults and the total model uncertainty set.

#### 5.3.2.1 Design of Speed Subsystem Controller Against Input Saturation

The error $z_{Vi} = V_i - \overline{x}_{vi} - \eta_{vi}$, $\eta_{vi}$ is the auxiliary signal to be designed introduced to deal with the input saturation problem. Derivation of the error yields

$$\dot{z}_{Vi} = f_{Vi} + \Delta f_{Vi} + d_{Vi} + g_{Vi}T_i^s - \dot{\overline{x}}_{vi} - \dot{\eta}_{vi} \tag{5.49}$$

The fixed-time disturbance observer is used to estimate $F_{Vi} = f_{Vi} + \Delta f_{Vi} + d_{Vi}$, and the fixed-time disturbance observer is [21] constructed as

$$\begin{cases} \dot{\hat{V}}_i = \hat{F}_{Vi} + g_{Vi}T_i^s + \theta_{V1}\operatorname{sig}(e_{V0})^{1/2} + \theta_{V2}\operatorname{sig}(e_{V0})^{3/2} \\ \dot{\hat{F}}_{Vi} = \theta_{V3}\operatorname{sig}(e_{V0})^{1/2} + \theta_{V4}\operatorname{sig}(e_{V0})^{3/2} \end{cases} \tag{5.50}$$

where $e_{V0} = V_i - \hat{V}_i$, $\hat{F}_{Vi}$ is the estimated value of $F_{Vi}$, and the parameters $\theta_{Vi}(i = 1, 2, 3, 4)$ are all positive parameters. Estimate error is defined as $e_{V1} = F_{Vi} - \hat{F}_{Vi}$. According to the fixed-time disturbance observer in Eq. (5.7), the error $e_{V0}$ and $e_{V1}$ converge to zero at fixed time $t_{V0}$, i.e., when $t > t_{V0}$, $\hat{V}_i = V_i$ and $\hat{F}_{Vi} = F_{Vi}$.

The auxiliary signal $\eta_{vi}$ is constructed, which is obtained by the following first-order dynamic model [22]

$$\dot{\eta}_{vi} = -\mu_{T0}\eta_{vi} - \mu_{T1}\operatorname{sig}(\eta_{vi})^c - \mu_{T2}\operatorname{sig}(\eta_{vi})^d + g_{Vi}(T_i^s - T_i) \tag{5.51}$$

where $\mu_{T0}$, $\mu_{T1}$ and $\mu_{T2}$ are positive parameters, $0 < c < 1$, $d > 1$.

Consider the following Lyapunov function

$$\bar{W}_{i1} = \frac{1}{2}z_{Vi}^2 \tag{5.52}$$

The derivative of $\bar{W}_{i1}$ is calculated to be

$$\begin{aligned} \dot{\bar{W}}_{i1} =& z_{Vi}(F_{Vi} + g_{Vi}T_i^s - \dot{\overline{x}}_{vi} + \mu_{T0}\eta_{vi} \\ &+ \mu_{T1}\operatorname{sig}(\eta_{vi})^c + \mu_{T2}\operatorname{sig}(\eta_{vi})^d - g_{Vi}\Delta T_i) \end{aligned} \tag{5.53}$$

Thus, the control input law of velocity subsystem is constructed as

$$\begin{aligned} T_{id} =& \frac{1}{g_{Vi}}(\dot{\bar{x}}_{vi} - \mu_{T1}\operatorname{sig}(\eta_{vi})^c - \mu_{T2}\operatorname{sig}(\eta_{vi})^d - \mu_{V1}\operatorname{sig}(z_{Vi})^a \\ & - \mu_{T0}\eta_{vi} - \mu_{V2}\operatorname{sig}(z_{Vi})^b - \hat{F}_{Vi}) - u_0\tanh(\frac{z_{Vi}g_{Vi}u_0}{\varepsilon_V}) \end{aligned} \tag{5.54}$$

where $\mu_{V0}$ and $\mu_{V1}$ are positive parameters, $0 < a < 1, b > 1, \varepsilon_V > 0$.

According to Lemma 5.2, the result is

$$|z_{Vi}g_{Vi}u_0| - z_{Vi}g_{Vi}u_0\tanh(\frac{z_{Vi}g_{Vi}u_0}{\varepsilon_V}) \le 0.2785\varepsilon_V \tag{5.55}$$

When $t > t_{V0}$, we have $\hat{F_{Vi}} = F_{Vi}$. Then according to Eqs. (5.53),(5.54) and (5.55), it can be derived as

$$\dot{\bar{W}}_{i1} \le -\mu_{V1}|z_{Vi}|^{\alpha+1} - \mu_{V2}|z_{Vi}|^{b+1} + 0.2785\varepsilon_V \tag{5.56}$$

#### 5.3.2.2 Fault-Tolerant Controller Design of Altitude Subsystem Based on ITSM

In this section, for the altitude subsystem is designed based on the backstepping structure, the fixed-time disturbance observer is used to estimate the lumped uncertainty information to deal with the uncertainty and actuator fault problems, and the ITSM control is used to design a controller with stronger robustness and anti-interference. The following four steps introduce the design of fault-tolerant controller.

Since the virtual control law needs to be derived many times in the back-stepping method, in order to avoid the problem of computational explosion caused by the virtual control law, the fixed-point method is introduced [23]

$$\zeta\dot{\kappa}_{ic} = -\operatorname{sgn}(e_{\kappa i})^{l_3} - e_{\kappa i} - \operatorname{sgn}(e_{\kappa i})^{l_4} \tag{5.57}$$

where $\zeta > 0, 0.5 < l_3 < 1, l_4 > 1$, defining $e_{\kappa i} = \kappa_{ic} - \kappa_{id}(\kappa = \gamma, \alpha, q)$. $\kappa_{ic}$ is the instruction of $\kappa_{id}$ after passing through the filter, and the first derivative of $\kappa_{id}$ is bounded, i.e., there exists $\sigma_\kappa > 0$ satisfying $|\dot{\kappa}_{id}| \le \sigma_\kappa$. According to reference [23], $e_{\kappa i}$ converges to zero at a fixed time $t_f$.

Step 1 : The error is defined as $z_{hi} = h_i - \bar{x}_{hi}$, and the ITSM surface of Step 1 is designed as

$$S_h = z_{hi} + \int_0^t \left(\mu_{h1}\operatorname{sig}(z_{hi})^a + \mu_{h2}\operatorname{sig}(z_{hi})^b\right)\mathrm{d}\tau. \tag{5.58}$$

where $h_1$ and $h_2$ are adjustable positive parameters, $1 < a < 1$ and $b > 1$. The Lyapunov function is considered as

$$\bar{W}_{i2} = \frac{1}{2V_M^2} S_h^2 \tag{5.59}$$

where $V_M$ is the maximum value of $V_i$.

Combined with the Eq. (5.57), the derivation of $\bar{W}_{i2}$ is calculated as

$$\dot{\bar{W}}_{i2} = \frac{S_h}{V_M^2}(g_{hi}(z_{\gamma i} + e_{\gamma i} + \gamma_{id}) - \bar{v}_{hi} + \mu_{h1}\operatorname{sig}(z_{hi})^a + \mu_{h2}\operatorname{sig}(z_{hi})^b) \tag{5.60}$$

The tracking error $z_{\gamma i} = \gamma_i - \gamma_{ic}$ and the filtering error $e_{\gamma i} = \gamma_{ic} - \gamma_{id}$ are defined.

Thus, the virtual control law of the first step is designed as

$$\begin{aligned} \gamma_{id} =& \frac{1}{g_{hi}}(\bar{v}_{hi} - \mu_{h1}\operatorname{sig}(z_{hi})^a - \mu_{h2}\operatorname{sig}(z_{hi})^b - \mu_{h3}\operatorname{sig}(S_h)^{l_3} \\ & - \mu_{h4}\operatorname{sig}(S_h)^{l_4} - g_{hi}z_{\gamma i}) \end{aligned} \tag{5.61}$$

where $\mu_{h3}$ and $\mu_{h4}$ are positive parameters.

When $t > t_f$, it derives that $e_{\gamma i} = 0$. By combining with Eq. (5.61), it follows that

$$\dot{\bar{W}}_{i2} = -\frac{\mu_{h3}}{V_M^2}|S_h|^{l_3+1} - \frac{\mu_{h4}}{V_M^2}|S_h|^{l_4+1} \tag{5.62}$$

Step 2: The integral terminal sliding surface of Step 2 is designed as

$$S_\gamma = z_{\gamma i} + \int_0^t \left(\mu_{\gamma 1}\operatorname{sig}\left(z_{\gamma i}\right)^a + \mu_{\gamma 2}\operatorname{sig}\left(z_{\gamma i}\right)^b\right)\mathrm{d}\tau \tag{5.63}$$

where $\mu_{\gamma 1}$ and $\mu_{\gamma 2}$ are adjustable positive parameters.

The fixed-time disturbance observer is constructed as

$$\begin{cases} \dot{\hat{\gamma}}_i = \hat{F}_{\gamma i} + g_{\gamma i}\alpha_i + \theta_{\gamma 1}\operatorname{sig}(e_{\gamma 0})^{1/2} + \theta_{\gamma 2}\operatorname{sig}(e_{\gamma 0})^{3/2} \\ \dot{\hat{F}}_{\gamma i} = \theta_{\gamma 3}\operatorname{sig}(e_{\gamma 0})^{1/2} + \theta_{\gamma 4}\operatorname{sig}(e_{\gamma 0})^{3/2} \end{cases} \tag{5.64}$$

where $e_{\gamma 0} = \gamma_i - \hat{\gamma}_i$, $\hat{F}_{\gamma i}$ is the estimated value of $F_{\gamma i}$, and the parameters $\theta_{\gamma i}(i = 1, 2, 3, 4)$ are all positive parameters. Estimate error is defined as $e_{\gamma 1} = F_{\gamma i} - \hat{F}_{\gamma i}$. According to the above fixed-time disturbance observer, the error $e_{\gamma 0}$ and $e_{\gamma 1}$ converge to zero at fixed time $t_{\gamma 0}$, i.e., when $t > t_{\gamma 0}$, $\hat{\gamma}_i = \gamma_i$ and $\hat{F}_{\gamma i} = F_{\gamma i}$.

The Lyapunov function is considered as

$$\bar{W}_{i3} = \frac{1}{2} S_\gamma^2 \tag{5.65}$$

Derivation of $\bar{W}_{i3}$ is calculated as

$$\dot{\bar{W}}_{i3} = S_\gamma (F_{\gamma i} + g_{\gamma i}(z_{\alpha i} + e_{\alpha i} + \alpha_{id}) - \dot{\gamma}_{ic} + \mu_{\gamma 1}\operatorname{sig}(z_{\gamma i})^a + \mu_{\gamma 2}\operatorname{sig}(z_{\gamma i})^b) \tag{5.66}$$

where the tracking error $z_{\alpha i} = \alpha_i - \alpha_{ic}$ and the filtering error $e_{\alpha i} = \alpha_{ic} - \alpha_{id}$ are defined. Thus, the virtual control law of the second step is designed as

$$\begin{aligned} \alpha_{id} =& \frac{1}{g_{\gamma i}}(\dot{\gamma}_{ic} - \mu_{\gamma 1}\operatorname{sig}(z_{\gamma i})^a - \mu_{\gamma 2}\operatorname{sig}(z_{\gamma i})^b - \mu_{\gamma 3}\operatorname{sig}(S_\gamma)^{l_3} \\ & - \mu_{\gamma 4}\operatorname{sig}(S_\gamma)^{l_4} - \hat{F}_{\gamma i} - g_{\gamma i} z_{\alpha i}) \end{aligned} \tag{5.67}$$

where $\mu_{\gamma 3} > 0$ and $\mu_{\gamma 4} > 0$.

When $t > t_{\gamma 0} + 2t_f$, we have $\hat{F}_{\gamma i} = F_{\gamma i}$ and $e_{\alpha i} = 0$. Then, we combine with Eq. (5.61), it follows that

$$\dot{\bar{W}}_{i3} = -\mu_{\gamma 3}\left|S_\gamma\right|^{l_3+1} - \mu_{\gamma 4}\left|S_\gamma\right|^{l_4+1} \tag{5.68}$$

Step 3: The integral terminal sliding surface of step 3 is designed as

$$S_\alpha = z_{\alpha i} + \int_0^t \left(\mu_{\alpha 1}\operatorname{sig}(z_{\alpha i})^a + \mu_{\alpha 2}\operatorname{sig}(z_{\alpha i})^b\right) \mathrm{d}\tau \tag{5.69}$$

where $\mu_{\alpha 1} > 0$ and $\mu_{\alpha 2} > 0$.

The fixed-time disturbance observer is constructed as

$$\begin{cases} \dot{\hat{\alpha}}_i = \hat{F}_{\alpha i} + g_{\alpha i} q_i + \theta_{\alpha 1}\operatorname{sig}(e_{\alpha 0})^{1/2} + \theta_{\alpha 2}\operatorname{sig}(e_{\alpha 0})^{3/2} \\ \dot{\hat{F}}_{\alpha i} = \theta_{\alpha 3}\operatorname{sig}(e_{\alpha 0})^{1/2} + \theta_{\alpha 4}\operatorname{sig}(e_{\alpha 0})^{3/2} \end{cases} \tag{5.70}$$

where $e_{\alpha 0} = \alpha_i - \hat{\alpha}_i$, $\hat{F}_{\alpha i}$ is the estimated value of $F_{\alpha i}$, and the parameters $\theta_{\alpha i} (i = 1, 2, 3, 4)$ are all positive parameters. Estimate error is defined as $e_{\alpha 1} = F_{\alpha i} - \hat{F}_{\alpha i}$. According to the above fixed-time disturbance observer, the error $e_{\alpha 0}$ and $e_{\alpha 1}$ converge to zero at fixed time $t_{\alpha 0}$, i.e., when $t > t_{\alpha 0}$, $\hat{\alpha}_i = \alpha_i$ and $\hat{F}_{\alpha i} = F_{\alpha i}$.

The Lyapunov function is considered as

$$\bar{W}_{i4} = \frac{1}{2} S_\alpha^2 \tag{5.71}$$

Taking the time-derivative of $\bar{W}_{i4}$, it has

$$\dot{\bar{W}}_{i4} = S_\alpha(F_{\alpha i} + g_{\alpha i}(z_{qi} + e_{qi} + q_{id}) - \dot{\alpha}_{ic} + \mu_{\alpha 1}\operatorname{sig}(z_{\alpha i})^a + \mu_{\alpha 2}\operatorname{sig}(z_{\alpha i})^b) \tag{5.72}$$

where the tracking error $z_{qi} = q_i - q_{ic}$ and the filtering error $e_{qi} = q_{ic} - q_{id}$ are defined. Thus, the virtual control law of the second step is designed as

$$\begin{aligned} q_{id} =& \frac{1}{g_{\alpha i}}(\dot{\alpha}_{ic} - \mu_{\alpha 1}\operatorname{sig}(z_{\alpha i})^a - \mu_{\alpha 2}\operatorname{sig}(z_{\alpha i})^b - \mu_{\alpha 3}\operatorname{sig}(S_\alpha)^{l_3} \\ & - \mu_{\alpha 4}\operatorname{sig}(S_\alpha)^{l_4} - \hat{F}_{\alpha i} - g_{\alpha i} z_{qi}) \end{aligned} \tag{5.73}$$

where $\mu_{\alpha 3} > 0$ and $\mu_{\alpha 4} > 0$.

When $t > t_{\gamma 0} + t_{\alpha 0} + 3t_f$, $\hat{F}_{\alpha i} = F_{\alpha i}$ and $e_{qi} = 0$. Then combined with Eq. (5.73), it follows that

$$\dot{\bar{W}}_{i4} = -\mu_{\alpha 3}\,|S_\alpha|^{l_3+1} - \mu_{\alpha 4}\,|S_\alpha|^{l_4+1} \tag{5.74}$$

Step 4: The integral terminal sliding surface of step 4 is designed as

$$S_q = z_{qi} + \int_0^t \left(\mu_{q1}\operatorname{sig}\left(z_{qi}\right)^a + \mu_{q2}\operatorname{sig}\left(z_{qi}\right)^b\right)\mathrm{d}\tau \tag{5.75}$$

where $\mu_{q1} > 0$ and $\mu_{q2} > 0$.

By combining the differential equation and the mathematical model of actuator fault in Eq. (5.10), what can be obtained is

$$\dot{q}_i = F_{qi} + g_{qi}(\rho_i\delta_{ei} + \overline{\delta}_i) = \overline{F}_{qi} + g_{qi}\delta_{ei} \tag{5.76}$$

where $\bar{F}_{qi} = F_{qi} + g_{qi}((\rho_i - 1)\delta_{ei} + \overline{\delta}_i)$.

The fixed-time disturbance observer is constructed as

$$\begin{cases} \dot{\hat{q}}_i = \hat{\overline{F}}_{qi} + g_{qi}\delta_{ei} + \theta_{q1}\operatorname{sig}(e_{q0})^{1/2} + \theta_{q2}\operatorname{sig}(e_{q0})^{3/2} \\ \dot{\overline{F}}_{qi} = \theta_{q3}\operatorname{sig}(e_{q0})^{1/2} + \theta_{q4}\operatorname{sig}(e_{q0})^{3/2} \end{cases} \tag{5.77}$$

where $e_{q0} = q_i - \hat{q}_i$, $\hat{F}_{qi}$ is the estimated value of $F_{qi}$, and the parameters $\theta_{qi}(i = 1, 2, 3, 4)$ are all positive parameters. Estimate error is defined as $e_{q1} = F_{qi} - \hat{F}_{qi}$. According to the above fixed-time disturbance observer, the error $e_{q0}$ and $e_{q1}$ converge to zero at fixed time $t_{q0}$, i.e., when $t > t_{q0}$, $\hat{q}_i = q_i$ and $\hat{F}_{qi} = F_{qi}$.

The Lyapunov function is considered as

$$\bar{W}_{i5} = \frac{1}{2} S_q^2 \tag{5.78}$$

Derivation of $\bar{W}_{i5}$ is calculated as

$$\dot{\bar{W}}_{i5} = S_q(\overline{F}_{qi} + g_{qi}\delta_{ei} - \dot{q}_{ic} + \mu_{q1}\,\mathrm{sig}(z_{qi})^a + \mu_{q2}\,\mathrm{sig}(z_{qi})^b) \tag{5.79}$$

Thus, the final actual control input is designed as

$$\begin{aligned}\delta_{ei} =& \frac{1}{g_{qi}}(\dot{q}_{ic} - \mu_{q1}\,\mathrm{sig}\left(z_{qi}\right)^a - \mu_{q2}\,\mathrm{sig}\left(z_{qi}\right)^b - \mu_{q3}\,\mathrm{sig}\left(S_q\right)^{l_3} \\ & - \mu_{q4}\,\mathrm{sig}(S_q)^{l_4} - \hat{\overline{F}}_{qi})\end{aligned} \tag{5.80}$$

where $\mu_{q3} > 0$ and $\mu_{q4} > 0$.

When $t > t_{\gamma 0} + t_{\alpha 0} + t_{q0} + 3t_f$, it has $\hat{F}_{qi} = F_{qi}$. Then, we combine with Eq. (5.80), it follows that

$$\dot{\bar{W}}_{i5} = -\mu_{q3}\left|S_q\right|^{l_3+1} - \mu_{q4}\left|S_q\right|^{l_4+1} \tag{5.81}$$

### 5.3.3 *Stability Analysis*

**Theorem 5.2** A longitudinal fixed-wing UAVs model (5.8)–(5.10) with input saturation, actuator faults, external disturbances, and model uncertainty problems is considered. If control laws for the velocity system (5.54) against input saturation and integral terminal sliding-mode control laws (5.61), (5.67), (5.73), and (5.80) for the altitude system are designed as well as fixed-time filters (5.57), based on fixed-time disturbance observers (5.50), (5.64), (5.70), and (5.77), then the system state converges to the region near the equilibrium point at fixed time and the tracking error converges to the region near zero at fixed time.

***Proof*** The Lyapunov function is chosen as

$$\bar{W}_{slide} = \bar{W}_{i2} + \bar{W}_{i3} + \bar{W}_{i4} + \bar{W}_{i5} \tag{5.82}$$

Combining Eqs. (5.62), (5.68), (5.74) and (5.81), the derivative of slide $\bar{W}_{slide}$ can be calculated as

$$
\begin{aligned}
\dot{\bar{W}}_{slide} =& -\frac{\mu_{h3}}{V_M^2}|S_h|^{l_3+1} - \frac{\mu_{h4}}{V_M^2}|S_h|^{l_4+1} \\
& - \mu_{\gamma 3}|S_\gamma|^{l_3+1} - \mu_{\gamma 4}|S_\gamma|^{l_4+1} - \mu_{\alpha 3}|S_\alpha|^{l_3+1} \\
& - \mu_{\alpha 4}\left|S_\alpha\right|^{l_4+1} - \mu_{q3}\left|S_q\right|^{l_3+1} - \mu_{q4}\left|S_q\right|^{l_4+1} \\
\leq & - 2^{(l_3+1)/2} V_M^{l_3-1} \mu_{h3} (\frac{S_h^2}{2V_M^2})^{(l_3+1)/2} \\
& - 2^{(l_4+1)/2} V_M^{l_4-1} \mu_{h4} (\frac{S_h^2}{2V_M^2})^{(l_4+1)/2} \\
& - 2^{(l_3+1)/2} \mu_{\gamma 3} (\frac{S_\gamma^2}{2})^{(l_3+1)/2} - 2^{(l_4+1)/2} \mu_{\gamma 4} (\frac{S_\gamma^2}{2})^{(l_4+1)/2} \\
& - 2^{(l_3+1)/2} \mu_{\alpha 3} (\frac{S_\alpha^2}{2})^{(l_3+1)/2} - 2^{(l_4+1)/2} \mu_{\alpha 4} (\frac{S_\alpha}{2})^{(l_4+1)/2} \\
& - 2^{(l_3+1)/2} \mu_{q3} (\frac{S_q}{2})^{(l_3+1)/2} - 2^{(l_4+1)/2} \mu_{q4} (\frac{S_q}{2})^{(l_4+1)/2}
\end{aligned} \tag{5.83}
$$

From Lemma 2.2, it can be inferred that

$$
\dot{\bar{W}}_{slide} \leq -2^{\frac{l_3+1}{2}} \lambda_3 \bar{W}_{slide}^{\frac{l_3+1}{2}} - 2^{\frac{3-l_4}{2}} \lambda_4 \bar{W}_{slide}^{\frac{l_4+1}{2}} \tag{5.84}
$$

where $\lambda_3 = \min\{\mu_{h3} V_M^{l_3-1}, \mu_{\gamma 3}, \mu_{\alpha 3}, \mu_{q3}\} > 0$, $\lambda_4 = \min\{\mu_{h4} V_M^{l_4-1}, \mu_{\gamma 4}, \mu_{\alpha 4}, \mu_{q4}\} > 0$. According to Lemma 2.2, it can be shown that the closed-loop system is stable in a fixed time, and the sliding surface $S_h$, $S_\gamma$, $S_\alpha$ and $S_q$ can converge to zero in a fixed time.

When the sliding surface reaches zero, it can be obtained that

$$
\begin{cases}
\dot{z}_{hi} = -\mu_{h1} \operatorname{sig}(z_{hi})^a - \mu_{h2} \operatorname{sig}(z_{hi})^b \\
\dot{z}_{\gamma i} = -\mu_{\gamma 1} \operatorname{sig}\left(z_{\gamma i}\right)^a - \mu_{\gamma 2} \operatorname{sig}\left(z_{\gamma i}\right)^b \\
\dot{z}_{\alpha i} = -\mu_{\alpha 1} \operatorname{sig}(z_{\alpha i})^a - \mu_{\alpha 2} \operatorname{sig}(z_{\alpha i})^b \\
\dot{z}_{qi} = -\mu_{q1} \operatorname{sig}\left(z_{qi}\right)^a - \mu_{q2} \operatorname{sig}\left(z_{qi}\right)^b
\end{cases} \tag{5.85}
$$

The Lyapunov function is chosen as

$$
W_z = \sum_{j \in \{V,h,\gamma,\alpha,q\}} \frac{1}{2} z_{ji}^2 \tag{5.86}
$$

Combining Eqs. (5.56) and (5.85), the derivative of slide $W_z$ can be calculated as

$$
\begin{aligned}
\dot{W}_z \leq & - \mu_{V1}\left|z_{Vi}\right|^{\alpha+1} - \mu_{V2}\left|z_{Vi}\right|^{b+1} + 0.2785\varepsilon_V \\
& - \mu_{\gamma 1}|z_{\gamma i}|^{a+1} - \mu_{\gamma 2}|z_{\gamma i}|^{b+1} - \mu_{\alpha 1}|z_{\alpha i}|^{a+1} - \mu_{\alpha 2}|z_{\alpha i}|^{b+1} \\
& - \mu_{q1}|z_{qi}|^{a+1} - \mu_{q2}\left|z_{qi}\right|^{b+1} - \mu_{h1}\left|z_{hi}\right|^{\alpha+1} - \mu_{h2}\left|z_{hi}\right|^{b+1}
\end{aligned} \tag{5.87}
$$

From Lemma 2.2 it follows that

$$\dot{W}_z \leq -2^{\frac{a+1}{2}}\lambda_5 W_z^{\frac{a+1}{2}} - 2^{\frac{b+1}{2}} 5^{\frac{1-b}{2}} \lambda_6 W_z^{\frac{b+1}{2}} + 0.2785\varepsilon_V \tag{5.88}$$

where $\lambda_5 = \min\{\mu_{V1}, \mu_{h1}, \mu_{\gamma 1}, \mu_{\alpha 1}, \mu_{q1}\} > 0$, $\lambda_6 = \min\{\mu_{V2}, \mu_{h2}, \mu_{\gamma 2}, \mu_{\alpha 2}, \mu_{q2}\} > 0$

Based on the above analysis and Lemma 3.2, it can be concluded that the tracking errors $z_{Vi}$, $z_{hi}$, $z_{\gamma i}$, $z_{\alpha i}$ and $z_{qi}$ of the fixed-wing UAVs state variables converge to the vicinity of zero in a fixed time. Theorem 5.2 is proven completely.

**Remark 5.2** The size of $0.2785\varepsilon_V$ in Eq. (5.88) depends on the parameter $\varepsilon_V$, and the $\varepsilon_V$ selected by simulation is very small, so it will not cause excessive error caused by $0.2785\varepsilon_V$ being too large.

## 5.4 Simulation Results

In order to verify the control effect of the FTC case of anti-input saturation formation based on fixed-time observer under switching topology proposed in this chapter, simulation experiments are carried out. During the simulation, a system composed of four fixed-wing UAV followers and a virtual leader is used, and its switching communication topology is shown in Fig. 5.1. The velocity reference signal and height reference signal are processed through the following two second-order filters to generate smooth reference signals $V_0$ and $h_0$

$$\frac{V_0}{V_d} = \frac{0.25}{s^2 + s + 0.25}, \quad \frac{h_0}{h_d} = \frac{0.36}{s^2 + 1.2\,s + 0.36}. \tag{5.89}$$

The speed reference signal given by the virtual leader is a step instruction from $V_d = 36\,\text{m/s}$ to $V_d = 40\,\text{m/s}$. The altitude reference signal given by the virtual leader is a step instruction from $h_d = 975\,\text{m}$ to $h_d = 1000\,\text{m}$. The velocity reference signal and the altitude reference signal are given by using two second-order filters as shown in Eq. (5.89) to generate smooth reference signals $V_0$ and $V_h$, respectively.

The expected formation of the formation system is that the expected relative height distance between each follower fixed-wing UAVs and the leader is $\chi_1 = 0$ m, $\chi_2 = 5$ m, $\chi_3 = -5$ m, $\chi_4 = 2$ m.

The configuration of body parameters and aerodynamic parameters of each fixed-wing UAV is the same, and the parameters are shown in Table 3.1 and Table 3.2.

The initial state of fixed-wing UAV1 is

$V_1(0) = 36$ m/s, $h_1(0) = 975$ m, $\gamma_1(0) = 0$ rad, $\alpha_1(0) = 0$ rad, $q_1(0) = 0$ rad/s;

The initial state of fixed-wing UAV2 is

$V_2(0) = 35$ m/s, $h_2(0) = 980$ m, $\gamma_2(0) = 0$ rad, $\alpha_2(0) = 0$ rad, $q_2(0) = 0$ rad/s;

The initial state of fixed-wing UAV3 is

$V_3(0) = 37\ \mathrm{m/s},\ h_3(0) = 970\ \mathrm{m},\ \gamma_3(0) = 0\ \mathrm{rad},\ \alpha_3(0) = 0\ \mathrm{rad},\ q_3(0) = 0\ \mathrm{rad/s}$;

The initial state of fixed-wing UAV4 is

$V_4(0) = 36\ \mathrm{m/s},\ h_4(0) = 977\ \mathrm{m},\ \gamma_4(0) = 0\ \mathrm{rad},\ \alpha_4(0) = 0\ \mathrm{rad},\ q_4(0) = 0\ \mathrm{rad/s}$;

The initial state quantities of the four first-order agents of the virtual speed system are

$\overline{x}_{v1} = 36\ \mathrm{m/s},\ \overline{x}_{v2} = 35\ \mathrm{m/s},\ \overline{x}_{v3} = 37\ \mathrm{m/s},\ \overline{x}_{v4} = 36\ \mathrm{m/s}.$

The initial state quantities of the four second-order agents of the virtual altitude system are

$$\begin{cases} \overline{x}_{h1} = 975\mathrm{m} \\ \overline{v}_{h1} = 0\mathrm{m/s} \end{cases},\ \begin{cases} \overline{x}_{h2} = 980\mathrm{m} \\ \overline{v}_{h2} = 0\mathrm{m/s} \end{cases},\ \begin{cases} \overline{x}_{h3} = 970\mathrm{m} \\ \overline{v}_{h3} = 0\mathrm{m/s} \end{cases},\ \begin{cases} \overline{x}_{h4} = 977\mathrm{m} \\ \overline{v}_{h4} = 0\mathrm{m/s} \end{cases}.$$

Aiming at the problem of actuator fault in fixed-wing UAVs formation, the design of simulation is that the followers fixed-wing UAV1 and fixed-wing UAV2 have rudder fault and the followers fixed-wing UAV3 and fixed-wing UAV4 have no fault. The fault mathematical model of fixed-wing UAV1 is

$$\delta_{e1}^{f} = \begin{cases} \delta_{e1},\ t \le 5\,s \\ (0.6 + 0.4\exp(-0.8(t-5)))\delta_{e1} + 0.01\sin(0.01(t-5)),\ t > 5\,s \end{cases}$$

The fault mathematical model of fixed-wing UAV2 is

$$\delta_{e2}^{f} = \begin{cases} \delta_{e2},\ t \le 10\,s \\ (0.7 + 0.3\exp(-1.2(t-10)))\delta_{e2} + 0.01\sin(0.01(t-10)),\ t > 10\,s \end{cases}$$

The uncertainty $\Delta f_{ji} (j = V, \gamma, \alpha, q)$ is between $-10\%$ and $10\%$ of $f_{ji}$. The interference are $d_{Vi} = 0.2\gamma_i \sin(0.1t)$, $d_{\gamma i} = 0.3\alpha_i \sin(0.05t)$, $d_{\alpha i} = 0.01\sin(0.1t)$ and $d_{qi} = 0.05\sin(0.02t)$. The input saturation is $T_m = 75$, then $u_0 = 10$.

The controller parameters of the upper virtual system are given in Table 5.1. The design parameters of the fixed-time disturbance observer are given in Table 5.2. The

**Table 5.1** Upper virtual system controller parameters

| Symbol | Value | Symbol | Value | Symbol | Value |
|---|---|---|---|---|---|
| $p_{11}$ | 1 | $p_{12}$ | 1.1 | $k_{11}$ | 1 |
| $p_{21}$ | 1 | $p_{22}$ | 2 | $k_{21}$ | 1 |
| $p_{31}$ | 6 | $p_{32}$ | 5 | $k_{31}$ | 0.8 |
| $l_1$ | 0.5 | $l_2$ | 1.5 | $k_{41}$ | 6 |
| $a$ | 0.5 | $b$ | 1.5 | $k_{12}$ | 1.1 |
| $k_{22}$ | 2 | $k_{32}$ | 1 | $k_{42}$ | 5 |

**Table 5.2** Fixed-time observer parameters

| Symbol | Value | Symbol | Value | Symbol | Value |
|---|---|---|---|---|---|
| $\theta_{r1}$ | 12 | $\theta_{r2}$ | 6 | $\theta_{a1}$ | 11 |
| $\theta_{r3}$ | 5 | $\theta_{r4}$ | 6 | $\theta_{a3}$ | 8 |
| $\theta_{r1}$ | 10 | $\theta_{r2}$ | 5 | $\theta_{q1}$ | 14 |
| $\theta_{r3}$ | 6 | $\theta_{r4}$ | 8 | $\theta_{q3}$ | 6 |
| $\theta_{a2}$ | 7 | $\theta_{a4}$ | 8 | $\theta_{q2}$ | 6 |
| $\theta_{q4}$ | 6 | | | | |

**Table 5.3** fixed-wing UAVs control law and filter parameters

| Symbol | Value | Symbol | Value | Symbol | Value |
|---|---|---|---|---|---|
| $\mu_{T1}$ | 5 | $\mu_{T2}$ | 3 | $\mu_{y3}$ | 0.1 |
| $\mu_{Y1}$ | 8 | $\mu_{Y2}$ | 5 | $\mu_{\alpha 1}$ | 10 |
| $c$ | 0.5 | $d$ | 2 | $\mu_{\alpha 3}$ | 0.5 |
| $\varepsilon_y$ | 3 | $l_3$ | 0.5 | $\mu_{q1}$ | 4 |
| $l_4$ | 1.2 | $\xi$ | 0.05 | $\mu_{q3}$ | 0.6 |
| $\mu_{h1}$ | 20 | $\mu_{h2}$ | 15 | $\mu_{y4}$ | 0.2 |
| $\mu_{h3}$ | 1 | $\mu_{h4}$ | 1 | $\mu_{\alpha 2}$ | 6 |
| $\mu_{y1}$ | 4 | $\mu_{y2}$ | 3 | $\mu_{\alpha 4}$ | 0.1 |
| $\mu_{q2}$ | 5 | $\mu_{q4}$ | 0.2 | $\mu_{r0}$ | 4 |

control law parameters and fixed-time filter parameters of the fixed-wing UAVs are given in Table 5.3.

The simulation curves of the anti-input saturation formation fault-tolerant scheme based on ITSM control for multiple fixed-wing UAVs are designed in this chapter are shown in Figs. 5.2, 5.3, 5.4, 5.5 and 5.6.

Figure 5.2 shows the output of each first-order agent of the upper virtual speed system. From the graph, it is found that the output signal can still be quickly consistent with the speed signal of leader although under switching communication. Its convergence is fast, and it can track consistently and reach the desired speed within 0.5 s at most. Figure 5.3 shows the output of each second-order agent of the virtual altitude system. It can be seen that under the switching communication graph, the MAS still completes the expected formation. It is obvious that the MAS can reach the expected formation within 6 s at most, and the convergence speed is fast. There is no chattering in the whole simulation, which can be seen that the designed integral sliding mode controller has good effect and overcomes the chattering problem of the sliding mode surface.

Figure 5.4 shows the flight speed response curves of the four fixed-wing UAVs. From the figure, it can be found that the four fixed-wing UAVs can reach the desired speed signal in about 6 s under the influence of input saturation. The speed response curve is smooth after 6 s, which shows that the designed control system based on

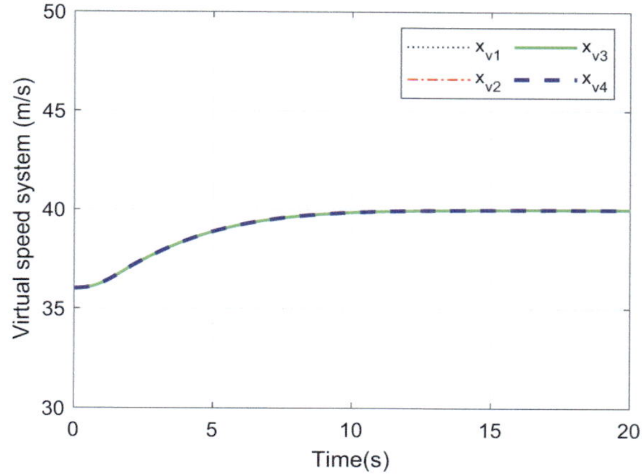


**Fig. 5.2** Output signal of virtual speed system

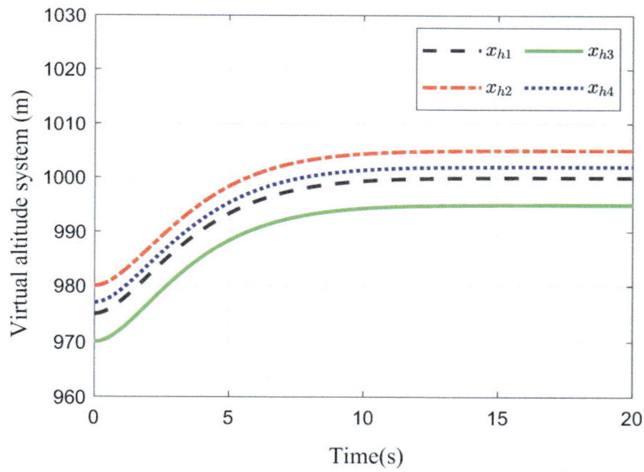


**Fig. 5.3** Output signal of virtual altitude system

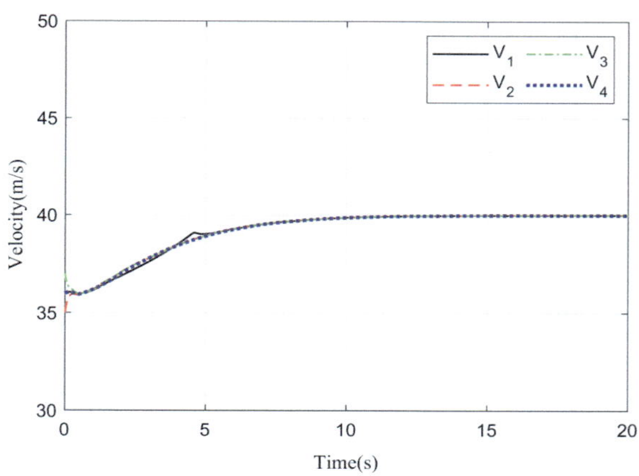


**Fig. 5.4** Flight speed of fixed-wing UAVs

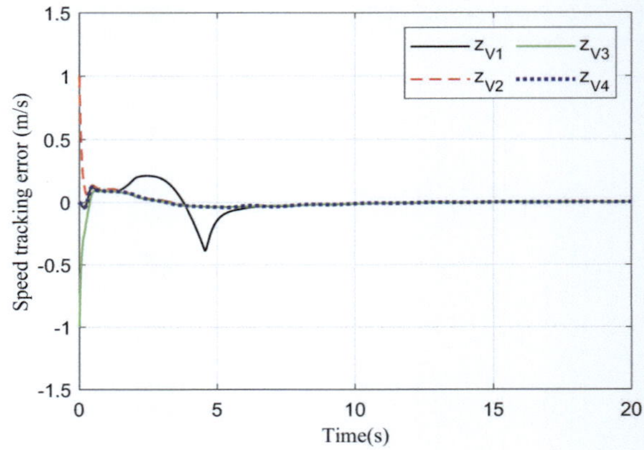


**Fig. 5.5** Speed tracking error of fixed-wing UAVs

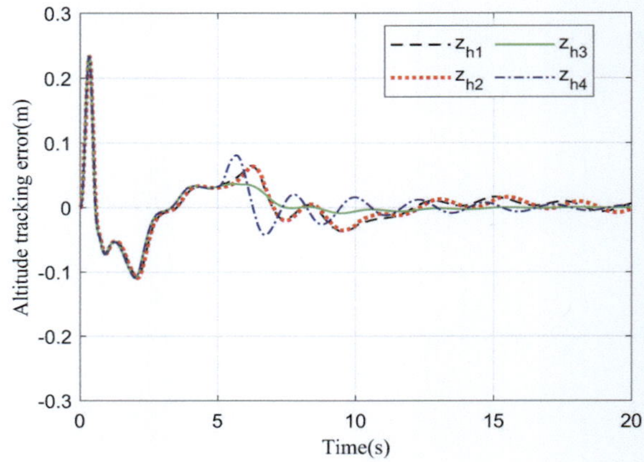


**Fig. 5.6** Altitude tracking error of fixed-wing UAVs

fixed time observer has good anti-saturation performance and strong anti-interference performance. Figure 5.5 shows the flight speed tracking error curves of the four fixed-wing UAVs. As seen from the figure, the speed tracking error of each fixed-wing UAV converges to the region near zero within 6 s, which is a fast convergence speed.

Figure 5.6 shows the altitude tracking error curves of the four fixed-wing UAVs. From the figure, it can be seen that the tracking error converges to the region near zero in about 10 s.

## 5.5 Conclusions

This chapter focuses on the development and design of a FTC strategy for formation under switching communication topologies for a vertical formation system of fixed-wing UAVs with model uncertainty, input saturation, and actuator faults. This chapter follows the fixed-time consistency principle and hierarchical control structure. In the upper layer formation controller design, considering the switching topology problem, an ITSM surface with higher-order sliding mode form is used in conjunction with the inverse step method to design the formation fault-tolerant controller, which not only requires the introduction of the communication topology relationship in the last step of the control law design, but also effectively attenuates the chattering problem of the sliding mode control. In the lower fixed-wing UAV tracking control design, a fixed-time auxiliary system is introduced to solve the control input saturation problem. Meanwhile, the fixed-time disturbance observer is used to estimate the aggregate information of model uncertainty and actuator fault. Finally, the effectiveness of the designed control scheme is verified by simulation experiments.

## References


1. Shi Y, Li J, Lv M, Wang N, Zhang B (2025) Distributed consensus control for 6-DOF fixed-wing multi-UAVs in asynchronously switching topologies. IEEE Trans Veh Technol 74(4):5649–5663
2. Zhang Z, Zheng L, Zhou Y, Guo Q (2022) A novel finite-time-gain-adjustment controller design method for UAVs tracking time-varying targets. IEEE Trans Intell Transp Syst 23(8):12531–12543
3. Wu J, Luo C, Luo Y, Li K (2021) Distributed UAV swarm formation and collision avoidance strategies over fixed and switching topologies. IEEE Trans Cybern 52(10):10969–10979
4. Wu J, Luo C, Min G, McClean S (2024) Formation control algorithms for multi-UAV systems with unstable topologies and hybrid delays. IEEE Trans Veh Technol 73(9):12358–12369
5. Feng H (2020) Distributed finite-time stochastic control for spatially interconnected markovian jump systems. J Franklin Inst 357(14):9195–9219
6. Liu K, Wang R, Zheng S, Dong S, Sun G (2022) Fixed-time disturbance observer-based robust fault-tolerant tracking control for uncertain quadrotor UAV subject to input delay. Nonlinear Dyn 107(6):2363–2390
7. Liu D, Bao W, Zhu X, Fei B, Men T, Xiao Z (2022) Cooperative path optimization for multiple UAVs surveillance in uncertain environment. IEEE Internet Things J 9(13):10676–10692
8. Dutta R, Sun L, Pack D (2018) A decentralized formation and network connectivity tracking controller for multiple unmanned systems. IEEE Trans Control Syst Technol 26(6):2206–2213
9. Kamdar R, Paliwal P, Kumar Y (2018) A state of art review on various aspects of multi-agent system. J Circuits Syst Comput 27(11):1830006
10. Zou Y, Xia K, He W (2022) Adaptive fault-tolerant distributed formation control of clustered vertical takeoff and landing UAVs. IEEE Trans Aerosp Electron Syst 58(2):1069–1082
11. Chen J, Li J, Yuan X (2020) Global fuzzy adaptive consensus control of unknown nonlinear multiagent systems. IEEE Trans Fuzzy Syst 28(3):510–522
12. Tripathi VK, Kamath AK, Behera L, Verma NK, Nahavandi S (2022) An adaptive fast terminal sliding-mode controller with power rate proportional reaching law for quadrotor position and altitude tracking. IEEE Trans Syst Man Cybern Syst 52(6):3612–3625
13. Wang R (2020) Adaptive output-feedback time-varying formation tracking control for multi-agent systems with switching directed networks. J Franklin Inst 357(1):551–568
14. Jiao R, Chou W, Rong Y, Dong M (2020) Anti-disturbance control for quadrotor UAV manipulator attitude system based on fuzzy adaptive saturation super-twisting sliding mode observer. Appl Sci 10(11):3719
15. Wang J, Wei Q, Sun L, Li Z (2022) Fault-tolerant adaptive pid switched control of robot manipulator based on average dwell time. Trans Inst Meas Control 44(16):3165–3175

16. Raoufi M, Habibi H, Yazdani A, Wang H (2022) Robust prescribed trajectory tracking control of a robot manipulator using adaptive finite-time sliding mode and extreme learning machine method. Robotics 11(5):111
17. Wang H, Cui G, Li H (2023) Fixed-time adaptive tracking control for a quadrotor unmanned aerial vehicle with input saturation. Actuators 12(3):130
18. Ni J, Shi P (2020) Adaptive neural network fixed-time leader-follower consensus for multiagent systems with constraints and disturbances. IEEE Trans Cybern 51(4):1835–1848
19. Baldi S, Roy S, Yang K, Liu D (2022) An underactuated control system design for adaptive autopilot of fixed-wing drones. IEEE/ASME Trans Mechatron 27(5):4045–4056
20. Zhang C, Zhang G, Dong Q (2023) Adaptive dynamic programming-based adaptive-gain sliding mode tracking control for fixed-wing unmanned aerial vehicle with disturbances. Int J Robust Nonlinear Control 33(2):1065–1097
21. Sun K, Liu L, Qiu J, Feng G (2021) Fuzzy adaptive finite-time fault-tolerant control for strict-feedback nonlinear systems. IEEE Trans Fuzzy Syst 29(4):786–796
22. Zhang W, Yue H, Chen Q, Jiang Y (2022) Fraction dynamic-surface-based adaptive neural finite-time control for stochastic nonlinear systems subject to unknown control directions, time-varying input delay and state delay. J Franklin Inst 359(18):10355–10391
23. Li J, Yang Y, Hua C, Guan X (2017) Fixed-time backstepping control design for high-order strict-feedback non-linear systems via terminal sliding mode. IET Control Theory Appl 11(8):1184–1193

# Chapter 6
# Fixed-Time FTC for UAVs-UGVs Under Switching Topologies

## 6.1 Introduction

Recently, numerous FTC methods have been developed to improve the safety, reliability, and maintainability of MASs. Among these approaches, sliding mode control stands out due to its inherent advantages, including robustness against uncertainties and external disturbances, as well as fault tolerance for unknown faults [6, 7, 9]. However, conventional sliding mode control has several drawbacks that limit its practical application in FTC. For example, traditional sliding mode control restricts system motion to an asymptotically stable sliding surface. To solve this problem, the terminal sliding mode control, which provides some outstanding performances such as finite-time convergence and high tracking precision, was developed in [16, 24]. Subsequently, fixed-time terminal sliding mode control was developed in [29, 33], featuring a settling time that is assignable regardless of initial conditions. While effective in theory, practical implementation faces challenges from control chattering, which induces undesirable mechanical oscillations in physical systems. To mitigate this issue, several solutions have been proposed, including boundary layer methods [10] and high-order sliding mode control (HOSMC) [17, 20]. Nevertheless, the inherent heterogeneity in UAV-UGV systems presents significant challenges in designing continuous fixed-time controllers for formation tracking applications.

In practical scenarios, communication topologies often become unreliable due to multiple factors including channel faults, dynamic link establishment, limited sensor ranges, and environmental disturbances [14, 22, 23]. This inherent unreliability makes the investigation of formation control under switching topologies both theoretically significant and practically valuable. Existing approaches utilizing Laplacian matrix and average dwell-time techniques have demonstrated capability in achieving multi-agent formation tracking under switching topologies [25, 28, 32]. However, these methods are fundamentally limited to asymptotic or exponential convergence of formation errors. To the best of our knowledge, the critical problem of fixed-time

 
K. Zhang et al., *Fixed-Time/Prescribed-Time Fault-Tolerant Control for Multiple Unmanned Systems*, Unmanned System Technologies,
https://doi.org/10.1007/978-981-95-2913-1_6

formation control for heterogeneous UAV-UGV systems under switching communication topologies remains unresolved in current literature.

Building upon the preceding analysis, this chapter investigates continuous fixed-time fault-tolerant formation tracking for heterogeneous UAV-UGV systems under both fixed and switching communication topologies. To guarantee successful formation execution, we propose a dual-control strategy: (1) a distributed formation controller for coordination in the $xy$ axis, and (2) a decentralized tracking controller for altitude regulation along the $z$ axis. The principal contributions of this chapter can be summarized as follows

(1) A novel continuous fixed-time formation control protocol is proposed for heterogeneous UAV-UGV systems subject to actuator faults and external disturbances. The developed controller maintains three key advantages: (i) it generates continuous control signals despite actuator faults, (ii) ensures fixed-time convergence of formation errors, and (iii) allows direct computation and adjustment of the convergence time through proper gain selection. While prior works [18, 30, 31] have addressed asymptotic formation tracking, this represents the first investigation of fixed-time FTC for aerial-ground vehicle systems.

(2) For each follower agent, a fixed-time disturbance observer is developed to simultaneously estimate actuator faults and external disturbances. Based on these estimates, we first introduce an innovative integral fixed-time sliding mode surface and establish a third-order dynamics model. Subsequently, by integrating the observer with backstepping control methodology, we design dual-mode fixed-time controllers: a distributed formation controller for planar ($xy$-axis) coordination and a decentralized tracking controller for altitude ($z$-axis) regulation, which collectively guarantee fixed-time convergence of the third-order system states to equilibrium despite actuator faults.

(3) Unlike existing formation control studies limited to fixed communication topologies [8, 12, 26], this chapter advances the field by addressing the more challenging and practical case of switching networks while achieving fixed-time convergence. Crucially, since fixed topologies represent a special case of switching topologies (where the switching set contains only one element), our results not only generalize previous work but also offer broader applicability to real-world scenarios where network connectivity may vary.

## 6.2 Problem Formulation and Preliminaries

### 6.2.1 Problem Formulation

Building upon the modeling framework established in Chap. 4, we formulate the composite dynamics for the $i$th ($i \in F_1 \cup F_2$) follower operating in the $xy$-plane under both actuator faults and external disturbances $d_i(t) \in R^{p_i}$ as follows

$$\ddot{\chi}_i(t) = A_i \left(I - \Lambda_i^{xy}\right) u_i^{xy}(t) + \tau_i^{xy}(t) + f_i^{xy}(t) + d_i^{xy}(t) \tag{6.1}$$

where $\chi_i(t) = [x_i, y_i]^{\mathrm{T}} \in R^2$ and $u_i^{xy}(t) \in R^2$ denote the position and control input in the $xy$ axis, respectively; $\Lambda_i \in R^2$ is the loss efficiency factor; $\tau_i^{xy}(t)$ is the bias fault parameter, $f_i^{xy}(t) \in R^2$ denotes the unknown non-linear vector function and $d_i^{xy}(t) \in R^2$ represents the external disturbance. The diagonal matrix $A_i = \mathrm{diag}\{1/m_i, 1/m_i\}$ if and only if $i \in U_2$, otherwise $A_i = I_2$.

Similarly, the dynamics of the $i$th follower ($i \in F_1$) in the $z$ axis can be redesigned as

$$\ddot{z}_i(t) = (1 - \rho_{iz})\, u_{iz}(t)/m_i + \tau_{iz}(t) + f_{iz}(t) + d_{iz}(t) - g \tag{6.2}$$

where $\rho_{iz} \in R$, $\tau_{iz}(t) \in R$, and $d_{iz}(t) \in R$ are the loss efficiency factor, bias fault parameter, and disturbance of the $i$th follower in the $z$ axis, respectively.

Combining the above process, the control-oriented model for the $i$th follower is established as

$$\begin{cases} \ddot{\chi}_i(t) = A_i u_i^{xy}(t) + \Delta_i^{xy}(t), i \in F_1 \cup F_2 \\ \ddot{z}_i(t) = u_{iz}(t)/m_i + \Delta_{iz}(t), i \in F_2 \end{cases} \tag{6.3}$$

where $\Delta_i^{xy}(t) = \tau_i^{xy}(t) + f_i^{xy}(t) + d_i^{xy}(t) - A_i \Lambda_i^{xy} u_i^{xy}(t)$, and $\Delta_{iz}(t) = \tau_{iz}(t) + f_{iz}(t) + d_{iz}(t) - \rho_i u_{iz}(t) - g$ are the lumped uncertainties in the $xy$ and $z$ axis, respectively.

**Remark 6.1** Figure 6.1 presents the schematic configuration of the proposed heterogeneous UAV-UGV formation system. The inherent differences in operational workspace and dynamic characteristics between aerial and ground vehicles introduce significant challenges in achieving synchronized formation behavior, rendering conventional distributed formation FTC strategies inadequate for such heterogeneous systems. To address these challenges, this chapter employs the kinematic models described by Eqs. (6.2) and (6.3) to characterize the heterogeneous UAV-UGV system, and develops distinct cooperative FTC protocols for motion coordination in the horizontal $xy$ and vertical $z$ dimensions. This dual-axis control approach enhances both the design flexibility of the control system and the cooperative performance of the entire formation, while maintaining practical implementation feasibility.

**Remark 6.2** In practical applications, obtaining precise parameter measurements for unmanned vehicles proves extremely challenging due to unavoidable environmental factors such as aerodynamic effects, wind disturbances, and actuator faults in dynamic operating conditions. Consequently, the terms $\Delta_i^{xy}(t)$ and $\Delta_{iz}(t)$ in Eqs. (6.2) and (6.3) are treated as composite uncertainties encompassing multiple complex components, including parametric uncertainties, external disturbances, and unknown actuator faults. To address this challenge, we develop fixed-time observers specifically designed to estimate these lumped uncertainty terms with guaranteed convergence properties.

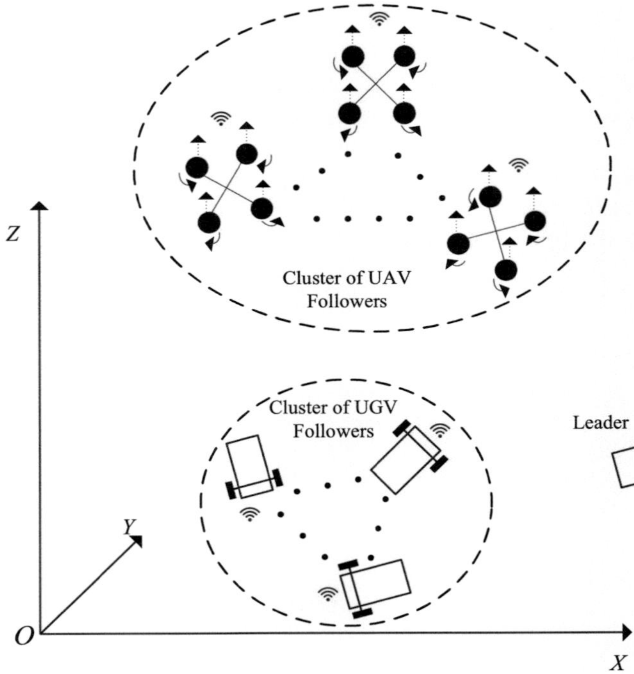


**Fig. 6.1** Schematic diagram of the studied UAVs-UGVs

### *6.2.2 Preliminaries*

This chapter studies a heterogeneous MASs comprising a leader agent labeled 0 and $N$ follower agents, whose inter-follower communication topology is modeled by an undirected graph $G = (S, E, A)$. Here, $S = \{s_1, s_2, ..., s_N\}$ represents the node set corresponding to the followers, $E = \{(s_i, s_j) : s_i, s_j \in S\}$ defines the edge set capturing their communication links, and $A = [a_{ij}] \in R^{N\times N}$ is the associated non-negative adjacency matrix where each element $a_{ij}$ quantifies the connection strength between agents $i$ and $j$. The adjacency element $a_{ij} > 0$ if and only if $(s_i, s_j) \in E$, otherwise, $a_{ij} = 0$. Let $D = \text{diag}\{\text{deg}_{\text{in}}(s_1), ..., \text{deg}_{\text{in}}(s_N)\}$ with $\text{deg}_{\text{in}}(s_i) = \sum_{j=1}^{N} a_{ij}$ denote the in-degree matrix of $G$. The Laplacian matrix $L = [l_{ij}] \in R^{N\times N}$ of $G$ is $L = D - A$. The undirected graph $G$ contains a spanning tree, if there is a root node which has path from it to every other node. The communication topology between the leader and the followers is denoted by $B = \text{diag}\{b_1, b_2, ..., b_N\}$. We assume $b_i > 0$, indicating that the $i$th follower can receive the leader's information in our framework. The complete communication network connecting all $N + 1$ agents is modeled by a directed graph $\bar{G}$, where we define $H = L + B$. An important property is that $H$ becomes invertible when $\bar{G}$ contains a spanning tree.

The interactive communication connection of the $N+1$ agents will be indicated by a directed switching graph. First, a switching signal $\sigma(t)$ is defined by $\sigma(t) : [0, +\infty) \rightarrow \Theta$ with $\Theta$ being a finite set of all possible topologies. Then, it is assumed that the switching topologies over the given directed graph set $\hat{\bar{G}}$, where $\hat{\bar{G}} = \{\bar{G}^1, \bar{G}^2, ..., \bar{G}^\Theta\}$. The graph $\bar{G}_{\sigma(t)} \in \hat{\bar{G}}$ is the communication topology graph at time $t$ and the time sequence $t_1, t_2, ...$ describes the switching sequence. Therefore, the switching topologies can be denoted by $\bar{G}_{\sigma(t)}$, and the $\bar{G}_{\sigma(t)}$ is fixed at any time interval.

**Assumption 6.1** The graph $\bar{G}$ contains a spanning tree, and the leader is the root node. Furthermore, the communication graph without leader is undirected connected.

**Assumption 6.2** For each time interval, the graph $\bar{G}_{\sigma(t)}$ $\left(t \in [t_k, t_{k+1}), k = 1, 2, ...\right)$ always contains a spanning tree rooted at the leader.

**Assumption 6.3** The UAV coefficients of $I_x$, $I_y$, and $I_z$, and $\xi_\phi$, $\xi_\theta$, and $\xi_\psi$ are unknown but bounded.

**Assumption 6.4** The parameters of fault severity and the control signal $u_0$ are time-varying and unmeasurable, but bounded, that is, there are several unknown constants $\bar{\rho}$, $\bar{\tau}$, and $\bar{u}_0$ such that $0 < \rho_{ik}(t) \le \bar{\rho} < 1$, $|\tau_{ik}| \le \bar{\tau} < \infty$, and $\|u_0\| \le \bar{u}_0 < \infty$.

**Assumption 6.5** The external disturbances $d_i^{xy}(t)$, and $d_i^z(t)$ are bounded and satisfy $\|d_i^{xy}(t)\| \le \bar{d}_1$, $\|d_i^z(t)\| \le \bar{d}_2$ with $\bar{d}_1$ and $\bar{d}_2$ being unknown constants.

**Lemma 6.1** ([1]) *Consider the following dynamic system*

$$\begin{cases} \dot{x}(t) = -o_1 \frac{x(t)}{\|x(t)\|^{1/2}} - o_2 x(t) \|x(t)\|^{m-1} + y(x) \\ y(x) = -o_3 \frac{x(t)}{\|x(t)\|} + \varsigma(t), \end{cases} \tag{6.4}$$

*where $\varsigma(t)$ is the disturbance. Moreover, the norm of $\varsigma(t)$ is bounded by a constant $W$, i.e., $\|\varsigma(t)\| \le W$. Then, $x(t), y(t) \in R^n$ can achieve fixed-time convergence within the following settling time*

$$T \le \left( \frac{1}{o_2(m-1)\nu^{m-1}} + \frac{2\left(\sqrt{n}\nu\right)^{1/2}}{o_1} \right) \left( 1 + \frac{o_3 + W}{(o_3 - W)\left(1 - \frac{\sqrt{2o_3}}{o_1}\right)} \right) \tag{6.5}$$

*where $\nu > 0, m > 1$. The gains $o_1$, $o_2$ and $o_3$ satisfy the conditions: $o_1 > \sqrt{2o_3}, o_2 > 0, o_3 > 4W$.*

**Control Objective**: To realize the desired fixed-time formation tracking, two user-defined time-varying vectors $h^{xy}(t) = \left[h_1^{xy}(t), h_2^{xy}(t), ..., h_N^{xy}(t)\right]^{\mathrm{T}}$ for $xy$ axis and $h^z(t) = \left[h_{N_1+1}^z(t), h_{N_1+2}^z(t), ..., h_{N_1+N_2}^z(t)\right]^{\mathrm{T}}$ for $z$ axis are introduced, where $h_i^{xy}(t)$ $(i = 1, 2, ..., N)$, and $h_{N_1+j}^z(t) \in R$ $(j = 1, 2, ..., N_2)$ are piecewise continuously differentiable. Then, for any bound initial conditions, the heterogeneous UAVs-UGVs need to realize the fixed-time formation tracking specified by $h^{xy}(t)$ and $h^z(t)$ within a fixed time, i.e.,

$$\begin{cases} \lim\limits_{t \to T_{\max}} \left(\chi_i(t) - x_0(t) - h_i^{xy}(t)\right) = 0, & i \in F_1 + F_2 \\ \lim\limits_{t \to T_{\max}} \left(z_i(t) - h_i^z(t)\right) = 0, & i \in F_1 \end{cases} \tag{6.6}$$

where $T_{\max} > 0$ is the convergence time that is independent of initial conditions.

**Remark 6.3** In practical scenarios like area mapping and obstacle avoidance, the heterogeneous UAVs-UGVs system must adapt its formation shape according to mission requirements and environmental conditions. This necessitates addressing the time-varying formation problem, where the followers' formation is governed by two piecewise continuously differentiable vectors $h^{xy}(t)$ and $h^z(t)$. These vectors simultaneously characterize: (1) the relative formation geometry among all agents, and (2) the overall formation configuration relative to the leader. Notably, when $h^{xy}(t) = 0$ and $h^z(t) = 0$, the time-varying formation tracking problem simplifies to a consensus problem. Therefore, the proper design of $h^{xy}(t)$ and $h^z(t)$ enables the formation system to achieve diverse macroscopic movement patterns, making this approach both theoretically sound and practically essential.

## 6.3 Main Results

### 6.3.1 Fixed-Time Fault-Disturbance Estimation Observer Design

**Theorem 6.1** *Consider the ith follower system of Eq. (6.3), and suppose that the lumped uncertainties $\Delta_i^{xy}(t)$ and $\Delta_{iz}(t)$ satisfy $\left\|\dot{\Delta}_i^{xy}(t)\right\| \leq W_1$ and $\left\|\dot{\Delta}_{iz}(t)\right\| \leq W_2$, where $W_1$ and $W_2$ are known constants. Let $\sigma_{i1}, \sigma_{i2}, \sigma_{i3}$ and $\sigma_{i4}$ as the states of the proposed fixed-time observer. Then, if the observers are designed as*

$$\begin{cases} \dot{\sigma}_{i1} = -o_{i1}\frac{s_{i1}}{\|s_{i1}\|^{\frac{1}{2}}} - o_{i2}s_{i1}\|s_{i1}\|^{m_{i1}-1} + \sigma_{i2} + A_i u_i^{xy} \\ \dot{\sigma}_{i2} = -o_{i3}\frac{s_{i1}}{\|s_{i1}\|} \end{cases} \tag{6.7}$$

*and*

$$\begin{cases} \dot{\sigma}_{i3} = -o_{i4}\frac{s_{i2}}{\|s_{i2}\|^{\frac{1}{2}}} - o_{i5}s_{i2}\|s_{i2}\|^{m_{i2}-1} + \sigma_{i4} + u_{iz}/m_i \\ \dot{\sigma}_{i4} = -o_{i6}\frac{s_{i2}}{\|s_{i2}\|} \end{cases} \tag{6.8}$$

*under the conditions*

$$o_{i1} > \sqrt{2o_{i3}}, o_{i2} > 0, o_{i3} > 4W_1 \tag{6.9}$$

*and*

$$o_{i4} > \sqrt{2o_{i5}}, o_{i5} > 0, o_{i6} > 4W_2 \tag{6.10}$$

*where* $s_{i1} = \sigma_{i1} - \dot{\chi}_i$, $s_{i2} = \sigma_{i3} - \dot{z}_i$, *and* $m_{i1} > 1$. *Then,* $\sigma_{i2}$ *and* $\sigma_{i4}$ *can estimate* $\Delta_i^{xy}$ *and* $\Delta_{iz}$ *within a fixed time, respectively.*

***Proof*** Differentiating $s_{i1}$ along Eq. (6.3) and Eq. (6.7) obtains

$$\begin{aligned}\dot{s}_{i1} &= \dot{\sigma}_{i1} - \ddot{\chi}_i \\ &= -o_{i1}\frac{s_{i1}}{\|s_{i1}\|^{\frac{1}{2}}} - o_{i2}s_{i1}\|s_{i1}\|^{m_{i1}-1} + \sigma_{i2} - \Delta_i^{xy}\end{aligned} \tag{6.11}$$

Letting $s_{i1}^* = \sigma_{i2} - \Delta_i^{xy}$ and taking the derivative of it gives

$$\dot{s}_{i1}^* = -o_{i3}\frac{s_{i1}}{\|s_{i1}\|} - \dot{\Delta}_i^{xy} \tag{6.12}$$

As a result, the error dynamics of $\Delta_i^{xy}$ can be obtained as

$$\begin{cases}\dot{s}_{i1} = -o_{i1}\frac{s_{i1}}{\|s_{i1}\|^{\frac{1}{2}}} - o_{i2}s_{i1}\|s_{i1}\|^{m_{i1}-1} + s_{i1}^* \\ \dot{s}_{i1}^* = -o_{i3}\frac{e_{i1}}{\|s_{i1}\|} - \dot{\Delta}_i^{xy}\end{cases} \tag{6.13}$$

In view of Lemma 6.1, when the gains $o_{i1}$, $o_{i2}$, and $o_{i3}$ satisfy condition Eq. (6.9), $s_{i1}$ and $s_{i1}^*$ can arrive at the origin within the following settling time

$$T_{1i} \leq \left(\frac{1}{o_{2i}(m_{i1}-1)\nu_{i1}^{m_{i1}-1}} + \frac{2(\sqrt{2}\nu_{i1})^{1/2}}{o_{i1}}\right) \times \left(1 + \frac{o_{i3}+M_1}{(o_{i3}-M_1)\left(1-\frac{\sqrt{2o_{i3}}}{o_{i1}}\right)}\right) \tag{6.14}$$

where $\nu_{i1} > 0$. Moreover, when $\nu_{i1} = \left(\frac{2^{1/4}o_{i1}}{o_{i2}}\right)^{\frac{1}{m_{i1}+1/2}}$, the minimum value of $T_{1i}$ can be obtained. Consequently, $\sigma_{i2}$ converges to $\Delta i^{xy}$ within the fixed time $T_{1i}$. Through analogous analysis, the fixed-time convergence of $\Delta_{iz}$ can be similarly demonstrated, though omitted here for brevity.

**Remark 6.4** Unlike existing finite-time formation control strategies that assume the lumped uncertainty is bounded, this chapter considers the more practical scenario where the derivative of the lumped uncertainty is bounded. The FTC methods for MASs presented in [5, 11] can eliminate estimation errors either asymptotically (as time approaches infinity) or within a finite time. In contrast, the proposed observers achieve fixed-time estimation of $\Delta_i^{xy}$ and $\Delta_{iz}$ is a crucial capability for safety-critical unmanned systems

### 6.3.2 Fixed-Time HOSMC Under Actuator Faults and Fixed Topologies

This section presents: (1) a distributed fixed-time fault-tolerant formation control scheme for the $xy$-plane, and (2) a decentralized tracking strategy for the $z$-axis, both developed under fixed communication topologies. First addressing the formation control problem, we design an integral fixed-time sliding mode surface to guarantee convergence within a fixed time interval.

$$S_{i1}=\int_0^t \left(\dot{e}_i + b_1 \text{sig}^{\alpha}(e_i) + b_2 \text{sig}^{\beta}(e_i)\right)\mathrm{d}\tau \tag{6.15}$$

where $e_i = \left[e_{ix}, e_{iy}\right]^{\mathrm{T}}$, and $\text{sig}^{l}(e_i) = [|e_{ix}|^{l}\text{sgn}(e_{ix}), \left|e_{iy}\right|^{l}\text{sgn}\left(e_{iy}\right)]^{\mathrm{T}}$. $\alpha$ and $\beta$ are positive constants satisfying the relation $0 < \alpha < 1$ and $\beta > 1$, $b_1 > 0$ and $b_2 > 0$. $e_i$ denotes the tracking error, which is defined as

$$e_i = \chi_i - x_0 - h_i^{xy}. \tag{6.16}$$

**Theorem 6.2** *For the system of Eq. (6.3), if the formation states reach the sliding mode surface of Eq. (6.15), the formation error $e_i$ converges within a fixed settling time.*

***Proof*** When the sliding mode surface $S_{i1} = 0$ is arrived, from Eq. (6.15), one obtains

$$\dot{e}_i = -b_1 \text{sig}^{\alpha}(e_i) - b_2 \text{sig}^{\beta}(e_i) \tag{6.17}$$

Consider the following Lyapunov function

$$V_1 = \frac{1}{2}\sum_{i=1}^{N} e_i^{\mathrm{T}} e_i \tag{6.18}$$

Then, its time derivative can be expressed as

$$\begin{aligned}
\dot{V}_1 &= \sum_{i=1}^{N} e_i^{\mathrm{T}} \dot{e}_i \\
&= \sum_{i=1}^{N} \left(-b_1 e_i^{\mathrm{T}} \text{sig}^{\alpha}(e_i) - b_i e_i^{\mathrm{T}} \text{sig}^{\beta}(e_i)\right) \\
&\le \sum_{i=1}^{N} \left(-b_1 \left(e_i^{\mathrm{T}} e_i\right)^{\frac{\alpha+1}{2}} - b_2 M^{\frac{1-\beta}{2}} \left(e_i^{\mathrm{T}} e_i\right)^{\frac{\beta_i+1}{2}}\right) \\
&\le -b_1 2^{\frac{\alpha+1}{2}} V_1^{\frac{\alpha+1}{2}} - b_2 2^{\frac{\beta+1}{2}} (MN)^{\frac{1-\beta}{2}} V_1^{\frac{\beta+1}{2}}.
\end{aligned} \tag{6.19}$$

where $M$ is the dimension of $e_i$. From Lemma 2.1, it can be easily concluded that the formation error $e_i$ are fixed-time stabilized. Then, the upper bound of convergence time $T_2$ satisfies

$$T_2 = \frac{1}{b_1 2^{\frac{\alpha-1}{2}}(1-\alpha_i)} + \frac{1}{b_2 2^{\frac{\beta-1}{2}}(MN)^{\frac{1-\beta}{2}}(\beta-1)} \tag{6.20}$$

**Remark 6.5** Following the same proof methodology as Eqs. (6.17)–(6.19) for the $xy$-axis case, the fixed-time stability of $z$-axis tracking error can be guaranteed. It should be noted that the formation error here is defined as $e_i = z_i - h_i^z$, where $h_i^z$ represents the desired trajectory for the $i$th ($i \in F_2$) follower in the $z$-axis direction.

Then, taking the first and second derivatives of the sliding surface of Eq. (6.15), one has

$$\begin{cases} \dot{S}_{i1} = \dot{e}_i + b_1 \mathrm{sig}^{\alpha}(e_i) + b_2 \mathrm{sig}^{\beta_i}(e_i) \\ \ddot{S}_{i1} = \ddot{e}_i + \lambda_{i1}\dot{e}_i + \lambda_{i2}\dot{e}_i \end{cases} \tag{6.21}$$

where

$$\lambda_{i1} = \mathrm{diag}\left\{\alpha b_1 |e_{ix}|^{\alpha-1}, \alpha b_1 \left|e_{iy}\right|^{\alpha-1}\right\} \tag{6.22}$$

and

$$\lambda_{i2} = \mathrm{diag}\left\{\beta b_2 |e_{ix}|^{\beta-1}, \beta b_2 \left|e_{iy}\right|^{\beta-1}\right\} \tag{6.23}$$

According to Eqs. (6.15), (6.21), and (6.22), the second-order nonlinear system of the $i$th follower can be transformed into the following third-order system

$$\begin{cases} \dot{S}_{i1} = S_{i2} \\ \dot{S}_{i2} = S_{i3} \\ \dot{S}_{i3} = \frac{d}{dt}\left(\lambda_{i1}\dot{e}_i + \lambda_{i2}\dot{e}_i + A_i u_i^{xy} + \Delta_i^{xy} - \ddot{h}_i^{xy} - \ddot{x}_0\right) \end{cases} \tag{6.24}$$

Then a backstepping-based method will be designed for system of Eq. (6.24). Firstly, the coordinate changes are given as

$$\begin{cases} \zeta_{i1} = S_{i1} \\ \zeta_{i2} = S_{i2} - \gamma_{i1} \\ \zeta_{i3} = S_{i3} - \gamma_{i2} \end{cases} \tag{6.25}$$

where $\gamma_{i1}$ and $\gamma_{i2}$ are the virtual control signals to be designed.

**Remark 6.6** Analysis of Eq. (6.15) reveals that the designed sliding mode surface combines characteristics of both integral sliding mode and fixed-time sliding mode surfaces. This hybrid structure provides superior transient response, steady-state performance, and guaranteed fixed-time stability. These features enable the developed

FTC system to effectively compensate for actuator faults while achieving faster formation system stabilization. Furthermore, Eq. (6.24) demonstrates a HOSMC structure, allowing the reconstructed controller to: (1) eliminate chattering effects, and (2) generate continuous control outputs similar to conventional HOSMC approaches.

**Theorem 6.3** *For the heterogeneous formation system described by Eq. (6.3) satisfying Assumption 6.1, when employing the fixed-time sliding mode manifold defined in Eq. (6.15), continuous fixed-time fault-tolerant formation tracking for the UAVs-UGVs system in the xy-plane can be achieved. This is accomplished through the combined action of the fixed-time observer in Eq. (6.7) and the subsequent distributed formation control protocol.*

$$u_i^{xy} = A_i^{-1}\left(\ddot{x}_0 + \ddot{h}_i^{xy} - \sigma_{i2} - \lambda_{i1}\dot{e}_i - \lambda_{i2}\dot{e}_i + u_{is}^{xy}\right) \tag{6.26}$$

*and*

$$u_{is}^{xy} = \int_0^t \left(\dot{\gamma}_{i2} - \chi_1 \mathrm{sig}(\zeta_{i3})^{c_1} - \delta_1 \mathrm{sig}(\zeta_{i3})^{d_1}\right)\mathrm{d}\tau \tag{6.27}$$

*where $\chi_1$ and $\delta_1$ are design positive constants. Besides, $c_1 \in (0, 1)$, and $d_1 > 1$. The virtual control inputs are designed as*

$$\gamma_{i2} = \dot{\gamma}_{i1} - \chi_2 \mathrm{sig}(\zeta_{i2})^{c_2} - \delta_2 \mathrm{sig}(\zeta_{i2})^{d_2} \tag{6.28}$$

*and*

$$\gamma_1 = -\chi_3 \mathrm{sig}((H \otimes I_2)\,\zeta_1)^{c_3} - \delta_3 \mathrm{sig}((H \otimes I_2)\,\zeta_1)^{d_3} \tag{6.29}$$

*where $\gamma_1 = \left[\gamma_{11}^{\mathrm{T}}, \gamma_{21}^{\mathrm{T}}, \dots, \gamma_{N1}^{\mathrm{T}}\right]^{\mathrm{T}}$, and $\zeta_1 = \left[\zeta_{11}^{\mathrm{T}}, \zeta_{21}^{\mathrm{T}}, \dots, \zeta_{N1}^{\mathrm{T}}\right]^{\mathrm{T}}$. Moreover, $c_2, c_3, d_2$, and $d_3$ are the controller parameters and satisfy $c_2, c_3 \in (0, 1)$, and $d_2, d_3 > 1$, $\chi_2, \delta_2, \chi_3$ and $\delta_3$ are positive constants to be chosen by the purpose of reaching a desired settling time.*

***Proof*** The stability analysis of the system in Eq. (6.25) is conducted through a three-step proof. The demonstration will reveal three finite convergence times $T_3$, $T_4$, and $T_5$ that remain independent of the error system's initial conditions, such that $\lim_{t\to T_3} \zeta_{i3} \to 0$, $\lim_{t\to T_4} \zeta_{i2} \to 0$, and $\lim_{t\to T_5} \zeta_{i1} \to 0$ for $i = 1, 2, \dots, N$, respectively.

*Step 1*: Select the following Lyapunov function

$$V_2 = \frac{1}{2}\sum_{i=1}^{N} \zeta_{i3}^{\mathrm{T}}\zeta_{i3} \tag{6.30}$$

Taking the derivative of $V_2$ along Eq. (6.24) with Eqs. (6.26) and (6.27) gives

$$\begin{aligned}\dot{V}_2 &= \sum_{i=1}^{N} \zeta_{i3}^{\mathrm{T}} \dot{\zeta}_{i3} \\ &= \sum_{i=1}^{N} \zeta_{i3}^{\mathrm{T}} \left( \frac{d}{dt} \begin{pmatrix} \lambda_{i1}\dot{e}_i + \lambda_{i2}\dot{e}_i + \\ A_i u_i^{xy} + \Delta_i^{xy} - \ddot{h}_i^{xy} - \ddot{x}_o \end{pmatrix} - \dot{\gamma}_{i2} \right) \\ &= \sum_{i=1}^{N} \zeta_{i3}^{\mathrm{T}} \left( \frac{d}{dt} \left( \Delta_i^{xy} - \sigma_{i2} + u_{is}^{xy} \right) - \dot{\gamma}_{i2} \right)\end{aligned} \tag{6.31}$$

Let $T_1 = \max\{T_{11}, T_{12}, ..., T_{1N}\}$. Then, according to Theorem 6.1, the observer guarantees that the estimation error $\Delta_i - \sigma_{i2}$ converges to zero in fixed time $T_1$, which implies that $\|\Delta_i - \sigma_{i2}\|$ is bounded at any finite time interval. Then, when $t \geq T_1$, one has

$$\begin{aligned}\dot{V}_2 &= \sum_{i=1}^{N} \zeta_{i3}^{\mathrm{T}} \left( \frac{d}{dt} u_{is}^{xy} - \dot{\gamma}_{i2} \right) \\ &= \sum_{i=1}^{N} \left( -\chi_1 \zeta_{i3}^{\mathrm{T}} \mathrm{sig}(\zeta_{i3})^{c_1} - \delta_1 \zeta_{i3}^{\mathrm{T}} \mathrm{sig}(\zeta_{i3})^{d_1} \right) \\ &\leq \sum_{i=1}^{N} \left( -\chi_1 \left( \zeta_{i3}^{\mathrm{T}} \zeta_{i3} \right)^{\frac{c_1+1}{2}} - \delta_1 M^{\frac{1-d_1}{2}} \left( \zeta_{i3}^{\mathrm{T}} \zeta_{i3} \right)^{\frac{d_1+1}{2}} \right) \\ &\leq -\chi_1 \left( \sum_{i=1}^{N} \zeta_{i3}^{\mathrm{T}} \zeta_{i3} \right)^{\frac{c_1+1}{2}} - \delta_1 (MN)^{\frac{1-d_1}{2}} \left( \sum_{i=1}^{N} \zeta_{i3}^{\mathrm{T}} \zeta_{i3} \right)^{\frac{d_1+1}{2}} \\ &\leq -\chi_1 2^{\frac{c_1+1}{2}} V_2^{\frac{c_1+1}{2}} - \delta_1 2^{\frac{d_1+1}{2}} (MN)^{\frac{1-d_1}{2}} V_2^{\frac{d_1+1}{2}}\end{aligned} \tag{6.32}$$

Then, the equilibrium point of $V_2 = 0$ is fixed-time stable within $T_1 + T_3$, and a fixed time of $T_3 = 1/\left[\chi_1 2^{\frac{c_1-1}{2}} (1 - c_1)\right] + 1/\left[\delta_1 2^{\frac{d_1-1}{2}} (MN)^{\frac{1-d_1}{2}} (d_1 - 1)\right]$ is obtained. Obviously, $T_3$ is independent of initial conditions.

*Step 2*: The following Lyapunov function is designed

$$V_3 = \frac{1}{2} \sum_{i=1}^{N} \zeta_{i2}^{\mathrm{T}} \zeta_{i2} \tag{6.33}$$

Based on Eq. (6.25), one has

$$\dot{\zeta}_{i2} = \zeta_{i3} + \gamma_{i2} - \dot{\gamma}_{i1} \tag{6.34}$$

After time $T_1 + T_3$, one has $\Delta_i - \sigma_{i2} = 0$ and $\zeta_{i3} = 0$. Then, for $t > T_1 + T_3$, by differentiating $V_3$ with respect to time and combining with protocols in Eqs. (6.28) and (6.34), the following results can be obtained

$$\begin{aligned}
\dot{V}_3 &= \sum_{i=1}^{N} \zeta_{i2}^{\mathrm{T}} \dot{\zeta}_{i2} \\
&= \sum_{i=1}^{N} \zeta_{i2}^{\mathrm{T}} \left( \zeta_{i3} - \chi_2 \mathrm{sig}(\zeta_{i2})^{c_2} - \delta_2 \mathrm{sig}(\zeta_{i2})^{d_2} \right) \\
&= \sum_{i=1}^{N} \left( -\chi_2 \zeta_{i2}^{\mathrm{T}} \mathrm{sig}(\zeta_{i2})^{c_2} - \delta_2 \zeta_{i2}^{\mathrm{T}} \mathrm{sig}(\zeta_{i2})^{d_2} \right) \\
&\le \sum_{i=1}^{N} \left( -\chi_2 \left( \zeta_{i2}^{\mathrm{T}} \zeta_{i2} \right)^{\frac{c_2+1}{2}} - \delta_2 M^{\frac{1-d_2}{2}} \left( \zeta_{i2}^{\mathrm{T}} \zeta_{i2} \right)^{\frac{d_2+1}{2}} \right) \\
&\le -\chi_2 \left( \sum_{i=1}^{N} \zeta_{i2}^{\mathrm{T}} \zeta_{i2} \right)^{\frac{c_2+1}{2}} - \delta_2 (MN)^{\frac{1-d_2}{2}} \left( \sum_{i=1}^{N} \zeta_{i2}^{\mathrm{T}} \zeta_{i2} \right)^{\frac{d_2+1}{2}} \\
&\le -\chi_2 2^{\frac{c_2+1}{2}} V_3^{\frac{c_2+1}{2}} - \delta_2 2^{\frac{d_2+1}{2}} (MN)^{\frac{1-d_2}{2}} V_3^{\frac{d_2+1}{2}}
\end{aligned} \tag{6.35}$$

Similarly, it can be concluded that the equilibrium point of $V_3 = 0$ is fixed-time stable and the convergence time is less than $T_1 + T_3 + T_4$, and $T_4 = 1/\left[ \chi_2 2^{\frac{c_2-1}{2}} (1 - c_2) \right] + 1/\left[ \delta_2 2^{\frac{d_2-1}{2}} (MN)^{\frac{1-d_2}{2}} (d_2 - 1) \right]$.

*Step 3:* Construct the following Lyapunov function

$$V_4 = \frac{1}{2} \zeta_1^{\mathrm{T}} (H \otimes I_2) \zeta_1 \tag{6.36}$$

Let $\zeta_2 = \left[ \zeta_{12}^{\mathrm{T}}, \zeta_{22}^{\mathrm{T}}, \ldots, \zeta_{N2}^{\mathrm{T}} \right]^{\mathrm{T}}$, then the following equation can be obtained

$$\dot{\zeta}_1 = \zeta_2 + \gamma_1 \tag{6.37}$$

Moreover, after time $T_1 + T_3 + T_4$, it can be obtained that $\Delta_i - \sigma_{i2} = 0$, $\zeta_{i3} = 0$ and $\zeta_{i2} = 0$. Then, for $t > T_1 + T_3 + T_4$, based on Eq. (6.29) and Eq. (6.37), the time derivative of $V_4$ is

$$\begin{aligned}
\dot{V}_4 &= \zeta_1^{T} (H \otimes I_2) \dot{\zeta}_1 \\
&= \zeta_1^{\mathrm{T}} (H \otimes I_2) \zeta_2 + \zeta_1^{\mathrm{T}} (H \otimes I_2) \gamma_1 \\
&= -\chi_3 [(H \otimes I_2) \zeta_1]^{\mathrm{T}} \mathrm{sig}((H \otimes I_2) \zeta_1)^{c_3} - \delta_3 [(H \otimes I_2) \zeta_1]^{\mathrm{T}} \mathrm{sig}((H \otimes I_2) \zeta_1)^{d_3} \\
&\le -\chi_3 \left\| (H \otimes I_2) \zeta_1 \right\|_{c_3+1}^{c_3+1} - \delta_3 \left\| (H \otimes I_2) \zeta_1 \right\|_{d_3+1}^{d_3+1} \\
&\le -\chi_3 \left( \frac{2\lambda_{\min^2}(H)}{\lambda_{\max}(H)} \right)^{\frac{c_3+1}{2}} V_4^{\frac{c_3+1}{2}} - \delta_3 \left( \frac{2\lambda_{\min^2}(H)}{\lambda_{\max}(H)} \right)^{\frac{d_3+1}{2}} V_4^{\frac{d_3+1}{2}}
\end{aligned} \tag{6.38}$$

where $\lambda_{\min}(H)$ and $\lambda_{\max}(H)$ represent the minimum and maximum eigenvalues of $H$, respectively. Based on the Lemma 3.3, one can obtain that $V_4$ is fixed-time stable and a fixed time $T_5 = 2/\left[ \chi_3 \left( \frac{2\lambda_{\min}^2}{\lambda_{\max}} \right)^{\frac{c_3+1}{2}} (1 - c_3) \right] + 2/\left[ \delta_3 \left( \frac{2\lambda_{\min}^2}{\lambda_{\max}} \right)^{\frac{d_3+1}{2}} (d_3 - 1) \right]$ can be obtained, which means that $\zeta_1$ will converge to zero in a fixed time of

$T_1 + T_3 + T_4 + T_5$. The analysis demonstrates that the proposed backstepping-based fixed-time sliding mode controller achieves guaranteed fixed-time convergence regardless of actuator faults and disturbances. Consequently, the heterogeneous UAVs-UGVs system described by Eq. (6.3) can accomplish fixed-time fault-tolerant formation tracking in the $xy$-plane within the predetermined settling time $T_1 + T_2 + T_3 + T_4 + T_5$.

**Remark 6.7** This chapter proposes a novel third-order state space model that fundamentally differs from prior fixed-time formation approaches with discontinuous control outputs [2, 21, 27]. The proposed design effectively eliminates chattering phenomena while maintaining continuous control signals similar to HOSMC implementations. The core innovation involves shifting the discontinuous characteristics generated by the sig(.) function to higher-order derivatives, thereby ensuring continuous control outputs. Furthermore, the controller architecture strategically incorporates the advantages of both integral sliding mode control and fixed-time observers to enhance overall performance. This design paradigm specifically addresses the critical requirements for FTC systems operating under unknown nonlinearities and external disturbances.

**Remark 6.8** The convergence times derived in Theorems 6.1–6.3 exhibit complete independence from the system's initial conditions. This crucial property ensures that the developed control protocol maintains consistent convergence performance across varying operating environments without requiring parameter readjustment. Furthermore, the controller design incorporates an explicit computational mechanism for determining and adjusting the convergence time upper bound to meet diverse performance specifications.

**Theorem 6.4** *The proposed decentralized FTC protocol comprising Eqs. (6.39)–(6.42), when combined with the fixed-time sliding mode surface in Eq. (6.15) and implemented using the observer from Eq. (6.8) under Assumption 6.1, ensures fixed-time convergence to zero for the follower-UAVs' z-axis tracking errors.*

$$u_{iz} = m_i \left( \ddot{h}_i^z - \sigma_{i4} - \lambda_{1i} \dot{e}_i - \lambda_{i2} \dot{e}_i + u_{is}^z \right) \tag{6.39}$$

*and*

$$u_{is}^z = \int_0^t \dot{\gamma}_{i2}^z - \mu_1 \text{sig}(\zeta_{i3})^{n_1} - \omega_1 \text{sig}(\zeta_{i3})^{m_1} \mathrm{d}\tau \tag{6.40}$$

*where* $e_i = z_i - h_i^z$ *is the tracking error of the* $i$*th follower* ($i \in U_2$), $\mu_1 > 0$ *and* $\omega_1 > 0$ *are the design parameters,* $n_1 \in (0, 1)$, $m_1 > 1$. *Moreover, the virtual inputs are designed to be of form*

$$\gamma_{i2}^z = \dot{\gamma}_{i1}^z - \mu_2 \text{sig}(\zeta_{i2})^{n_2} - \omega_2 \text{sig}(\zeta_{i2})^{m_2} \tag{6.41}$$

*and*

$$\gamma_{i1}^{z} = -\mu_3 \mathrm{sig}(\zeta_{i1})^{n_3} - \omega_3 \mathrm{sig}(\zeta_{i1})^{m_3} \tag{6.42}$$

*where* $\mu_2, \mu_3, \omega_2$ *and* $\omega_3$ *are preselected positive constants, and* $n_2, n_3 \in (0, 1)$, $m_2, m_3 > 1$.

***Proof*** Since $\gamma_{i1}^{z}$ represents the local virtual input of the $i$th agent, the Lyapunov function in Step 3 can be designed as $V_4 = 1/2 \left( \sum_{i=1}^{N_2} \zeta_{i1}^{\mathrm{T}} \zeta_{i1} \right)$, then the rest of the proof can be established by following the proof procedure in Theorem 6.3. For the space limitation, we omit the detailed proof herein.

**Remark 6.9** For simplicity, the study in [15] maintains a uniform and fixed height for all follower UAVs during formation. However, in practical applications such as obstacle navigation and cooperative surveillance, the ability to adjust UAV altitude while maintaining stability is crucial for task completion. The proposed decentralized time-varying formation FTC algorithm in the $z$-axis significantly enhances the system's practical applicability by addressing this need.

### 6.3.3 Fixed-Time FTC Design Under Actuator Faults and Switching Topologies

In practical applications, communication links between agents may be intermittently established or unexpectedly lost due to unreliable network connections. Therefore, investigating fixed-time FTC for MASs with time-varying topologies becomes essential. For convenience, let $L_{\sigma(t)}$ and $B_{\sigma(t)}$ denote the Laplacian matrix and the interconnection relationship between the leader and followers at time $t$, respectively. Then, a matrix $H_{\sigma(t)} = L_{\sigma(t)} + B_{\sigma(t)}$ is defined. Obviously, under Assumption 6.2, $H_{\sigma(t)}$ is positive-definite and symmetric at any time interval.

**Theorem 6.5** *For heterogeneous UAVs-UGVs systems operating under switching topologies* $\bar{G}_{\sigma(t)}$, *the fixed-time fault-tolerant formation tracking problem along all three axes* $(xyz)$ *can be effectively addressed when Assumption 6.2 is satisfied. The solution framework combines: the fixed-time observers in Eqs. (6.7)–(6.8), the distributed FTC protocol specified in Eqs. (6.26)–(6.28) and (6.43), along with the decentralized FTC law given by Eqs. (6.39)–(6.42) that incorporates the fixed-time sliding mode surface defined in Eq. (6.15).*

$$\gamma_1 = -\chi_3 \mathrm{sig}\left(\left(H_{\sigma(t)} \otimes I_2\right) \zeta_1\right)^{c_3} - \delta_3 \mathrm{sig}\left(\left(H_{\sigma(t)} \otimes I_2\right) \zeta_1\right)^{d_3} \tag{6.43}$$

***Proof*** Following the proof procedure in Steps 1–2 of Theorem 6.3, one concludes that $\zeta_{i3}$ and $\zeta_{i2}$ can converge to zero when $t \geq T_1 + T_3 + T_4$ under switching topologies $\bar{G}_{\sigma(t)}$. Then, for the error system $\zeta_1$, a topology-dependent Lyapunov function is defined as

$$V_4 = \frac{1}{2}\zeta_1^{\mathrm{T}}\left(H_{\sigma(t)} \otimes I_2\right)\zeta_1 \tag{6.44}$$

For $t \geq T_1 + T_3 + T_4$, one obtains that

$$\begin{aligned}
\dot{V}_4 &= \zeta_1^{\mathrm{T}}\left(H_{\sigma(t)} \otimes I_2\right)\dot{\zeta}_1 \\
&= \zeta_1^{\mathrm{T}}\left(H_{\sigma(t)} \otimes I_2\right)\zeta_2 + \zeta_1^{\mathrm{T}}\left(H_{\sigma(t)} \otimes I_2\right)\gamma_1 \\
&= -\chi_3\left[\left(H_{\sigma(t)} \otimes I_2\right)\zeta_1\right]^{\mathrm{T}}\mathrm{sig}\left(\left(H_{\sigma(t)} \otimes I_2\right)\zeta_1\right)^{c_3} \\
&\quad - \delta_3\left[\left(H_{\sigma(t)} \otimes I_2\right)\zeta_1\right]^{\mathrm{T}}\mathrm{sig}\left(\left(H_{\sigma(t)} \otimes I_2\right)\zeta_1\right)^{d_3} \\
&\leq -\chi_3\left\|\left(H_{\sigma(t)} \otimes I_2\right)\zeta_1\right\|^{c_3+1} - \delta_3\left\|\left(H_{\sigma(t)} \otimes I_2\right)\zeta_1\right\|^{d_3+1} \\
&\leq -\chi_3\left(\frac{2\lambda_{\mathrm{min}^2}\left(H_\sigma\right)}{\lambda_{\max}\left(H_\sigma\right)}\right)^{\frac{c_3+1}{2}} V_4^{\frac{c_3+1}{2}} - \delta_3\left(\frac{2\lambda_{\mathrm{min}^2}\left(H_\sigma\right)}{\lambda_{\max}\left(H_\sigma\right)}\right)^{\frac{d_3+1}{2}} V_4^{\frac{d_3+1}{2}}
\end{aligned} \tag{6.45}$$

where $\lambda_{\min}\left(H_{\sigma(t)}\right)$ and $\lambda_{\max}\left(H_{\sigma(t)}\right)$ are the minimum and maximum eigenvalues of matrix $H_{\sigma(t)}$, and are defined as $\lambda_{\min}\left(H_\sigma\right) = \min\left\{\lambda_{\min}\left(H_{\sigma(t_1)}\right), \lambda_{\min}\left(H_{\sigma(t_2)}\right), \ldots\right\}$, and $\lambda_{\max}\left(H_\sigma\right) = \max\left\{\lambda_{\max}\left(H_{\sigma(t_1)}\right), \lambda_{\max}\left(H_{\sigma(t_2)}\right), \ldots\right\}$.

Then, fixed-time theorem ensures that $\zeta_1$ is fixed-time stable, and a fixed settling time can be obtained as

$$T_5 = \frac{2}{\chi_3\left(\frac{2\lambda_{\mathrm{min}^2}(H_\sigma)}{\lambda_{\max}(H_\sigma)}\right)^{\frac{c_3+1}{2}}\left(1 - c_3\right)} + \frac{2}{\delta_3\left(\frac{2\lambda_{\mathrm{min}^2}(H_\sigma)}{\lambda_{\max}(H_\sigma)}\right)^{\frac{d_3+1}{2}}\left(d_3 - 1\right)} \tag{6.46}$$

Consequently, by integrating the outcomes of Theorem 6.3, we can establish that the desired formation tracking for the leader-follower UAVs-UGVs system is achievable under connected switching topologies, thereby completing the proof.

**Remark 6.10** To limit the amplitude of singularity terms $\alpha b_1|e_{ix}|^{\alpha-1}$ and $\alpha b_1\left|e_{iy}\right|^{\alpha-1}$ in control signal, the following saturation function is employed.

$$\mathrm{sat}(x, y) = \begin{cases} x, & \text{if } |x| < y, \\ y\mathrm{sign}(x), & \text{if } |x| \geq y. \end{cases} \tag{6.47}$$

Using $e_{ix}$ as a representative example for detailed analysis, its solution can be directly expressed as $e_{ix}(t) = e_{ix}(0) + \int_0^t \dot{e}_{ix}(\tau)\mathrm{d}\tau$. When $\dot{e}_{ix}(t) > 0$, the error $e_{ix}(t)$ exhibits monotonic growth until exiting the singularity region. Conversely, when $\dot{e}_{ix}(t) < 0$, $e_{ix}(t)$ decreases monotonically until departure from the singularity region. Both scenarios demonstrate that system states only temporarily reside within the singularity area. Consequently, the singularity region's presence does not affect the stability analysis outcomes, and the settling time's upper bound remains computable through the fixed-time theorem.

**Remark 6.11** This chapter advances beyond existing UAVs-UGVs formation systems [15, 19] by considering a more general switching topology framework, where previous results emerge as special cases. While prior studies [4, 13, 28] have made significant contributions to asymptotic formation control under switching networks, Theorem 6.5 demonstrates fixed-time formation tracking achievement despite switching topologies and actuator faults—a crucial advancement for safety-critical MASs requiring enhanced safety, reliability, and maneuverability. The current approach, however, imposes stringent eigenvalue requirements on switching topologies for fixed-time convergence. This limitation motivates future research directions, particularly exploring integration of emerging techniques like translation-adaptive strategies to achieve fixed-time convergence while relaxing eigenvalue constraints on convergence time.

**Remark 6.12** The stability of the proposed heterogeneous UAVs-UGVs system is primarily threatened by two categories of faults: individual component faults and interconnection faults. Without proper FTC design, these faults could compromise both the global control objectives and overall system stability. The theoretical framework developed in this chapter guarantees fixed-time stability despite actuator faults, thereby enhancing the system's maneuverability to reach stable operating conditions. However, our current analysis assumes a directed spanning tree network topology for FTC strategy development, which remains vulnerable to interconnection faults including sensor range limitations, communication link faults, and cyber attacks. Additionally, signal degradation caused by network attenuation or excessive inter-agent distances may introduce faulty coupling signals, potentially destabilizing the entire system.

**Remark 6.13** The proposed control scheme significantly reduces communication and computational load once the sliding surface is attained, with the protocol's time complexity primarily occurring during the pre-sliding phase. Leveraging fixed-time control theory in our observer-based controller design enables parameter tuning to maintain processor operation within industrial computer capabilities, preventing prolonged processor occupancy. Simulation results demonstrate the algorithm's execution time remains below 0.0001 s. Per Shannon's sampling theorem, successful signal reconstruction requires the harmonic signal frequency to exceed 5kHz. Given the operational characteristics of UAVs and UGVs, where formation signal frequencies are substantially below 5kHz, the proposed algorithm proves both theoretically valid and practically implementable.

**Remark 6.14** The inherent nonlinear dynamics, abrupt actuator faults, and unknown system characteristics of heterogeneous UAVs-UGVs systems present significant challenges for conventional approaches using gradient descent methods and

Hamilton-Jacobi-Bellman (HJB) equation solutions through Jacobi stability analysis. These difficulties in achieving desired cooperative performance motivate our adoption of Lyapunov stability theory for designing fault-tolerant formation control protocols. By employing the fixed-time Lyapunov stability theorem, we develop a formation controller that simultaneously achieves faster response and enhanced fault tolerance—capabilities currently unattainable through Jacobi stability methods. This work consequently identifies an important research gap: the development of fixed-time controllers for heterogeneous UAVs-UGVs systems based on Jacobi stability frameworks, which represents a promising direction for future investigation.

## 6.4 Simulation Experiments

Two simulation experiments validate the designed controller's effectiveness. The heterogeneous UAVs-UGVs system comprises one leader (agent 0), two UGV followers (agents 1–2), and two UAV followers (agents 3–4), with each agent's model parameters identical to those in Chap. 4. Following Theorems 6.1–6.5, we configure the gains for the proposed fixed-time observer as $o_{i1} = 5$, $o_{i2} = 2$, and $o_{i3} = 4$, and the controller parameters are given by $b_1 = 1$, $b_2 = 1.1$, $\alpha = 0.5$, $\beta = 1.1$, $c_1 = c_2 = c_3 = 0.5$, $d_1 = d_2 = d_3 = 1.4$, $\chi_1 = 6$, $\delta_1 = 5$, $\chi_2 = 0.8$, $\delta_2 = 0.7$, $\chi_3 = 0.9$, and $\delta_3 = 1.8$. To confirm the performance of fault tolerance, actuator faults are considered, and the fault scenario is provided as follows

$$\rho_{1x} = \begin{cases} 0, t \le 5\,s \\ 0.2, t > 5\,s, \end{cases} \rho_{1y} = \begin{cases} 0, t \le 5\,s \\ 0.3, t > 5\,s, \end{cases}$$

$$\rho_{2x} = \begin{cases} 0, t \le 7.5\,s \\ 0.35, t > 7.5\,s, \end{cases} \rho_{2y} = \begin{cases} 0, t \le 7.5\,s \\ 0.27, t > 7.5\,s, \end{cases}$$

$$\rho_{3x} = \begin{cases} 0, t \le 6\,s \\ 0.15 - 0.1\sin(t), 6\,s < t \le 10\,s \\ 0.25, t > 10\,s, \end{cases}$$

$$\rho_{3y} = \begin{cases} 0, t \le 6\,s \\ 0.18, t > 6\,s, \end{cases} \rho_{3z} = \begin{cases} 0, t \le 6\,s \\ 0.2, t > 6\,s, \end{cases}$$

$$\rho_{4x} = \begin{cases} 0, t \le 8\,s \\ 0.17, t > 8\,s, \end{cases} \rho_{4z} = \begin{cases} 0, t \le 8\,s \\ 0.4, t > 8\,s, \end{cases}$$

$$\rho_{4y} = \begin{cases} 0, t \le 8\,s \\ 0.18 - 0.05\cos(t), 8\,s < t \le 11\,s \\ 0.15, t > 11\,s, \end{cases}$$

and

$$\tau_1 = [-0.2\cos(t), -\sin(0.5t)]^{\mathrm{T}},$$

$$\tau_2 = [-0.15\sin(1.2t), -0.45\sin(2t)]^{\mathrm{T}},$$

$$\tau_3 = [0.2\sin(2t), 0.4\cos(2t), -0.2\cos(3t)]^{\mathrm{T}},$$

$$\tau_4 = [0.1\sin(1.7t), 0.3\cos(1.8t), -0.3\sin(3t)]^{\mathrm{T}}.$$

Additionally, the simulation accounts for unknown uncertainties in all follower-UAVs' aerodynamic damping parameters (with 20% variation) while incorporating external disturbances for follower agents, specified as $d_1 = [0.1\sin(t), -0.3\cos(0.7t)]^{\mathrm{T}}$, $d_2 = [0.45\cos(t), 0.7\sin(2t)]^{\mathrm{T}}$ for UGV followers, and the disturbances of UAV followers are designed as $d_3 = [-0.3\cos(0.5t), -0.2\sin(1.5t), 0.27\sin(2t)]^{\mathrm{T}}$, and $d_4 = [0.5\sin(0.25t), 0.55\cos(1.5t), 0.3\sin(2t)]^{\mathrm{T}}$. For comparison, a simulation on "distributed fixed-time observer based fault-tolerant formation control for UAVs-UGVs [3]" is conducted.

### 6.4.1 Fixed-Time Formation Tracking Under Fixed Topologies

In this section, the fixed-time fault-tolerant formation tracking for studied UAVs-UGVs under fixed topologies is considered. The initial conditions are selected as $x_0(0) = [0, 0]^{\mathrm{T}}$, $x_1(0) = [5, -0.5]^{\mathrm{T}}$, $x_2(0) = [6, -2.5]^{\mathrm{T}}$, $x_3(0) = [-1, 3.7, 1]^{\mathrm{T}}$, $x_4(0) = [-5, 1, 0]^{\mathrm{T}}$. The formation vectors in $xy$ axis are designed as $h_1^{xy} = [1, -1]^{\mathrm{T}}$, $h_2^{xy} = [5, -5]^{\mathrm{T}}$, $h_3^{xy} = [-5, 5]^{\mathrm{T}}$, and $h_4^{xy} = [-7, 7]^{\mathrm{T}}$, and the desired paths in $z$ axis are selected as $h_3^z = 0.5 + 1.5t$, and $h_4^z = e^{0.2t}$. The control signal of the leader is designed as $u_0 = [0.5, \sin(t)]^{\mathrm{T}}$. The adjacency matrix for the followers shown in Fig. 6.2 is described as $L = \begin{bmatrix} 0.9 & -0.4 & 0 & -0.5 \\ -0.4 & 1.1 & -0.7 & 0 \\ 0 & -0.7 & 0.9 & -0.2 \\ -0.5 & 0 & -0.2 & 0.7 \end{bmatrix}$.

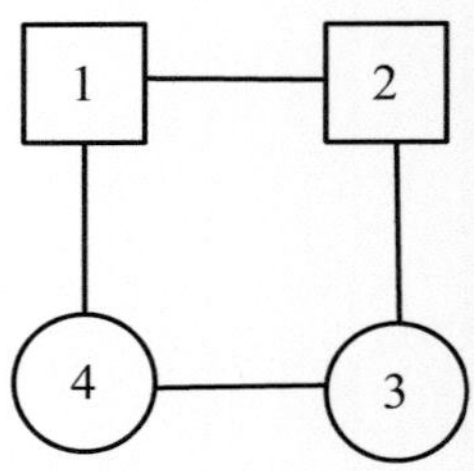


**Fig. 6.2** Communication topology

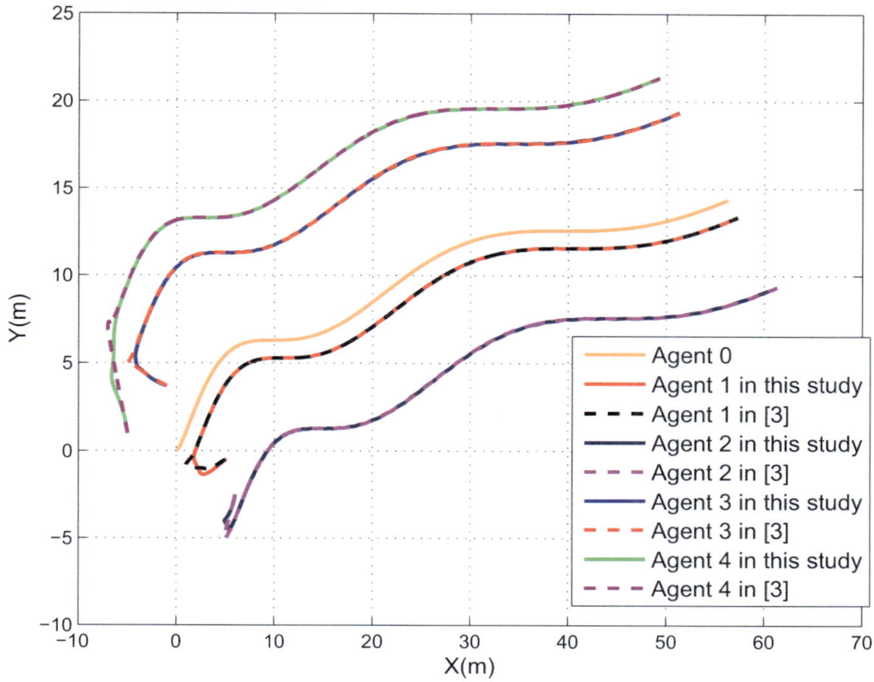


**Fig. 6.3** Position trajectories of the UAVs-UGVs in the $xy$ axis

The comparative results under fixed topology conditions are illustrated in Figs. 6.3, 6.4, 6.5, 6.6, 6.7 and 6.8. Figures 6.4 and 6.7 demonstrate that both our proposed method and the observer-based fixed-time controller [3] achieve rapid transient response and excellent tracking performance for heterogeneous MASs subject to actuator faults and disturbances. However, as evidenced in Figs. 6.5, 6.6 and 6.8, our approach distinguishes itself from [3] by delivering uniformly bounded and continuous control inputs—a characteristic enabled by the HOSMC technique that aligns perfectly with our theoretical predictions.

The results demonstrate successful formation tracking achievement within 5 s. While observer implementation introduces some convergence delay, the complete heterogeneous MAS convergence time remains shorter than that of the transformed third-order system in Eq. (6.24), satisfying $T < T_1 + T_2 + T_3 + T_4 + T_4 = 32.0114$ s. Both theoretical analysis and simulation results reveal an inverse relationship between convergence time and energy consumption—faster convergence demands greater energy expenditure. This trade-off necessitates careful balancing between response speed and control input during controller structure optimization.

Figures 6.7 and 6.8 present the position errors and control inputs in the $z$-axis, respectively. Despite abrupt actuator faults, the system maintains accurate trajectory tracking within a fixed time frame, demonstrating the superior fault-tolerant

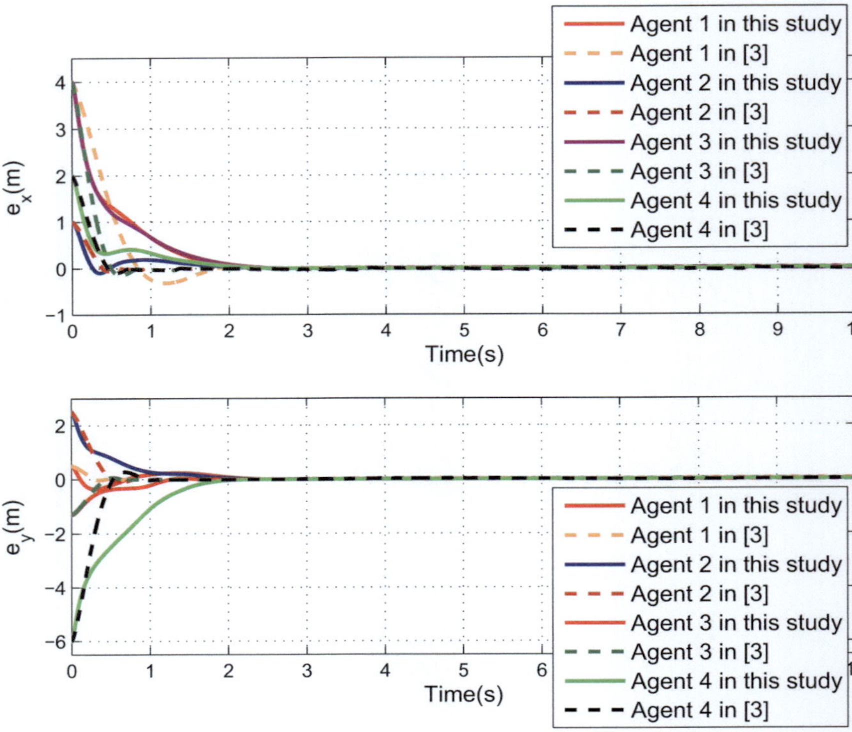


**Fig. 6.4** Formation errors in the $xy$ axis

capability of our decentralized tracking controller. Figure 6.8 reveals distinct characteristics in the control input magnitudes: specifically, $u_{3_z}$ for $t > 6$ s exhibits larger amplitudes compared to $t < 6$ s. This behavior stems from actuator faults occurring after 6 s, necessitating increased control effort to compensate for the fault-induced performance degradation—a key advantage of the proposed algorithm.

### 6.4.2 *Fixed-Time Formation Tracking Under Switching Topologies*

This section investigates fixed-time fault-tolerant time-varying formation control for heterogeneous UAVs-UGVs systems operating under switching topologies. Suppose that the communication graphs are shown in Fig. 6.9, and we can get $\lambda_{\max}(H_\sigma) = 3.1927$ and $\lambda_{\min}(H_\sigma) = 1$. The formation system begins from topology (1) and switches to topology (2) at $t = 5$ s, and switches to topology (3) at $t = 10$ s, then switches to topology (4) at $t = 15$ s.

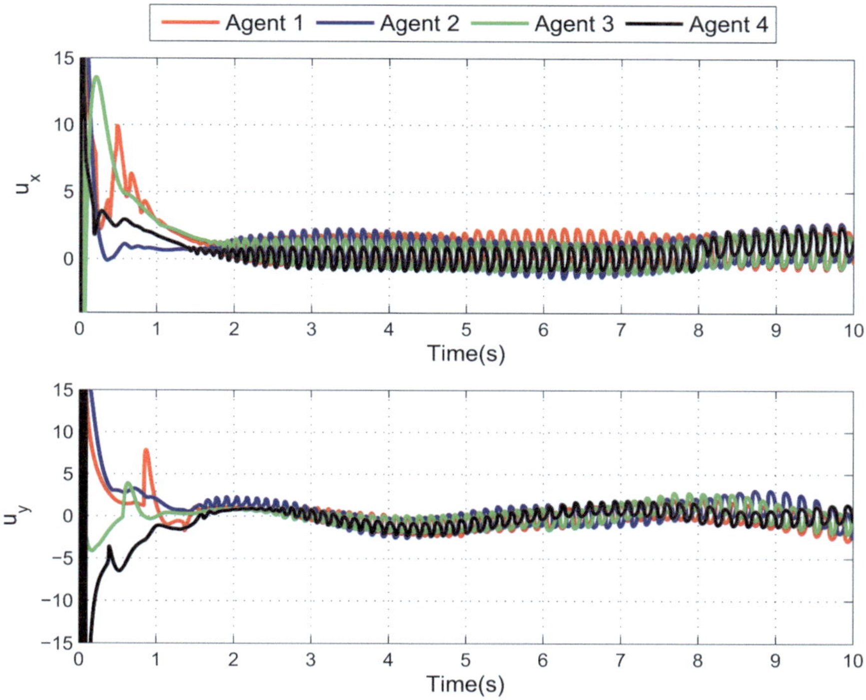


**Fig. 6.5** Control inputs by using the proposed method

The leader-follower formation systems are required to realize a time-varying circular formation configuration, and the expected formation vectors for follower are denoted as $h_i^{xy}(t) = [0.3i \sin(0.25i * t), 0.3i \cos(0.25i * t)]^{\mathrm{T}}$. To validate the fixed-time property of the developed algorithm, the initial states are $x_1(0) = [-0.1, 0.2]^{\mathrm{T}}$, $x_2(0) = [0.5, -0.3]^{\mathrm{T}}$, $x_3(0) = [-0.4, 0]^{\mathrm{T}}$, and $x_4(0) = [0, 0.6]^{\mathrm{T}}$. The control signal of the leader is redesigned as $u_0 = [0.02, 0.02]^{\mathrm{T}}$.

Simulation results in Figs. 6.10, 6.11 and 6.12 demonstrate the system's performance. Figure 6.10 reveals that the heterogeneous UAVs-UGVs system successfully achieves the desired time-varying circular formation around the leader using our proposed algorithm, despite switching topologies and actuator faults. Figure 6.11 indicates nearly identical convergence speed to Fig. 6.4, confirming the fixed-time stability's effectiveness in reducing initial condition dependence. Furthermore, Fig. 6.12 demonstrates smooth control inputs enabled by the HOSMC method. These results collectively verify that our approach delivers satisfactory performance across all critical metrics: tracking accuracy, fixed-time convergence, fault tolerance, and control input quality.

The proposed method, while ensuring control input continuity, typically generates harmonic control signals with specific frequencies. Although increasing con-

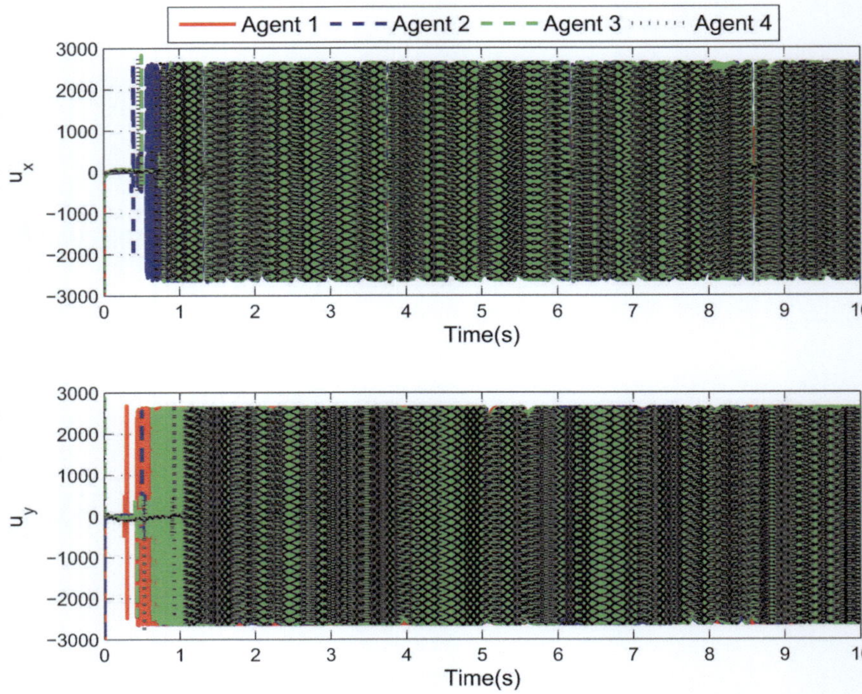


Fig. 6.6 Control inputs by using the method in [3]

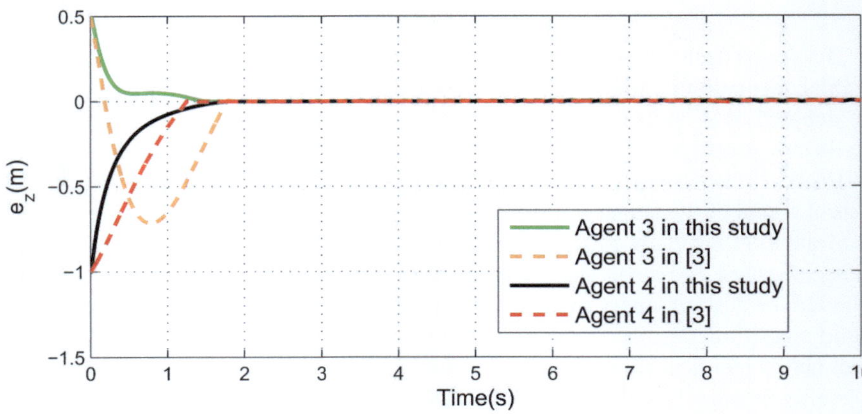


Fig. 6.7 Tracking errors in the $z$ axis

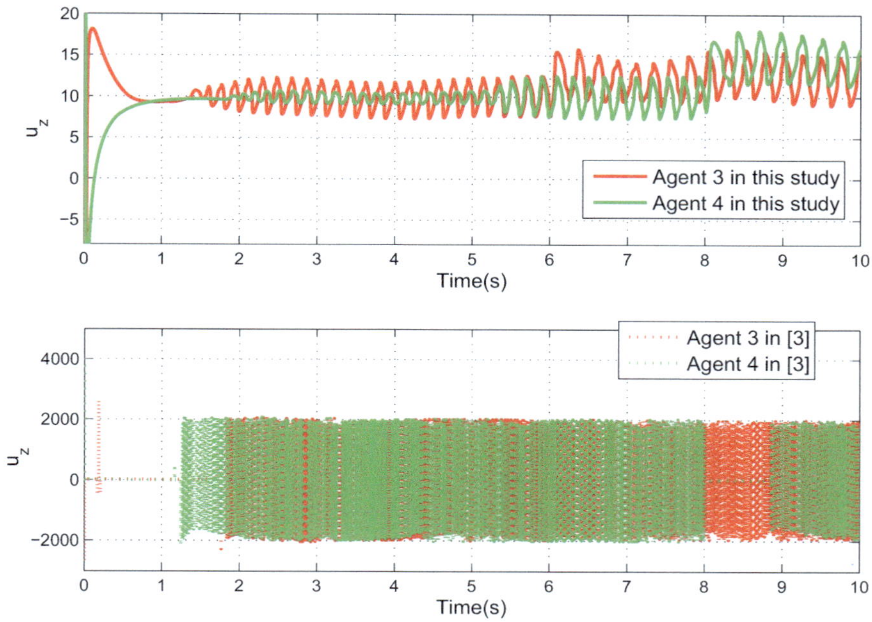


**Fig. 6.8** Control inputs in the $z$ axis

**Fig. 6.9** Switching topologies

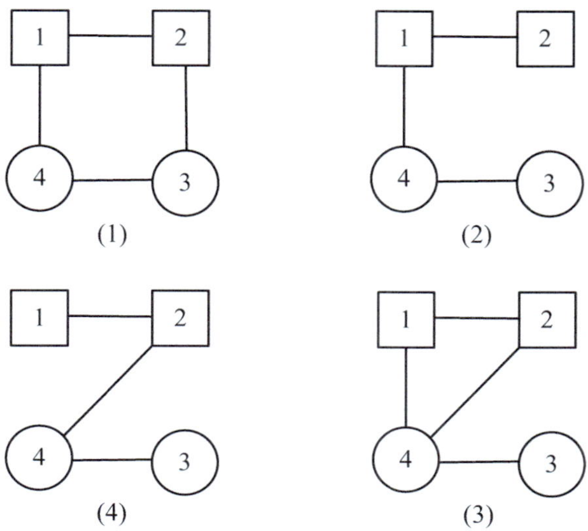

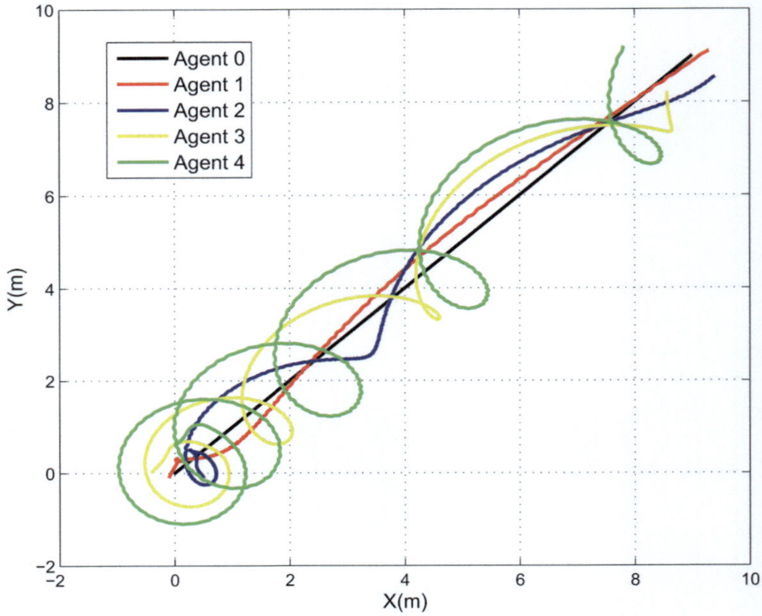


**Fig. 6.10** Position trajectories of the UAVs-UGVs in the $xy$ axis

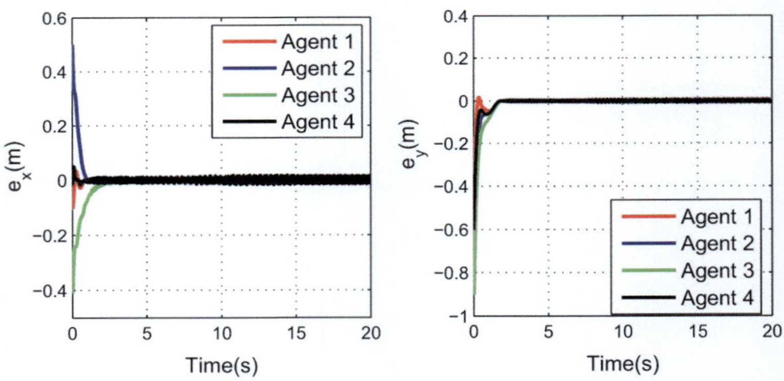


**Fig. 6.11** Formation errors in the $xy$ axis

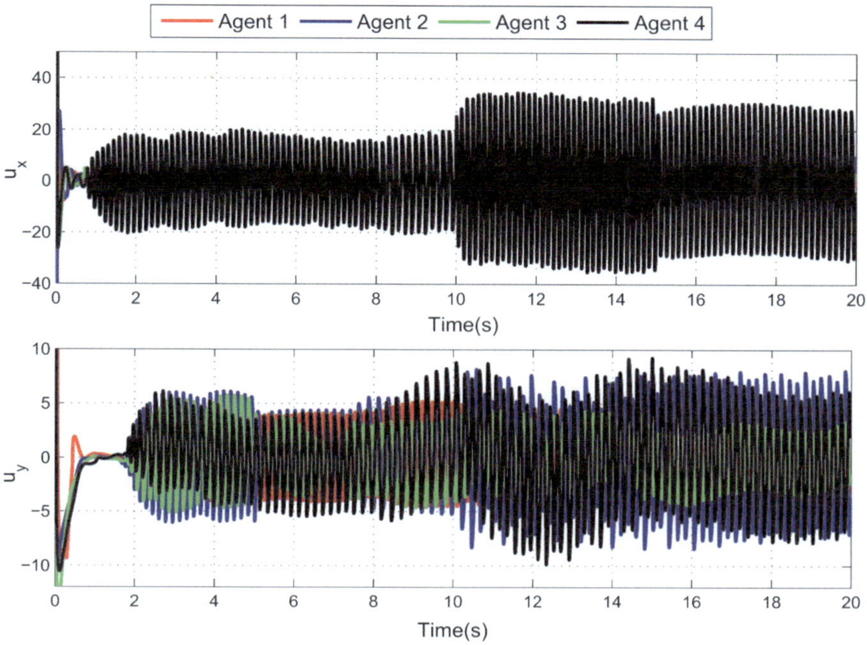


**Fig. 6.12** Control inputs in the $xy$ axis

troller parameters improves tracking accuracy and convergence rate, this enhancement comes at the cost of undesirable high-amplitude, high-frequency control signals. Consequently, practical implementation requires careful parameter tuning to balance convergence performance with acceptable control input characteristics.

## 6.5 Conclusions

This chapter addresses continuous fixed-time fault-tolerant formation tracking for heterogeneous MASs under both fixed and switching topologies. The solution framework involves: (1) developing a fixed-time observer for simultaneous actuator fault and disturbance estimation; (2) integrating HOSMC, fixed-time stability theory, and backstepping techniques to create a novel observer-based formation protocol that achieves leader-follower coordination with chattering suppression, fault tolerance, fixed-time convergence, and enhanced tracking precision; (3) deriving an explicit expression for the convergence time upper bound. Two comprehensive formation simulations validate the theoretical advances, demonstrating all claimed capabilities including the predetermined convergence timeframe.

## References


1. Basin M, Bharath Panathula C, Shtessel Y (2017) Multivariable continuous fixed-time second-order sliding mode control: design and convergence time estimation. IET Control Theory Appl 11(8):1104–1111
2. Chang S, Wang Y, Zuo Z, Yang H (2022) Fixed-time formation control for wheeled mobile robots with prescribed performance. IEEE Trans Control Syst Technol 30(2):844–851
3. Cheng W, Zhang K, Jiang B, Ding SX (2022) Fixed-time formation tracking for heterogeneous multi-agent systems under actuator faults and directed topologies. IEEE Trans Aerosp Electron Syst 58(4):3063–3077
4. Deng C, Che W-W (2021) Fault-tolerant fuzzy formation control for a class of nonlinear multiagent systems under directed and switching topology. IEEE Trans Syst Man Cybern Syst 51(9):5456–5465
5. Deng C, Che W-W, Shi P (2020) Cooperative fault-tolerant output regulation for multiagent systems by distributed learning control approach. IEEE Trans Neural Netw Learn Syst 31(11):4831–4841
6. Dong H, Lin X, Gao S, Cai B, Ning B (2020) Neural networks-based sliding mode fault-tolerant control for high-speed trains with bounded parameters and actuator faults. IEEE Trans Veh Technol 69(2):1353–1362
7. Hua L, Zhong Q, Soh YC, Shi K, Kwon O-M, Yan L, Zhong S (2024) Distributed sliding-mode consensus tracking control for fuzzy delayed multiagent systems under hybrid cyber-attacks. IEEE Trans Fuzzy Syst 32(6):3471–3483
8. Huang Y, Meng Z, Sun J (2024) Distributed constrained optimization for second-order multiagent systems via event-based communication. IEEE Trans Syst Man Cybern Syst 54(9):5317–5326
9. Hussein M, Zhang Y, Liu Z (2024) Distributed fault-tolerant formation control design via high-order sliding mode for a team of car-like vehicles. IEEE Trans Intell Veh 9(2):4221–4231
10. Incremona GP, Rubagotti M, Ferrara A (2017) Sliding mode control of constrained nonlinear systems. IEEE Trans Autom Control 62(6):2965–2972
11. Li H, Wu Y, Chen M (2021) Adaptive fault-tolerant tracking control for discrete-time multiagent systems via reinforcement learning algorithm. IEEE Trans Cybern 51(3):1163–1174
12. Li Y-Y, Li Y-X (2024) Resilient distributed fixed-time tracking of heterogeneous uavs-ugvs systems against dos attacks. IEEE Trans Syst Man Cybern Syst 54(9):5780–5790
13. Liu H, Ma T, Lewis FL, Valavanis KP (2020) Robust time-varying formation control for a set of quad-copters with switching interaction communication topology. IEEE Trans Veh Technol 69(7):6880–6890
14. Liu Z, Huang H, Park JH, Huang J, Wang X, Lv M (2023) Adaptive fuzzy control for unknown nonlinear multiagent systems with switching directed communication topologies. IEEE Trans Fuzzy Syst 31(7):2487–2494
15. Rahimi R, Abdollahi F, Naqshi K (2014) Time-varying formation control of a collaborative heterogeneous mult-agent system. Robot Auton Syst 62(12):1799–1805
16. Shen W, Yao D, Shen C, Li H (2023) Finite-time fault-tolerant control for swashplate of integrated hydraulic transformer with uncertain mismatched interference. IEEE Trans Industr Electron 70(6):6347–6355
17. Shi S, Zhang G, Min H, Xu S (2024) Prescribed-time high-order sliding mode control subject to mismatched terms with unknown gain functions. IEEE Trans Syst Man Cybern Syst 54(5):3184–3194
18. Shi Y, Hua Y, Yu J, Dong X, Lu J, Ren Z (2024) Cooperative fault-tolerant formation tracking control for heterogeneous air-ground systems using a learning-based method. IEEE Trans Aerosp Electron Syst 60(2):1505–1518
19. Vallejo-Alarcon M, Castro-Linares R, Velasco-Villa M (2015) Unicycle-type robot & quadrotor leader-follower formation backstepping control. IFAC-PapersOnLine 48(19):51–56

20. Van M (2018) An enhanced robust fault tolerant control based on an adaptive fuzzy pid-nonsingular fast terminal sliding mode control for uncertain nonlinear systems. IEEE/ASME Trans Mechatron 23(3):1362–1371
21. Wang C, Tnunay H, Zuo Z, Lennox B, Ding Z (2019) Fixed-time formation control of multirobot systems: Design and experiments. IEEE Trans Industr Electron 66(8):6292–6301
22. Wang F, Hu Z, Zhou C, Lin X (2025) Dynamic event-triggered consensus control of multiagent systems under markov switching topology. IEEE Trans Autom Sci Eng 22:11163–11175
23. Wang Y, Yan H, Chang Y, Liu X, Wang M (2024) Dynamic event-triggered control for persistent dwell-time switched nonlinear multiagent systems with random packet loss. IEEE Trans Syst Man Cybern Syst 54(4):2045–2054
24. Xu SS-D, Chen C-C, Wu Z-L (2015) Study of nonsingular fast terminal sliding-mode fault-tolerant control. IEEE Trans Industr Electron 62(6):3906–3913
25. Yang Y, He W, Li S (2024) Refined dynamic event-triggering cluster consensus of multiagent systems with fixed/switching topology. IEEE Trans Cybern 54(3):1377–1390
26. Yin T, Gu Z, Park JH (2024) Event-based intermittent formation control of multi-uav systems under deception attacks. IEEE Trans Neural Netw Learn Syst 35(6):8336–8347
27. Zhang Z, Song H, Wang H, Wang X, Dong H (2021) Cooperative multi-scenario departure control for virtual coupling trains: a fixed-time approach. IEEE Trans Veh Technol 70(9):8545–8555
28. Zhao H, Dai X, Zhang Q, Ding J (2020) Robust event-triggered model predictive control for multiple high-speed trains with switching topologies. IEEE Trans Veh Technol 69(5):4700–4710
29. Zhao J, Lu P, Du C, Cao F (2024) Active fault-tolerant strategy for flight vehicles: Transfer learning-based fault diagnosis and fixed-time fault-tolerant control. IEEE Trans Aerosp Electron Syst 60(1):1047–1059
30. Zhao W, Liu H, Valavanis KP, Lewis FL (2022) Fault-tolerant formation control for heterogeneous vehicles via reinforcement learning. IEEE Trans Aerosp Electron Syst 58(4):2796–2806
31. Zhao W, Liu H, Wan Y, Lin Z (2022) Data-driven formation control for multiple heterogeneous vehicles in air-ground coordination. IEEE Trans Control Netw Syst 9(4):1851–1862
32. Zhou H, Zuo Y, Tong S (2024) Observer-based fuzzy event-triggered consensus fault-tolerant control for nonlinear multiagent systems under switching topologies. IEEE Trans Fuzzy Syst 32(8):4660–4670
33. Zuo Z (2015) Non-singular fixed-time consensus tracking for second-order multi-agent networks. Automatica 54(4):305–309

# Part III
# Prescribed-Time FTC for Multiple Unmanned Systems

# Chapter 7
# Fully Actuated System Approach-Based Prescribed-Time FTC for Unmanned Helicopters

## 7.1 Introduction

As a distinct characterization of practical systems, the HOFAS is defined by the invertibility of its control matrix. Consequently, a methodology termed the HOFAS approach was proposed to efficiently resolve various control challenges [6]. This approach offers two key merits: (1) it streamlines controller design and overcomes limitations in model reduction, particularly when coupling terms exist within such systems; (2) it ensures full-actuation properties. Along this research trajectory, a series of relevant investigations have been conducted. For instance, the authors in [4, 5, 13] devised an adaptive passive FTC scheme for nonlinear HOFAS to maintain stability, while scholars in [1, 2] explored an active FTC scheme for HOFAS through the use of a nonlinear extended state observer. Meanwhile, the HOFAS approach has been applied to convert under-actuated systems into fully-actuated models, thereby enabling the attainment of desired control objectives. Theoretical studies in this field include [14], where a 3-DOF quadrotor was transformed into two fully-actuated subsystems, with a tracking control method designed accordingly. The authors in [18] proposed a HOFAS-based model predictive control and implemented it in both a numerical and a mechanical system. For additional system types, refer to [6, 7, 10, 27]. Notably, the characteristic of the HOFAS approach also contributes to enhancing the formation control performance of under-actuated multiple 6-DOF unmanned helicopters, serving as another motivation for this chapter.

Inspired by the preceding discussions, this chapter aims to address the prescribed-time fault-tolerant formation control problem for collision-avoidant multi-unmanned helicopters through the HOFAS approach. The core contributions of this chapter are highlighted as follows:

(1) Unlike existing studies in [9, 15–17] that decompose systems into outer-loop and inner-loop subsystems, this work treats under-actuated multi-unmanned helicopters with both actuator gain and bias faults as an integrated system. To facilitate control design, three high-order fully actuated subsystems are constructed.

 
K. Zhang et al., *Fixed-Time/Prescribed-Time Fault-Tolerant Control for Multiple Unmanned Systems*, Unmanned System Technologies,
https://doi.org/10.1007/978-981-95-2913-1_7

(2) Based on the decentralized principle, a prescribed-time fault estimation observer is proposed, which is capable of approximating lumped faults within a user-specified time/rate. Contrary to finite/fixed-time observers, the estimation time/rate of this observer can be known in advance, and relevant parameters can be explicitly determined.
(3) Following the estimation of fault information, a prescribed-time fault-tolerant formation control protocol is developed by integrating the HOFAS approach and artificial repulsive potential functions. This protocol not only compensates for actuator fault redundancy but also ensures high-precision formation tracking with collision avoidance within a preassigned time, while reducing the complexity of control design procedures.

## 7.2 Problem Formulation and Preliminaries

### *7.2.1 Problem Formulation*

To enable the configuration of flight dynamics and aerodynamics for multi-unmanned helicopter formations, the $i$th unmanned helicopter is characterized within the body-fixed frame $\mathrm{R}_i = (\mathrm{O}_i, x_i, y_i, z_i)$ and the inertial frame $\mathrm{R}_e = (\mathrm{O}_e, x_e, y_e, z_e)$, and a leader-follower formation diagrammatic sketch are shown in Fig. 7.1.

Leveraging the frameworks in [9, 15–17], an under-actuated 6-DOF rigid-body model of the $i$th unmanned helicopter under partial actuator faults is constructed as follows

$$\begin{cases} \dot{P}_i = V_i, \ i = 1, 2 \ldots, N \\ \dot{V}_i = ge_3 + r_i e_3(-g + Z^i_{\omega}\omega_i + Z^i_{col}\delta^i_{col}) \\ \dot{\Theta}_i = \Pi_i(\Theta_i)\Omega_i \\ \dot{\Omega}_i = -J_i^{-1}S(\Omega_i)^{\times}J_{ii} + A_i\Omega_i + B_iU_i^{\mathrm{F}} + e_3N^i_{col}\delta^i_{col} \end{cases} \tag{7.1}$$

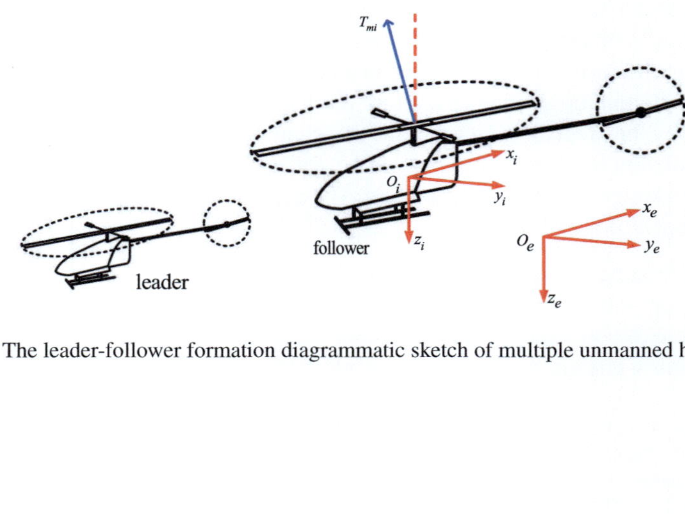


**Fig. 7.1** The leader-follower formation diagrammatic sketch of multiple unmanned helicopters

where the vectors $P_i = [x_i, y_i, z_i]^{\mathrm{T}}$, $V_i = [u_i, v_i, \omega_i]^{\mathrm{T}}$ and $\Theta_i = [\phi_i, \theta_i, \psi_i]^{\mathrm{T}}$ represent the position, velocity, and Euler angles in $\mathrm{R}_i$; and $\Omega_i = [p_i, q_i, r_i]^{\mathrm{T}}$ denotes the Euler angular velocity defined in $\mathrm{R}_e$. Gravitational acceleration is denoted by $g$, with $e_3 = [0, 0, 1]^{\mathrm{T}}$; $z_{\omega}^i$, $Z_{\mathrm{col}}^i$, and $N_{\mathrm{col}}^i$ are constants linked to the main rotor rotational speed, main rotor collective pitch angle, and total main rotor coupling coefficient. $J_i = \mathrm{diag}\{J_{xi}, J_{yi}, J_{zi}\}$ characterizes the diagonal inertia matrix; $\delta_{\mathrm{col}}^i$ denotes the control torque associated with the main rotor collective pitch, while $U_i^{\mathrm{F}} = [\delta_{\mathrm{lo}i}^{\mathrm{F}}, \delta_{\mathrm{la}i}^{\mathrm{F}}, \delta_{\mathrm{t}i}^{\mathrm{F}}]^{\mathrm{T}}$ represents the actuator control output for the longitudinal flapping angle, lateral cyclic pitch, and tail rotor collective pitch, respectively.

Additionally, the rotation matrix $R_i$, skew-symmetric matrix $S(\Omega_i)^{\times}$, attitude kinematic matrix $\Pi_i(\Theta_i)$, and parameter matrices $A_i$, $B_i$ are defined as below

$$R_i = \begin{bmatrix} \cos\theta_i\cos\psi_i & \sin\phi_i\sin\theta_i\cos\psi_i - \cos\phi_i\sin\psi_i & \cos\phi_i\sin\theta_i\cos\psi_i + \sin\phi_i\sin\psi_i \\ \cos\theta_i\sin\psi_i & \sin\phi_i\sin\theta_i\sin\psi_i + \cos\phi_i\cos\psi_i & \cos\phi_i\sin\theta_i\sin\psi_i - \sin\phi_i\cos\psi_i \\ -\sin\theta_i & \sin\phi_i\cos\theta_i & \cos\phi_i\cos\theta_i \end{bmatrix}$$

$$S(\Omega_i)^{\times} = \begin{bmatrix} 0 & -r_i & q_i \\ r_i & 0 & -p_i \\ -q_i & p_i & 0 \end{bmatrix}, \quad \Pi_i(\Theta_i) = \begin{bmatrix} 1 & \sin\phi_i\tan\theta_i & \cos\phi_i\tan\theta_i \\ 0 & \cos\phi_i & -\sin\phi_i \\ 0 & \sin\phi_i/\cos\theta_i & \cos\phi_i/\cos\theta_i \end{bmatrix}$$

$$A_i = \begin{bmatrix} -\iota_{mi}L_{bi} & -\iota_{mi}L_{Ai} & 0 \\ -\iota_{mi}m_{bi} & -\iota_{mi}m_{Ai} & 0 \\ 0 & 0 & N_{ri} \end{bmatrix}, \quad B_i = \begin{bmatrix} L_{loi} & L_{lai} & 0 \\ m_{loi} & m_{lai} & 0 \\ 0 & 0 & N_{Ti} \end{bmatrix}$$

where $\iota_{m_i}$ denotes the main rotor time constant; $N_{r_i}$ represents the yaw angle damping coefficient; $N_{T_i}$ depends on the tail rotor rotational speed and blade radius. The coefficients $L_{\mathrm{lo}i} = L_{A_i}A_{\mathrm{lo}i} + L_{b_i}B_{\mathrm{lo}i}$, $L_{\mathrm{la}i} = L_{A_i}A_{\mathrm{la}i} + L_{b_i}b_{\mathrm{la}i}$, $m_{\mathrm{lo}i} = m_{A_i}A_{\mathrm{lo}i} + m_{b_i}b_{\mathrm{lo}i}$, and $m_{\mathrm{la}i} = m_{A_i}A_{\mathrm{la}i} + m_{b_i}b_{\mathrm{la}i}$ are defined, where $L_{A_i}$, $L_{b_i}$, $m_{A_i}$, $m_{b_i}$ denote the main rotor's lateral and longitudinal wave motion coefficients, while $L_{\mathrm{lo}i}$, $L_{\mathrm{la}i}$, $m_{\mathrm{lo}i}$, $m_{\mathrm{la}i}$ represent cross-drive coefficients.

Subsequently, the $i$th under-actuated unmanned helicopter model in Eq. (7.1) is decomposed into three subsystems.

Subsystem 1

$$\ddot{z}_i = g - m_i^{-1}T_{m_i}\cos\phi_i\cos\theta_i \tag{7.2}$$

Subsystem 2

$$\begin{cases} \ddot{x}_i = -m_i^{-1}T_{m_i}(\cos\phi_i\sin\theta_i\cos\psi_i + \sin\phi_i\sin\psi_i) \\ \ddot{y}_i = -m_i^{-1}T_{m_i}(\cos\phi_i\sin\theta_i\sin\psi_i - \sin\phi_i\cos\psi_i) \\ \dot{\phi}_i = p_i + \sin\phi_i\tan\theta_i q_i + \cos\phi_i\tan\theta_i r_i \\ \dot{\theta}_i = \cos\phi_i q_i - \sin\phi_i r_i \\ \dot{p}_i = \bar{f}_{i1} + L_{loi}\delta_{loi}^f + L_{lai}\delta_{lai}^f \\ \dot{q}_i = \bar{f}_{i2} + m_{loi}\delta_{loi}^f + m_{lai}\delta_{lai}^f \end{cases} \tag{7.3}$$

Subsystem 3

$$\begin{cases} \dot{\psi}_i = q_i \sin\phi_i / \cos\theta_i + r_i \cos\phi_i / \cos\theta_i \\ \dot{r}_i = \bar{f}_{i3} + N_{Ti}\delta_{ti}^{f} \end{cases} \tag{7.4}$$

where $T_{m_i}$ denotes the main rotor thrust controlled by collective pitch, with $m_i$ as the total mass of the $i$th unmanned helicopter. $\bar{f}_{i1} = J_{xi}^{-1}(J_{yi} - J_{zi})q_i r_i - \iota_{mi} L_{bi} p_i - \iota_{mi} L_{Ai} q_i$, $\bar{f}_{i2} = J_{yi}^{-1}(J_{zi} - J_{xi})p_i r_i - \iota_{mi} m_{bi} p_i - \iota_{mi} m_{Ai} q_i$ and $\bar{f}_{i3} = J_{zi}^{-1}(J_{xi} - J_{yi})p_i q_i + N_{ri} r_i + N_{mi}\delta_{mi}$.

Gain faults and bias faults are common in aircraft actuators. Building on [11, 28], the control torque under these faults is modeled as

$$\delta_{Li}^{\mathrm{F}} = [\rho_{i1}\delta_{loi} + \zeta_{i1}, \rho_{i2}\delta_{lai} + \zeta_{i2}]^{\mathrm{T}} \triangleq \delta_{Li} + \bar{\Theta}_{i2} \tag{7.5}$$

$$\delta_{ti}^{\mathrm{F}} = \rho_{i3}\delta_{ti} + \zeta_{i3} \triangleq \delta_{ti} + \bar{\Theta}_{i3} \tag{7.6}$$

where $\delta_{Li}^{\mathrm{F}} = [\delta_{loi}^{\mathrm{F}}, \delta_{lai}^{\mathrm{F}}]^{\mathrm{T}}$. where $0 < \rho_{\min}^{i} \le \rho_{im} < 1$ $(m = 1, 2, 3)$ characterizes partial actuator effectiveness loss (with $\rho_{\min}^{i}$ as the minimum of a positive continuous function), and $\zeta_{im}$ denotes fault-induced additional bias torque. The lumped faults to be estimated and compensated are

$$\begin{aligned} \bar{\Theta}_{i2} &= [(\rho_{i1} - 1)\delta_{\mathrm{lo}i} + \zeta_{i1}, (\rho_{i2} - 1)\delta_{\mathrm{la}i} + \zeta_{i2}]^{\mathrm{T}}, \\ \bar{\Theta}_{i3} &= (\rho_{i3} - 1)\delta_{t_i} + \zeta_{i3}. \end{aligned} \tag{7.7}$$

**Remark 7.1** To reduce the time and space complexity of the FTC algorithm, only the actuator input $u_i = [\delta_{\mathrm{lo}i}, \delta_{\mathrm{la}i}, \delta_{t_i}]^{\mathrm{T}}$ is assumed to be fault-prone. This scenario is validated in references [12, 19, 25].

**Remark 7.2** The actuator fault model (Eqs. (7.5) and (7.6)) includes mainly four main cases, as discussed in [11, 28]. The gain factor $\rho_{im}$ may arise from partial torque/electronics loss or connection degradation, while the bias factor $\zeta_{im}$ could stem from rotor/electronics damage or connection deviation.

For simplicity, define $S_{i1} = z_i$, $S_{i2} = [x_i, y_i]^{\mathrm{T}}$, $S_{i2,1} = [\phi_i, \theta_i]^{\mathrm{T}}$, $S_{i2,2} = [p_i, q_i]^{\mathrm{T}}$, $S_{i3} = \psi_i$, $S_{i3,1} = r_i$, and control inputs $u_{i1} = m_i^{-1} T_{m_i}$, $u_{i2} = [\delta_{\mathrm{lo}i}, \delta_{\mathrm{la}i}]^{\mathrm{T}}$, $u_{i3} = \delta_{t_i}$. The faulty subsystems (Eqs. (7.2)–(7.4)) are then converted into three high-order fully-actuated subsystems

$$S_{i1}^{(2)} = f_{i1} + g_{i1}u_{i1} \tag{7.8}$$

$$S_{i2}^{(4)} = f_{i2} + g_{i2}u_{i2} + \Theta_{i2} \tag{7.9}$$

$$S_{i3}^{(2)} = f_{i3} + g_{i3}u_{i3} + \Theta_{i3} \tag{7.10}$$

where $f_{i1} = g$, $g_{i1} = -\cos\phi_i \cos\theta_i$, $f_{i2} = g_{i2,1}^{(1)} f_{i2,1} + g_{i2,1} f_{i2,1}^{(1)} + g_{i2,1}^{(1)} g_{i2,2} S_{i2,2} + g_{i2,1} g_{i2,2}^{(1)} S_{i2,2} + g_{i2,1} g_{i2,2} f_{i2,2} + f_{i2,3}^{(1)}$, $g_{i2} = g_{i2,1} g_{i2,2} g_{i2,3}$, $\Theta_{i2} = g_{i2}\bar{\Theta}_{i2}$ and $f_{i3} = f_{i3,1}^{(1)} + g_{i3,1}^{(1)} S_{i3,1} + g_{i3,1} f_{i3,2}$, $g_{i3} = N_{Ti} g_{i3,1}$, $\Theta_{i3} = g_{i3}\bar{\Theta}_{i3}$, in which the parameters $f_{i2,1}$, $f_{i2,2}$, $f_{i2,3}$, $f_{i3,1}$, $f_{i3,2}$, $g_{i2,1}$, $g_{i2,2}$, $g_{i2,3}$, $g_{i3,1}$ are separately listed as follows

$$f_{i2,1} = \begin{bmatrix} \cos\phi_i \tan\theta_i r_i \\ -\sin\phi_i r_i \end{bmatrix},\ f_{i2,2} = \begin{bmatrix} \bar{f}_{i1} \\ \bar{f}_{i2} \end{bmatrix},\ g_{i2,2} = \begin{bmatrix} 1 & \sin\phi_i \tan\theta_i \\ 0 & C_{\phi_i} \end{bmatrix}$$

$$g_{i2,1} = u_{i1} \begin{bmatrix} \sin\phi_i \sin\theta_i \cos\psi_i - \cos\phi_i \sin\psi_i & -\cos\phi_i C_{\theta_i} \cos\psi_i \\ \sin\phi_i \sin\theta_i \sin\psi_i + \cos\phi_i \sin\psi_i & -\cos\phi_i C_{\theta_i} \sin\psi_i \end{bmatrix}$$

$$g_{i2,3} = \begin{bmatrix} L_{loi} & L_{lai} \\ m_{loi} & m_{lai} \end{bmatrix},\ g_{i2,4} = \begin{bmatrix} -\cos\phi_i \sin\theta_i & -\sin\phi_i \\ \sin\phi_i & -\cos\phi_i \sin\theta_i \end{bmatrix}$$

$$f_{i2,3} = g_{i2,4} h_{i2}^{(1)} u_{i1} + g_{i2,4} h_{i2} u_{i1}^{(1)},\ h_{i2}^{\mathrm{T}} = \left[\cos\psi_i{}^{\mathrm{T}}\ \sin\psi_i{}^{\mathrm{T}}\right]^{\mathrm{T}}$$

$$f_{i3,1} = \sin\phi_i \cos\theta_i^{-1} q_i,\ f_{i3,2} = \bar{f}_{i3},\ g_{i3,1} = \cos\phi_i \cos\theta_i^{-1}$$

**Assumption 7.1** ([9]) The Euler angle vectors (i.e., the roll angle $\phi_i$, the pitch angle $\theta_i$, and the yaw angle $\psi_i$) are bounded and satisfy the ranges of $\phi_i \in (-\frac{\pi}{2}, \frac{\pi}{2})$, $\theta_i \in (-\frac{\pi}{2}, \frac{\pi}{2})$ and $\psi_i \in (-\pi, \pi)$, respectively.

**Assumption 7.2** ([6]) For any $T > 0$, $\det(g_{i1}) \neq 0$, $\det(g_{i2}) \neq 0$ and $\det(g_{i3}) \neq 0$, this is the conception of the fully-actuated condition.

**Remark 7.3** Unlike the existing results in [9, 15–17] dividing the $i$th under-actuated unmanned helicopter in Eq. (7.1) into the outer and inner-loop subsystems, the HOFAS approach simplify the controller design through separating into three high-order fully-actuated subsystems as Eqs. (7.8)–(7.10). Therefore, it is possible to track position $[S_{i2}, S_{i1}]^{\mathrm{T}}$ and yaw angle $S_{i3}$ and make the desired trajectories of other two angles $\phi_i^d$ and $\theta_i^d$ to keep the concordance of the conditions

$$\phi_i^d = \arcsin\left(\frac{\cos\psi_i^d A_{iy} - \sin\psi_i^d A_{ix}}{u_{i1}}\right) \tag{7.11}$$

$$\theta_i^d = \arctan\left(\frac{\cos\psi_i^d A_{ix} + \sin\psi_i^d A_{iy}}{A_{iz} - g}\right) \tag{7.12}$$

where $\psi_i^d$ denotes the desired yaw trajectory (i.e., the yaw angle of the virtual leader $S_{03}$). Here, $A_{iz} = S_{i1}^{(2)}$ and $[A_{ix}, A_{iy}]^{\mathrm{T}} = S_{i2}^{(2)}$, derived directly from Eqs. (7.8) and (7.9).

### 7.2.2 *Preliminaries*

Before presenting the main results, the following definitions and lemmas are introduced. Consider the general nonlinear system as

$$\dot{x}(t) = f(x(t), t),\ x(0) = x_0 \tag{7.13}$$

where $x(t) \in R^m$ is the state vector, and the origin is an equilibrium point.

**Definition 7.1** ([20, 21]) For a user-specified settling time $T_p \leq T_f \leq T_{\max}$ (independent of initial conditions, with $T_p$ denoting data transmission and processing time), the origin of system (7.13) is

1. Prescribed-time convergent if, as $T \to T_f^-$, the solution $x(t)$ converges to the origin.
2. Practically prescribed-time convergent if, as $T \to T_f^-$, $x(t)$ converges to a neighborhood of the origin, i.e., $|x(t)| \leq \epsilon$ for some $\epsilon > 0$.

**Lemma 7.1** ([3, 24]) *For system (7.13), if there exists a positive-definite, differentiable Lyapunov function $V(t)$ such that*

1. $\dot{V}(t) \leq -\kappa_1 V(t) - \kappa_2 \frac{\dot{\mu}(t)}{\mu(t)} V(t)$ *(with $\kappa_1, \kappa_2 > 0$), the origin is prescribed-time convergent with settling time $T_f$.*
2. $\dot{V}(t) \leq -\kappa_1 V(t) - \kappa_2 \frac{\dot{\mu}(t)}{\mu(t)} V(t) + \varepsilon$ *(with $\kappa_1, \kappa_2, \varepsilon > 0$), the origin is practically prescribed-time convergent with settling time $T_f$.*

The function $\mu(t)$ is a monotonically increasing piecewise function defined as

$$\mu(t) = \begin{cases} \left( \dfrac{T_f}{T_f + T_0 - T} \right)^n, & t \in [T_0, T_f) \\ 1, & t \in [T_f, \infty) \end{cases} \tag{7.14}$$

where $T_0$ is the initial time and $n$ is the system order. This function satisfies $\mu(T_0) = 1$ and $\lim_{t \to T_f^-} \mu(t) = \infty$.

**Remark 7.4** The lemmas in [8, 29] are special cases of Lemma 7.1 (with $\kappa_2 = 2$). This chapter generalizes these results. The term $\kappa_2 \frac{\dot{\mu}(t)}{\mu(t)}$ plays a critical role in bounding the convergence time/rate of estimation and control. If errors fail to converge to the origin (or a small neighborhood) within the user-specified $T_f$, this term may drive system states to infinity (consistent with simulation results).

**Remark 7.5** In the time-varying function $\mu(t)$, $T_f$ is a user-defined settling time (pre-assignable, not dependent on system parameters). $T_f$ should be set based on practical needs and computational constraints. Multiple forms of $\mu(t)$ exist for prescribed-time or practical prescribed-time convergence (e.g., [21, 26]).

**Lemma 7.2** ([6]) *For any $\varsigma > 0$, there exist matrices $A_r \in \mathbb{R}^{m \times m}$ ($r = 0, 1, \ldots, n-1$) such that $Re\{\lambda_r(\Phi(A^{0 \sim n-1}))\} \leq -\frac{\varsigma}{2}$ for $r = 1, 2, \ldots, nm$. There exists a matrix $P \triangleq P(A^{0 \sim n-1}) = [P_0, P_1, \ldots, P_{n-1}]$ ($P_r \in \mathbb{R}^{nm \times m}$) satisfying the Lyapunov inequality*

$$\Phi^T(A^{0 \sim n-1}) P + P \Phi(A^{0 \sim n-1}) \leq -\varsigma P \tag{7.15}$$

*where $I_{nm}$ is the nm-dimensional identity matrix. The notation below is also introduced as*

$$P_L \triangleq P_L(A^{0\sim n-1}) = P \begin{bmatrix} 0 \\ I_m \end{bmatrix} \tag{7.16}$$

**Lemma 7.3** ([6]) *For any Hurwitz matrix $F \in \mathbb{R}^{nm\times nm}$, matrices $A^{0\sim n-1}$ and non-singular $\mathrm{V} \in \mathbb{R}^{nm\times nm}$ satisfy*

$$\Phi(A^{0\sim n-1}) = \mathrm{V} f \mathrm{V}^{-1} \tag{7.17}$$

*where $A^{0\sim n-1} = -zF^n\mathrm{V}^{-1}(z, F)$, and*

$$\mathrm{V} = \mathrm{V}(z, F) = \begin{bmatrix} z \\ zF \\ \vdots \\ zF^{n-1} \end{bmatrix} \tag{7.18}$$

*where $z \in \mathbb{R}^{nm\times nm}$ is arbitrary (with $det(\mathrm{V}(Z, f)) \neq 0$).*

**Control Objective**. Using the HOFAS approach, the objectives are

1. Design a prescribed-time fault estimation observer to simultaneously estimate subsystem states and lumped faults $\Theta_{i2}$, $\Theta_{i3}$.
2. Devise a prescribed-time active fault-tolerant formation control protocol to ensure multi-helicopter tracking of the virtual leader, maintain inter-vehicle distances, and avoid collisions.

That is to say, when $T \to T_f^-$, there exist the following stability conditions hold

$$\begin{cases} \lim\limits_{T\to T_f^-} S_{i1} - S_{j1} = C_{iz}^d - C_{jz}^d, \ \lim\limits_{T\to T_f^-} S_{i3} - S_{j3} = 0 \\ \lim\limits_{T\to T_f^-} S_{i2} - S_{j2} = C_{i\mathrm{xy}}^d - C_{j\mathrm{xy}}^d, \ \lim\limits_{T\to T_f^-} S_{i1} - S_{01} = C_{iz}^d \\ \lim\limits_{T\to T_f^-} S_{i2} - S_{02} = C_{i\mathrm{xy}}^d, \ \lim\limits_{T\to T_f^-} S_{i3} = S_{03} \end{cases} \tag{7.19}$$

where $T_f$ is the user-assigned settling time, $[C_{i\mathrm{xy}}^d, C_{iz}^d]^\mathrm{T}$ denotes the desired relative formation distance of the $i$th helicopter from the virtual leader, and $[S_{02}, S_{01}]^\mathrm{T}$, $S_{03}$ represent the virtual leader's position and yaw angle.

## 7.3 Main Results

### 7.3.1 Prescribed-Time Fault Estimation Observer Design

To address the impact of lumped faults $\Theta_{i2}$ and $\Theta_{i3}$ in Eqs. (7.9) and (7.10), inspired by prior studies [3, 8], a prescribed-time fault estimation observer is designed as

$$\begin{cases} \hat{S}_{i2}^{(4)} = f_{i2} + g_{i2}u_{i2} + \hat{\Theta}_{i2} \\ \dot{\hat{\Theta}}_{i2} = -(\kappa_{i1} + \kappa_{i2}\frac{\dot{\mu}_{i1}}{\mu_{i1}})e_{i2} - \kappa_{i5}\text{sign}(e_{i2}) - \Gamma_{i1} \end{cases} \tag{7.20}$$

and

$$\begin{cases} \hat{S}_{i3}^{(2)} = f_{i3} + g_{i3}u_{i3} + \hat{\Theta}_{i3} \\ \dot{\hat{\Theta}}_{i3} = -(\kappa_{i3} + \kappa_{i4}\frac{\dot{\mu}_{i2}}{\mu_{i2}})e_{i3} - \kappa_{i6}\text{sign}(e_{i3}) - \Gamma_{i2} \end{cases} \tag{7.21}$$

where $\hat{S}_{i2}^{(4)}$, $\hat{S}_{i3}^{(2)}$ and $\hat{\Theta}_{i2}$, $\hat{\Theta}_{i3}$ are the estimations of subsystem states $S_{i2}^{(4)}$, $S_{i3}^{(2)}$ and lumped faults $\Theta_{i2}$, $\Theta_{i3}$, respectively. $e_{i2} = \hat{S}_{i2}^{(3)} - S_{i2}^{(3)}$, $e_{i3} = \dot{\hat{S}}_{i3} - \dot{S}_{i3}$, $\tilde{\Theta}_{i2} = \hat{\Theta}_{i2} - \Theta_{i2}$ and $\tilde{\Theta}_{i3} = \hat{\Theta}_{i3} - \Theta_{i3}$ represent the estimation errors. Additionally, $\kappa_{ik}$, $k = 1, 2, \ldots, 6$ is a designed positive constant, and $\mu_{i1}$, $\mu_{i2}$ denote the scaling functions as in Eq. (7.14) with the settling-times $T_{i1f}$, $T_{i2f}$. The terms $\Gamma_{i1}$ and $\Gamma_{i2}$ are $\Gamma_{i1} = \int_0^{\mathrm{T}} \kappa_{i7}\text{sign}(e_{i2}(\iota))d\iota$, $\Gamma_{i2} = \int_0^{\mathrm{T}} \kappa_{i8}\text{sign}(e_{i3}(\iota))d\iota$ with $\kappa_{i7} \geq |-\dot{\Theta}_{i2}|$ and $\kappa_{i8} \geq |-\dot{\Theta}_{i3}|$.

**Remark 7.6** During multi-unmanned helicopter formation flight, lumped faults $\Theta_{i2}$ and $\Theta_{i3}$ are bounded but not precisely known, making the designed fault estimation observer more practical. It is important that the observer in Eqs. (7.20) and (7.21) requires $-\dot{\Theta}_{i2}$ and $-\dot{\Theta}_{i3}$ to be bounded. Since control input $u_i$, gain factor $\rho_{im}$, and bias factor $\zeta_{im}$ $(m = 1, 2, 3)$ are bounded, $\Theta_{i2}$ and $\Theta_{i3}$ are also bounded.

**Remark 7.7** The decentralized observers in Eqs. (7.20) and (7.21) can be extended to a distributed form, where estimation errors become $\bar{e}_{i2} = \sum_{j \in N_i} a_{ij}(e_{i2} - e_{j2}) + b_i e_{i2}$ and $\bar{e}_{i3} = \sum_{j \in N_i} a_{ij}(e_{i3} - e_{j3}) + b_i e_{i3}$ ($a_{ij}$ and $b_i$ are coupling strengths). The decentralized form is adopted here due to its lower computational complexity and equivalent estimation performance.

**Theorem 7.1** *For the $i$th faulty subsystems in Eqs. (7.9) and (7.10), under the proposed prescribed-time fault estimation observer in Eqs. (7.20)–(7.23), the estimation errors of subsystem states $e_{i2}$, $e_{i3}$ and lumped faults $\tilde{\Theta}_{i2}$, $\tilde{\Theta}_{i3}$ converge to the origin within the user-specified time $T_{if}$.*

***Proof*** Taking derivatives of estimation errors with respect to Eqs. (7.20)–(7.23) gives

$$\begin{cases} \dot{e}_{i2} = \hat{S}_{i2}^{(4)} - S_{i2}^{(4)} = \hat{\Theta}_{i2} - \Theta_{i2} \\ \dot{e}_{i3} = \hat{S}_{i3}^{(2)} - S_{i3}^{(2)} = \hat{\Theta}_{i3} - \Theta_{i3} \end{cases} \tag{7.22}$$

Based on the Eqs. (7.20) and (7.21), the derivatives $\dot{e}_{i2}$ and $\dot{e}_{i3}$ can be rewritten as

$$\begin{cases} \dot{e}_{i2} = -(\kappa_{i1} + \kappa_{i2}\frac{\dot{\mu}_{i1}}{\mu_{i1}})e_{i2} - \kappa_{i5}\text{sign}(e_{i2}) + \bar{\Gamma}_{i1} \\ \dot{\bar{\Gamma}}_{i1} = -\kappa_{i7}\text{sign}(e_{i2}) - \dot{\Theta}_{i2} \end{cases} \tag{7.23}$$

and

$$\begin{cases} \dot{e}_{i3} = -(\kappa_{i3} + \kappa_{i4}\frac{\dot{\mu}_{i2}}{\mu_{i2}})e_{i3} - \kappa_{i6}\text{sign}(e_{i3}) + \bar{\Gamma}_{i2} \\ \dot{\bar{\Gamma}}_{i2} = -\kappa_{i8}\text{sign}(e_{i3}) - \dot{\Theta}_{i3} \end{cases} \tag{7.24}$$

where $\bar{\Gamma}_{i1}$ and $\bar{\Gamma}_{i2}$ denote the auxiliary variables.

Before moving on, we construct the candidate Lyapunov functionals as

$$\begin{cases} V_{i1} = \frac{1}{2}e_{i2}^2 \\ V_{i2} = \frac{1}{2}e_{i3}^2 \end{cases} \tag{7.25}$$

Taking the derivatives of $V_{i1}$ and $V_{i2}$ with respect of time, one derives that

$$\begin{cases} \dot{V}_{i1} = -(\kappa_{i1} + \kappa_{i2}\frac{\dot{\mu}_{i1}}{\mu_{i1}})e_{i2}^2 - \kappa_{i5}|e_{i2}| + e_{i2}\bar{\Gamma}_{i1} \\ \dot{V}_{i2} = -(\kappa_{i3} + \kappa_{i4}\frac{\dot{\mu}_{i2}}{\mu_{i2}})e_{i3}^2 - \kappa_{i6}|e_{i3}| + e_{i3}\bar{\Gamma}_{i2} \end{cases} \tag{7.26}$$

By the property of the sign function, $\text{sign}(\bar{\Gamma}_{i1})$, $\text{sign}(\bar{\Gamma}_{i2})$ in Eqs. (7.23) and (7.24) are opposite to $\text{sign}(e_{i2})$, $\text{sign}(e_{i3})$, satisfying $e_{i2}\bar{\Gamma}_{i1} \leq 0$ and $e_{i3}\bar{\Gamma}_{i2} \leq 0$ for all $T \geq 0$.

Hence, one can obtain the following inequalities hold

$$\begin{cases} \dot{V}_{i1} \leq -\bar{\kappa}_{i1}V_{i1} - \bar{\kappa}_{i2}\frac{\dot{\mu}_{i1}}{\mu_{i1}}V_{i1}, \ \forall\, t \to T_{i1f}^{-} \\ \dot{V}_{i2} \leq -\bar{\kappa}_{i3}V_{i2} - \bar{\kappa}_{i4}\frac{\dot{\mu}_{i2}}{\mu_{i2}}V_{i2}, \ \forall\, t \to T_{i2f}^{-} \end{cases} \tag{7.27}$$

where $\bar{\kappa}_{im} = 2\kappa_{im}$, $m = 1, 2, 3, 4$. According to Lemma 7.1, it follows that the approximation errors of subsystem states $e_{i2} \to 0$ and $e_{i3} \to 0$ when $t \to T_{i1f}^{-}$ and $t \to T_{i2f}^{-}$, i.e., $\lim_{T \to T_{i1f}^{-}} V_{i1} = 0$ and $\lim_{t \to T_{i2f}^{-}} V_{i2} = 0$ also hold.

From the relationship in above-mentioned equations, that is, $\dot{e}_{i2} = \tilde{\Theta}_{i2}$ and $\dot{e}_{i3} = \tilde{\Theta}_{i3}$, we can conclude that the approximation errors of lumped faults also converge to the origin when $t \to T_{i1f}^{-}$ and $t \to T_{i2f}^{-}$, respectively.

Since $V_{i1}$ and $V_{i2}$ are continuous, for $t \geq T_{i1f}$ and $t \geq T_{i2f}$, the inequalities $\dot{V}_{i1} \leq -\bar{\kappa}_{i1}V_{i1}$ and $\dot{V}_{i2} \leq -\bar{\kappa}_{i3}V_{i2}$ ensure $e_{i2}, e_{i3}, \tilde{\Theta}_{i2}, \tilde{\Theta}_{i3} = 0$. Thus, the observer achieves fault estimation within $T_{if} = \max\{T_{i1f}, T_{i2f}\}$, which completes the proof.

**Remark 7.8** The above theorem focuses on individual estimation errors, but the overall convergence for $N$ helicopters holds within $T_1 = \max_{1 \leq i \leq N} T_{if}$, as each observer provides accurate estimates.

### 7.3.2 Prescribed-Time FTC Design

Using the proposed observer in Eqs. (7.20) and (7.21), this section presents an active fault-tolerant formation control design to ensure tracking within a user-specified time.

Before presenting the designed control protocol, some concepts about graph theory need to be introduced.

The communication topology among $N$ followers is modeled by a fixed graph $\mathrm{G} = (V, \mathrm{E}, \mathrm{A})$, where

- $V = \{\upsilon_1, \upsilon_2, \ldots, \upsilon_N\}$ is the node set,
- $\mathrm{E} = \{(\upsilon_i, \upsilon_j)\}$ is the edge set with $j \in N_i$ if $(\upsilon_i, \upsilon_j) \in \mathrm{E}$,
- $\mathrm{A} = [a_{ij}] \in \mathbb{R}^{N\times N}$ is the adjacency matrix ($a_{ij} = 1$ if connected, else 0).

For the virtual leader-follower network, a directed graph uses $\mathrm{B} = \mathrm{diag}\{b_1, \ldots, b_N\}$ with $b_i = 1$ if follower $i$ connects to the leader.

**Assumption 7.3** ([20, 24]) The undirected and directed graphs are connected, and at least one follower links to the virtual leader.

On this foundation, a prescribe-time active fault-tolerant formation control protocol is proceeding in the next, aiming at preserving the target geometric pattern during the formation flying.

Define the formation position tracking errors $\bar{e}_{i1}$ (i.e., $z$-direction), $\bar{e}_{i2}$ (i.e., $x$ and $y$-directions), and the yaw angle tracking error $\bar{e}_{i3}$ between the relative distances and the virtual leader for each unmanned helicopter as below

$$\begin{cases} \bar{e}_{i1} = \sum\limits_{j\in N_i} a_{ij}((S_{i1} - S_{j1}) - (C^d_{iz} - C^d_{jz})) + \Upsilon_{i1} \\ \bar{e}_{i2} = \sum\limits_{j\in N_i} a_{ij}((S_{i2} - S_{j2}) - (C^d_{ixy} - C^d_{jxy})) + \Upsilon_{i2} \\ \bar{e}_{i3} = \sum\limits_{j\in N_i} a_{ij}(S_{i3} - S_{j3}) + b_i(S_{i3} - S_{03}) \end{cases} \tag{7.28}$$

where the terms are $\Upsilon_{i1} = b_i(S_{i1} - S_{01} - C^d_{iz})$ and $\Upsilon_{i2} = b_i(S_{i2} - S_{02} - C^d_{ixy})$.

On the basis of the defined formation tracking errors Eq. (7.28), the formation control protocols $u_{i1}, u_{i2}$ and $u_{i3}$ in the $i$th high-order fully-actuated subsystems Eqs. (7.8)–(7.10) are designed as follows

$$\begin{cases} u_{i1} = -(g_{i1}\varpi_i)^{-1}\big(A_{i1}^{(0\sim1)}\bar{e}_{i1}^{(0\sim1)} + \varpi_i f_{i1} \\ +C_{i1}\frac{\dot{\mu}_{i3}}{\mu_{i3}}p_{i1L}^{\mathrm{T}}\bar{e}_{i1}^{(0\sim1)} - \sum\limits_{j\in N_i} a_{ij}S_{j1}^{(2)} - b_i S_{01}^{(2)}\big) \\ u_{i2} = -(g_{i2}\varpi_i)^{-1}\big(A_{i2}^{(0\sim3)}\bar{e}_{i2}^{(0\sim3)} + \varpi_i f_{i2} + \varpi_i\hat{\Theta}_{i2} + u_{i2,1} \\ +C_{i2}\frac{\dot{\mu}_{i4}}{\mu_{i4}}p_{i2L}^{\mathrm{T}}\bar{e}_{i2}^{(0\sim3)} - \sum\limits_{j\in N_i} a_{ij}\hat{S}_{j2}^{(4)} - b_i S_{02}^{(4)}\big) \\ u_{i3} = -(g_{i3}\varpi_i)^{-1}\big(A_{i3}^{(0\sim1)}\bar{e}_{i3}^{(0\sim1)} + \varpi_i f_{i3} + \varpi_i\hat{\Theta}_{i3} \\ +C_{i3}\frac{\dot{\mu}_{i5}}{\mu_{i5}}p_{i3L}^{\mathrm{T}}\bar{e}_{i3}^{(0\sim1)} - \sum\limits_{j\in N_i} a_{ij}\hat{S}_{j3}^{(2)} - b_i S_{03}^{(2)}\big) \end{cases} \tag{7.29}$$

where $\varpi_i = \sum_{j\in N_i} a_{ij} + b_i$. $C_{im}$, $m = 1, 2, 3$ reveals as a positive constant satisfying $C_{im} \geq \tilde{\kappa}_{im}/\lambda_{\min}(\frac{\bar{p}_{imL}}{p_{im}})$ and $\bar{p}_{imL} = p_{imL}p_{imL}^{\mathrm{T}}$. In addition, $\mu_{i3}$, $\mu_{i4}$ and $\mu_{i5}$ separately denote the scaling functions as in Eq. (7.14) with the settling-times $T_{i3f}$, $T_{i4f}$ and $T_{i5f}$.

Besides, the term $u_{i2,1} = \gamma_i \sum_{j\in N_i} \nabla_{\mathbf{s}_{i2}}\Psi_{ij}(\|d_{ij}\|)$ implies the repulsive force to avoid collisions, with $\gamma_i$ being a positive constant, and $\Psi_{ij}(\|d_{ij}\|)$ being an APF as below

$$\Psi_{ij}(\|d_{ij}\|) = \begin{cases} \|d_{ij}\|\exp^{(\|d_{ij}\|-r_{\mathrm{in}}^i)^{-2}}, & \|d_{ij}\| \in [r_{\mathrm{in}}^i, r_{\mathrm{out}}^i) \\ 0, & \text{otherwise} \end{cases} \tag{7.30}$$

where $\|d_{ij}\| = \|S_{i2} - S_{j2}\|$ is the actual relative distance between the $i$th and the $j$th unmanned helicopter on xoy plane. $r_{\mathrm{in}}^i$ and $r_{\mathrm{out}}^i$ denote the radii of collision avoidance region and detection region, respectively.

From Eq. (7.30), it can be easily concluded that: i) $\Psi_{ij}(\|d_{ij}\|)$ tends to zero when the $j$th unmanned helicopter lies outside of the detection region $r_{\mathrm{out}}^i$; and ii) $\Psi_{ij}(\|d_{ij}\|)$ tends to infinity when the $j$th unmanned helicopter lies on the surface of the collision avoidance region $r_{\mathrm{in}}^i$.

**Remark 7.9** The actual relative distance is $\|d_{ij}\| = \|P_i - P_j\|$ with $P_i = [S_{i2}, S_{i1}]^{\mathrm{T}}$, and collisions are avoided when $\|d_{ij}\| \geq r_{\mathrm{in}}^i$. Here, collision avoidance is considered only on the $xoy$ plane, assuming helicopters take off in formation with good $z$-axis tracking.

**Remark 7.10** The desired relative distance $[C_{ixy}^d, C_{iz}^d]^{\mathrm{T}}$ is assumed time-invariant, but extensions to time-varying patterns are straightforward. The results also extend to multiple-leader scenarios (containment or formation-containment control).

**Theorem 7.2** For the $i$th subsystems in Eqs. (7.8)–(7.10), under the proposed observer in Eqs. (7.20)–(7.23) and control protocol in Eq. (7.16), formation tracking errors $\bar{e}_{i1}$, $\bar{e}_{i2}$, $\bar{e}_{i3}$ converge to a neighborhood of the origin within user-specified time $T_{if}$.

***Proof*** The proof proceeds in three steps to analyze prescribed-time convergence.

**Step 1**. Taking the second derivative of $\bar{e}_{i1}$ and using Eq. (7.8), it has

$$\begin{aligned}\bar{e}_{i1}^{(2)} &= \sum_{j\in N_i} a_{ij}(S_{i1}^{(2)} - S_{j1}^{(2)}) + b_i(S_{i1}^{(2)} - S_{01}^{(2)}) \\ &= \varpi_i(f_{i1} + g_{i1}u_{i1}) - \sum_{j\in N_i} a_{ij}S_{j1}^{(2)} - b_i S_{01}^{(2)}\end{aligned} \tag{7.31}$$

Substituting control protocol Eq. (7.16), it derives that

$$\bar{e}_{i1}^{(2)} + A_{i1}^{(0\sim1)}\bar{e}_{i1}^{(0\sim1)} = -C_{i1}\frac{\dot{\mu}_{i3}}{\mu_{i3}}p_{i1L}^{\mathrm{T}}\bar{e}_{i1}^{(0\sim1)} \tag{7.32}$$

which rewrites as

$$\dot{\bar{e}}_{i1}^{(0\sim1)} = \Phi_{i1}(A_{i1}^{(0\sim1)})\bar{e}_{i1}^{(0\sim1)} + \begin{bmatrix} 0 \\ -C_{i1}\frac{\dot{\mu}_{i3}}{\mu_{i3}}p_{i1L}^{\mathrm{T}}\bar{e}_{i1}^{(0\sim1)}\end{bmatrix} \tag{7.33}$$

By Lemma 7.2, there exists a positive-definite matrix $p_{i1}$ satisfying

$$\Phi_{i1}^{\mathrm{T}}p_{i1} + p_{i1}\Phi_{i1} \le -\varsigma I. \tag{7.34}$$

Define the Lyapunov functional as

$$V_{i3} = \frac{1}{2}(\bar{e}_{i1}^{(0\sim1)})^{\mathrm{T}}p_{i1}\bar{e}_{i1}^{(0\sim1)} \tag{7.35}$$

Differentiating $V_{i2}$ and combining the Eq. (7.33), it has

$$\begin{aligned} l\dot{V}_{i3} &= \frac{1}{2}(\bar{e}_{i1}^{(0\sim1)})^{\mathrm{T}}\big(\Phi_{i1}(A_{i1}^{(0\sim1)})^{\mathrm{T}}p_{i1} + p_{i1}\Phi_{i1}(A_{i1}^{(0\sim1)})\big)\bar{e}_{i1}^{(0\sim1)} \\ &+ (\bar{e}_{i1}^{(0\sim1)})^{\mathrm{T}}p_{i1}\begin{bmatrix} 0 \\ -C_{i1}\frac{\dot{\mu}_{i3}}{\mu_{i3}}p_{i1L}^{\mathrm{T}}\bar{e}_{i1}^{(0\sim1)}\end{bmatrix} \\ &\le -\frac{\check{\kappa}_{i1}}{2}(\bar{e}_{i1}^{(0\sim1)})^{\mathrm{T}}p_{i1}\bar{e}_{i1}^{(0\sim1)} - C_{i1}\frac{\dot{\mu}_{i3}}{\mu_{i3}}(\bar{e}_{i1}^{(0\sim1)})^{\mathrm{T}}\bar{p}_{i1L}\bar{e}_{i1}^{(0\sim1)}\end{aligned} \tag{7.36}$$

implying $\dot{V}_{i3} \le -\check{\kappa}_{i1}V_{i3} - \tilde{\kappa}_{i1}\frac{\dot{\mu}_{i3}}{\mu_{i3}}V_{i3}$. By Lemma 7.1, $\bar{e}_{i1} \to 0$ as $t \to T_{i3f}^{-}$, and $\dot{V}_{i3} \le -\check{\kappa}_{i1}V_{i3}$ for $t \ge T_{i3f}$ ensures sustained convergence.

**Step 2**. Taking the fourth derivative of $\bar{e}_{i2}$ and using Eq. (7.9)

$$\begin{aligned}\bar{e}_{i2}^{(4)} &= \sum_{j\in N_i} a_{ij}(\hat{S}_{i2}^{(4)} - \hat{S}_{j2}^{(4)}) + b_i(\hat{S}_{i2}^{(4)} - S_{02}^{(4)}) \\ &= \varpi_i(f_{i2} + g_{i2}u_{i2} + \hat{\Theta}_{i2}) - \sum_{j\in N_i} a_{ij}\hat{S}_{j2}^{(4)} - b_i S_{02}^{(4)}\end{aligned} \tag{7.37}$$

To compensate for the estimated lumped fault $\hat{\Theta}_{i2}$ and prevent collisions on the xoy plane, the control protocol given in (7.29) is implemented. This ensures effective fault mitigation and safe operation, and we have

$$\bar{e}_{i2}^{(4)} + A_{i2}^{(0\sim3)}\bar{e}_{i2}^{(0\sim3)} = -u_{i2,1} - \frac{\dot{\mu}_{i4}}{\mu_{i4}} p_{i2L}^{\mathrm{T}}\bar{e}_{i2}^{(0\sim3)} \tag{7.38}$$

Based on the derived control protocol, the closed-loop system is obtained as follows

$$\dot{\bar{e}}_{i2}^{(0\sim3)} = \Phi_{i2}(A_{i2}^{(0\sim3)})\bar{e}_{i2}^{(0\sim3)} - \begin{bmatrix} 0_{3*1} \\ u_{i2,1} + \frac{\dot{\mu}_{i4}}{\mu_{i4}} p_{i2L}^{\mathrm{T}}\bar{e}_{i2}^{(0\sim3)} \end{bmatrix} \tag{7.39}$$

By Lemma 7.2, there exists a positive-definite matrix $p_{i2} \triangleq p_{i2}(A_{i2}^{0\sim3})$ satisfying the inequality (7.16). Accordingly, we choose the Lyapunov functional candidate as

$$V_{i4} = \frac{1}{2}(\bar{e}_{i2}^{(0\sim3)})^{\mathrm{T}} p_{i2}\bar{e}_{i2}^{(0\sim3)} \tag{7.40}$$

Differentiating $V_{i4}$ for $t \to T_{i4f}^{-}$, it has

$$\begin{aligned}\dot{V}_{i4} &= \frac{1}{2}(\bar{e}_{i2}^{(0\sim3)})^{\mathrm{T}}\big(\Phi_{i2}(A_{i2}^{(0\sim3)})^{\mathrm{T}} p_{i2} + p_{i2}\Phi_{i2}(A_{i2}^{(0\sim3)})\big)\bar{e}_{i2}^{(0\sim3)} \\ &- (\bar{e}_{i2}^{(0\sim3)})^{\mathrm{T}} p_{i2}\begin{bmatrix} 0_{3*1} \\ u_{i2,1} + \frac{\dot{\mu}_{i4}}{\mu_{i4}} p_{i2L}^{\mathrm{T}}\bar{e}_{i2}^{(0\sim3)} \end{bmatrix} \\ &\le -\frac{\check{\kappa}_{i2}}{2}(\bar{e}_{i2}^{(0\sim3)})^{\mathrm{T}} p_{i2}\bar{e}_{i2}^{(0\sim3)} - C_{i2}\frac{\dot{\mu}_{i4}}{\mu_{i4}}(\bar{e}_{i2}^{(0\sim3)})^{\mathrm{T}}\bar{p}_{i2L}\bar{e}_{i2}^{(0\sim2)} \\ &- \gamma_i(\bar{e}_{i2}^{(0\sim3)})^{\mathrm{T}} p_{i2L}\sum\nolimits_{j\in N_i}\nabla_{\mathbf{s}_{i2}}\Psi_{ij}(\|d_{ij}\|)\end{aligned} \tag{7.41}$$

where $\check{\kappa}_{i2}$ is a positive constant.

From the ARPF definition in Eq. (2.28), if $\|d_{ij}\| > r_{\text{out}}^i$, the repulsive force $\sum_{j\in N_i}\nabla_{\mathbf{s}_{i2}}\Psi_{ij}(\|d_{ij}\|) = 0$. Since the formation maintains the desired pattern, each helicopter spends minimal time in $\|d_{ij}\| \in (r_{\text{in}}^i, r_{\text{out}}^i)$, bounding $\bar{e}_{i2}$ during this period. Thus, the repulsive force satisfies

$$-\gamma_i(\bar{e}_{i2}^{(0\sim3)})^{\mathrm{T}} p_{i2L}\sum_{j\in N_i}\nabla_{\mathbf{s}_{i2}}\Psi_{ij}(\|d_{ij}\|) \le \varepsilon_i \tag{7.42}$$

where $\varepsilon_i > 0$ is a bound. This gives

$$\dot{V}_{i4} \leq -\check{\kappa}_{i2} V_{i4} + \varepsilon_i \tag{7.43}$$

meaning $\bar{e}_{i2}$ converges to $\{V_{i4} \mid V_{i4} \leq \frac{\varepsilon_i}{\check{\kappa}_{i2}}\}$ within $T_{i4f}$. By Lemma 7.1, for $t \geq T_{i4f}$, $\dot{V}_{i4} \leq -\check{\kappa}_{i2} V_{i4} + \varepsilon_i$ bounds $\bar{e}_{i2}$ in $\{V_{i4} \mid V_{i4} \leq \frac{2\varepsilon_i}{\check{\kappa}_{i2}}\}$.

Define the quadratic energy function

$$\mathrm{E}_i = \frac{1}{2} d_{ij}^{\mathrm{T}} d_{ij} + \frac{1}{2} \hat{S}_{i2}^{(3)\mathrm{T}} \hat{S}_{i2}^{(3)} \tag{7.44}$$

Differentiating $\mathrm{E}_i$ and substituting $u_{i2}$

$$\begin{aligned} \dot{\mathrm{E}}_i &= d_{ij}^{\mathrm{T}} (\dot{S}_{i2} - \dot{S}_{j2}) + \hat{S}_{i2}^{(3)\mathrm{T}} (f_{i2} + g_{i2} u_{i2} + \hat{\Theta}_{i2}) \\ &= d_{ij}^{\mathrm{T}} (\dot{S}_{i2} - \dot{S}_{j2}) - \varpi_i^{-1} \hat{S}_{i2}^{(3)\mathrm{T}} (u_{i2,1} + u_{i2,2}) \end{aligned} \tag{7.45}$$

where $u_{i2,2} = A_{i2}^{(0\sim3)} \bar{e}_{i2}^{(0\sim3)} + C_{i2} \frac{\dot{\mu}_{i4}}{\mu_{i4}} p_{i2L}^{\mathrm{T}} \bar{e}_{i2}^{(0\sim3)} - b_i S_{02}^{(4)} - \sum_{j \in N_i} a_{ij} \hat{S}_{j2}^{(4)}$, signals $\dot{S}_{i2}$, $\hat{S}_{i2}^{(3)}$, $u_{i2,1}$, $u_{i2,2}$ are bounded.

As $d_{ij} \to r_{\mathrm{in}}^i$, $-\varpi_i^{-1} \hat{S}_{i2}^{(3)\mathrm{T}} u_{i2,1} \to \infty$. For the proper parameters, it has

$$\begin{aligned} -(\varpi_i \gamma_i)^{-1} \hat{S}_{i2}^{(3)\mathrm{T}} u_{i2,1} &\geq \frac{1}{2} d_{ij}^{\mathrm{T}} d_{ij} + \frac{1}{2} \hat{S}_{i2}^{(3)\mathrm{T}} \hat{S}_{i2}^{(3)} \\ &+ (\varpi_i \gamma_i)^{-1} \hat{S}_{i2}^{(3)\mathrm{T}} u_{i2,2} - \gamma_i^{-1} d_{ij}^{\mathrm{T}} (\dot{S}_{i2} - \dot{S}_{j2}) \end{aligned} \tag{7.46}$$

Substituting Eq. (7.46) into Eq. (7.45) yields the inequality $\dot{\mathrm{E}}_i \geq \gamma_i \mathrm{E}_i$ holds. Following the analysis in [23], it follows that

$$d_{ij}^{\mathrm{T}} d_{ij} \geq 2\exp^{\gamma_i (T_{if} - T_{i0})} \mathrm{E}_i(T_{i0}) - \hat{S}_{i2}^{(3)\mathrm{T}} \hat{S}_{i2}^{(3)} \tag{7.47}$$

where $T_{i0}$ denote the initial time. Since the quadratic term $\hat{S}_{i2}^{(3)\mathrm{T}} \hat{S}_{i2}^{(3)}$ is bounded, for sufficiently large $\gamma_i$ and proper selection of the initial state $\mathrm{E}_i(T_{i0})$, the following inequality holds

$$2\exp^{\gamma_i (T_{if} - T_{i0})} \mathrm{E}_i(T_{i0}) - \hat{S}_{i2}^{(3)\mathrm{T}} \hat{S}_{i2}^{(3)} \geq (r_{\mathrm{in}}^i)^2 \tag{7.48}$$

i.e., the inequality $|d_{ij}| \geq r_{\mathrm{in}}^i$ verifies that the designed APF guarantees collision avoidance between neighboring unmanned helicopters in formation.

**Step 3**. Following Steps 1–2, the second derivative of $\bar{e}_{i3}$ is

$$\begin{aligned} \bar{e}_{i3}^{(2)} &= \sum_{j \in N_i} a_{ij} (\hat{S}_{i3}^{(2)} - \hat{S}_{j3}^{(2)}) + b_i (\hat{S}_{i3}^{(2)} - S_{03}^{(2)}) \\ &= \varpi_i (f_{i3} + g_{i3} u_{i3} + \hat{\Theta}_{i3}) - \sum_{j \in N_i} a_{ij} \hat{S}_{j3}^{(2)} - b_i S_{03}^{(2)} \end{aligned} \tag{7.49}$$

Using Eq. (2.26), we get

$$\bar{e}_{i3}^{(2)} + A_{i3}^{(0\sim1)}\bar{e}_{i3}^{(0\sim1)} = -C_{i3}\frac{\dot{\mu}_{i5}}{\mu_{i5}}p_{i3L}^{\mathrm{T}}\bar{e}_{i3}^{(0\sim1)} \tag{7.50}$$

Also, it can be rewritten the state-space form as

$$\dot{\bar{e}}_{i3}^{(0\sim1)} = \Phi_{i3}(A_{i3}^{(0\sim1)})\bar{e}_{i3}^{(0\sim1)} + \begin{bmatrix} 0 \\ -C_{i3}\frac{\dot{\mu}_{i5}}{\mu_{i5}}p_{i3L}^{\mathrm{T}}\bar{e}_{i3}^{(0\sim1)} \end{bmatrix} \tag{7.51}$$

For the Eq. (7.51), we define the Lyapunov functional

$$V_{i5} = \frac{1}{2}(\bar{e}_{i3}^{(0\sim1)})^{\mathrm{T}}p_{i3}\bar{e}_{i3}^{(0\sim1)} \tag{7.52}$$

where $p_{i3} \triangleq p_{i3}(A_{i3}^{0\sim1})$. Differentiating $V_{i5}$ and applying Lemma 7.2, it has

$$\begin{aligned}
\dot{V}_{i5} &= \frac{1}{2}(\bar{e}_{i3}^{(0\sim1)})^{\mathrm{T}}\big(\Phi_{i3}(A_{i3}^{(0\sim1)})^{\mathrm{T}}p_{i3} + p_{i3}\Phi_{i3}(A_{i3}^{(0\sim1)})\big)\bar{e}_{i3}^{(0\sim1)} \\
&+ (\bar{e}_{i3}^{(0\sim1)})^{\mathrm{T}}p_{i3}\begin{bmatrix} 0 \\ -C_{i3}\frac{\dot{\mu}_{i5}}{\mu_{i5}}p_{i3L}^{\mathrm{T}}\bar{e}_{i3}^{(0\sim1)} \end{bmatrix} \\
&\le -\frac{\check{\kappa}_{i3}}{2}(\bar{e}_{i3}^{(0\sim1)})^{\mathrm{T}}p_{i3}\bar{e}_{i3}^{(0\sim1)} - C_{i3}\frac{\dot{\mu}_{i5}}{\mu_{i5}}(\bar{e}_{i3}^{(0\sim1)})^{\mathrm{T}}\bar{p}_{i3L}\bar{e}_{i3}^{(0\sim1)}
\end{aligned} \tag{7.53}$$

leading to $\dot{V}_{i5} \le -\check{\kappa}_{i3}V_{i5} - \tilde{\kappa}_{i3}\frac{\dot{\mu}_{i5}}{\mu_{i5}}V_{i5}$. Following Step 1, $\bar{e}_{i3} \to 0$ as $t \to T_{i5f}^{-}$ and remains zero for $t \ge T_{i5f}$.

In summary, using neighbor information, tracking errors $\bar{e}_{im}$ $(m = 1, 2, 3)$ converge to a small neighborhood of the origin within $\bar{T}_{if} = \max\{T_{i3f}, T_{i4f}, T_{i5f}\}$, achieving formation tracking, which completes the proof.

**Remark 7.11** Similar to the fault observer design, the entire $N$-follower system achieves tracking within $T_2 = \max_{1\le i\le N}\bar{T}_{if}$.

**Remark 7.12** Unlike [9, 15–17] using backstepping or sliding mode control, the HOFAS approach directly designs control protocols with fewer steps, offsetting non-linearities to form a continuous linear closed-loop system.

**Remark 7.13** Matrix forms $A_{im}^{(0\sim N_i-1)}$, $p_{im}$, and $p_{imL}$ are critical for control performance, solvable via Lemmas 7.2 and 7.3. The design steps are:

Step 1. Determine observer parameters $\kappa_{ik}$, $\mu_{i1}$, $\mu_{i2}$ to estimate $\hat{\Theta}_{i2}$, $\hat{\Theta}_{i3}$, and compute $T_{if}$ and $T_1$.

Step 2. Set virtual leader trajectories to define $\bar{e}_{im}$.

Step 3. Choose Hurwitz matrices $F_{i1}, F_{i3} \in \mathbb{R}^{2\times2}$, $F_{i2} \in \mathbb{R}^{8\times8}$ and $Z_{im}$ to ensure $\mathrm{V}_{im}$ nonsingular.

Step 4. Calculate $A_{im}^{0\sim N_i-1} = -Z_{im}F_{im}^{N_i}\mathrm{V}_{im}^{-1}$ and $\Phi(A_{im}^{0\sim N_i-1}) = \mathrm{V}_{im}F_{im}\mathrm{V}_{im}^{-1}$.
Step 5. Select $\varsigma_{im}$ to derive $p_{im}$ and $p_{imL}$.
Step 6. Determine controller parameters $a_{ij}$, $b_i$, $C_{im}$, $\mu_{im}$ to compute $\bar{T}_{if}$ and $T_2$.

**Remark 7.14** Prescribed-time convergence offers a novel stability analysis framework compared to finite/fixed-time convergence. Conventional methods yield fault-dependent convergence sets, while prescribed-time FTC allows user-defined convergence regions, aligning with practical control requirements.

## 7.4 Simulation Results

Simulation experiments validate the proposed prescribed-time fault-tolerant formation control using a team of one virtual leader and four follower helicopters. The communication topology in Fig. 7.2 shows couplings: $a_{12} = a_{21} = 1, a_{13} = a_{31} = 0$, $a_{14} = a_{41} = 1$, $a_{23} = a_{32} = 1$, $a_{24} = a_{42} = 0$, $a_{34} = a_{43} = 1$, with leader connections $b_1 = 1$, $b_2 = b_3 = b_4 = 0$.

The physical parameters for the $i$th under-actuated 6-DOF unmanned helicopter is modified as: $m_i = 8.2\text{Kg}$, $g = 9.8m/S^2$, $z_{\omega i} = -0.7615S^{-1}$, $z_{mi} = -131.4125m/\text{rad}\cdot S^2$, $N_{mi} = -0.3705S^{-2}$, $J_i = \text{diag}\{0.18, 0.34, 0.28\}$, and the two parameter matrices

$$A_i = \text{diag}\{-48.1757, -25.5048, -0.9808\}s^{-1}$$

$$b_i = \begin{bmatrix} 0 & 1689.5 & 0 \\ 894.5 & 0 & 0 \\ 0 & 0 & 135.8 \end{bmatrix} s^{-2},\ i = 1, 2, 3, 4$$

Each helicopter tracks the virtual leader's trajectories

$$S_{01} = 0.1t,\ S_{02} = [20\sin(0.5t), 20\cos(0.5t)]^{\mathrm{T}},\ S_{03} = \sin(0.1t)$$

The desired relative distances are $C_{iz}^d = 0, C_{1\text{xy}}^d = [-a, -a]^{\mathrm{T}}, C_{2\text{xy}}^d = [0, 0]^{\mathrm{T}}, C_{3\text{xy}}^d = [a, -a]^{\mathrm{T}}$, $C_{4\text{xy}}^d = [-a, a]^{\mathrm{T}}$ $(a = 2)$.

Lumped faults are introduced at $t = 3$ s: $\bar{\Theta}_{i2} = [-0.01\delta_{\text{lo}i}, -0.01\delta_{\text{lai}}]^{\mathrm{T}}$ for $i = 1, 2$ and $\bar{\Theta}_{i2} = [-0.005\delta_{\text{lo}i}, -0.005\delta_{\text{lai}}]^{\mathrm{T}}$ for $i = 3, 4$; at $t = 4$ s: $\bar{\Theta}_{i3} = -0.7\delta_{t_i}$ for

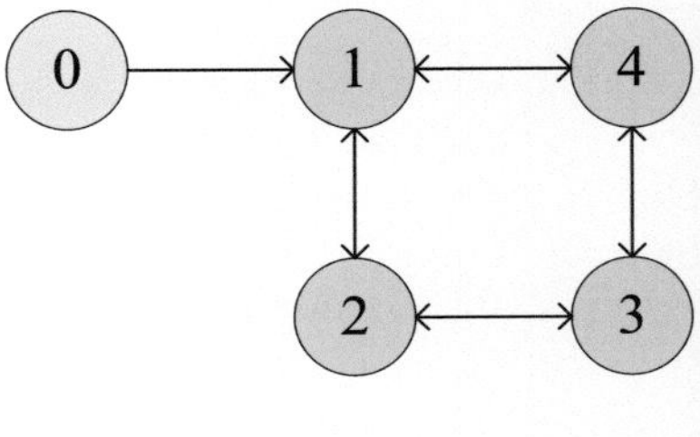


**Fig. 7.2** The communication network graph of one virtual leader and four follower unmanned helicopters

all $i$. Observer parameters: $\kappa_{i1} = \kappa_{i3} = 100, \kappa_{i2} = 4, \kappa_{i4} = 2, \kappa_{i5} = \kappa_{i6} = 0.1, \kappa_{i7} = \kappa_{i8} = 0.1$, $T_{i1f} = 3.2$ s, $T_{i2f} = 4.2$ s. Control parameters: $C_{im} = 4$ $(m = 1, 2, 3)$, $T_{i3f} = T_{i5f} = 0.5$ s, $T_{i4f} = 6$ s. ARPF parameters: $\gamma_i = 16, r^i_{\text{in}} = 1, r^i_{\text{out}} = 2$. Using Lemmas 7.2 and 7.3, matrices $A_{im}^{(0\sim N_i-1)}$, $p_{im}$, and $p_{imL}$ are derived as follows. Select:

$$Z_{i1} = \begin{bmatrix}1 & 1\end{bmatrix}, \quad Z_{i3} = \begin{bmatrix}1 & 1\end{bmatrix}, \quad F_{i1} = \begin{bmatrix}-4 & -2\\ 2 & -4\end{bmatrix},$$

$$F_{i3} = \begin{bmatrix}-4 & -2\\ 2 & -4\end{bmatrix}, \quad F_{i2} = \begin{bmatrix}F_{i21} & 0\\ 0 & F_{i22}\end{bmatrix},$$

$$F_{i21} = \begin{bmatrix}-2.25 & -0.1\\ 0.1 & -2.25\end{bmatrix}, \quad F_{i22} = \text{diag}\{-2.15, -2.45, -2.45, -2.35, -2.05, -2.25\},$$

$$Z_{i2} = \begin{bmatrix}1 & 0 & 1 & 0 & 1 & 0 & 1 & 0\\ 0 & 1 & 0 & 1 & 0 & 1 & 0 & 1\end{bmatrix}.$$

From Eq. (7.18), $\text{V}_{im}$ $(m = 1, 2, 3)$ is nonsingular: $\det\text{V}_{i1} \neq 0, \det\text{V}_{i2} \neq 0, \det\text{V}_{i3} \neq 0$. Compute:

$$A_{i1}^{(0\sim1)} = -Z_{i1}F_{i1}^2\text{V}_{i1}^{-1}, \quad A_{i2}^{(0\sim3)} = -Z_{i2}F_{i2}^4\text{V}_{i2}^{-1}, \quad A_{i3}^{(0\sim1)} = -Z_{i3}F_{i3}^2\text{V}_{i3}^{-1}.$$

Using Eq. (7.17), $\Phi_{im}(A_{im}^{(0\sim n_i-1)})$ is derived. Solving inequality (7.15) with $\varsigma_{im} = 0.001$ yields:

$$p_{i1L} = p_{i1}\begin{bmatrix}0\\ I\end{bmatrix}, \quad p_{i2L} = p_{i2}\begin{bmatrix}0_{6\times1}\\ I_{2\times1}\end{bmatrix}.$$

Matrix forms are omitted for brevity.

Also, the matrix $\Phi_{im}(A_{im}^{(0\sim n_i-1)})$ can be calculated in terms of the Eq. (7.17). Then, by solving the inequality (7.15) with $\varsigma_{im} = 0.001$, the matrices $p_{im}$ and $p_{i1L} = p_{i1}\begin{bmatrix}0\\ I\end{bmatrix}$, $p_{i2L} = p_{i2}\begin{bmatrix}0_{6*1}\\ I_{2*1}\end{bmatrix}$ can be derived separately. Here, the forms of these matrices are omitted.

Simulations use initial conditions: $S_{i1}(0) = 0$, $S_{i3}(0) = 0.1$, $S_{12}(0) = [0, 1.5]^{\text{T}}$, $S_{22}(0) = [2, 2]^{\text{T}}$, $S_{32}(0) = [2, 4]^{\text{T}}$, $S_{42}(0) = [1.5, 0]^{\text{T}}$. Results are shown in Figs. 7.3, 7.4, 7.5, 7.6, 7.7, 7.8 and 7.9. Figures 7.3, 7.4, 7.5 and 7.6 depict position trajectories $(x_i, y_i, z_i)$ and Euler angles $(\psi_i)$. Figure 7.7 shows 3D formation performance, confirming protocol (7.29) achieves the desired pattern within $T_2 = 6$ s, with unconstrained initial conditions. Figure 7.8 plots relative distances $\|d_{ij}\|$, verifying collision avoidance ($\|d_{ij}\| \geq r^i_{\text{in}}$). Figure 7.9 shows fault estimation errors $\|\tilde{\Theta}_{i2}\|$ and $\|\tilde{\Theta}_{i3}\|$ converging to zero within $T_1 = 4.2$ s.

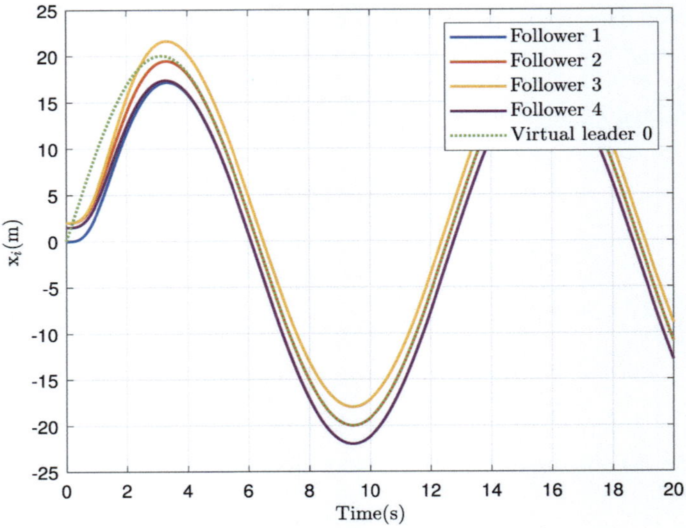


**Fig. 7.3** Tracking results of the position $x_i$ for each follower unmanned helicopter

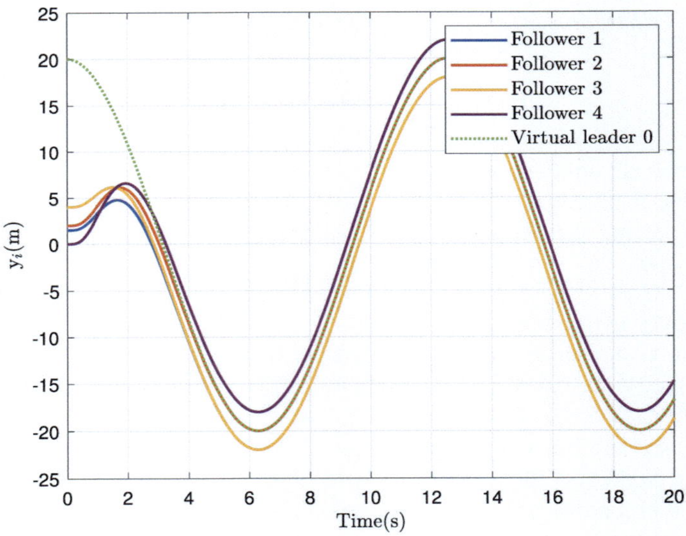


**Fig. 7.4** Tracking results of the position $y_i$ for each follower unmanned helicopter

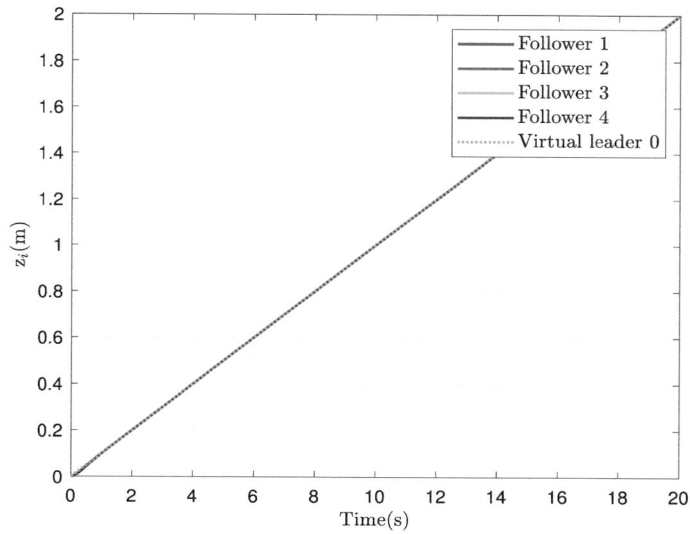


**Fig. 7.5** Tracking results of the position $z_i$ for each follower unmanned helicopter

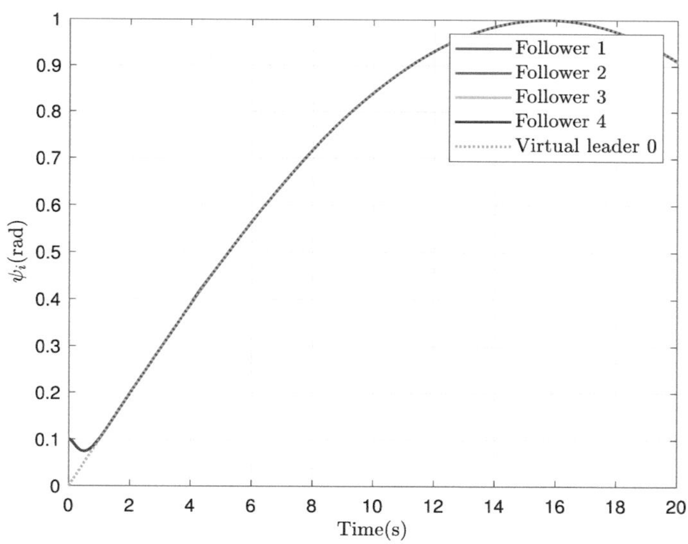


**Fig. 7.6** Tracking results of the Euler angle $\psi_i$ for each follower unmanned helicopter

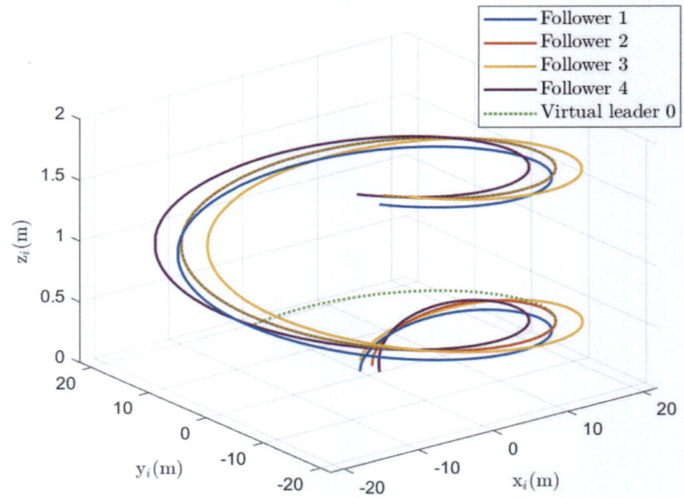


**Fig. 7.7** Formation trajectories in 3-orthogonal plane

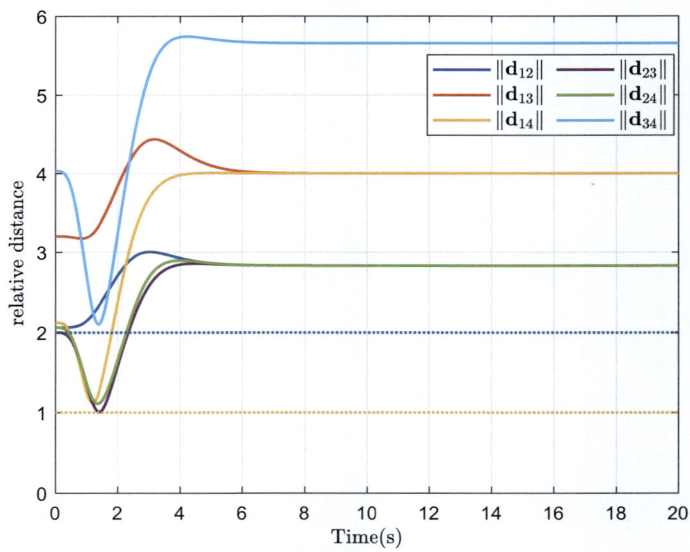


**Fig. 7.8** The actual relative distances $\|d_{ij}\|$ among unmanned helicopters with considering collision avoidance

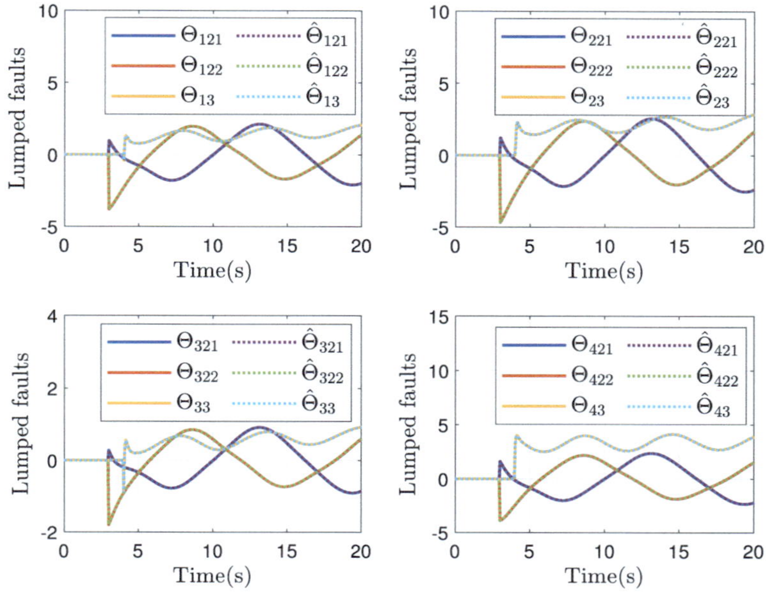


**Fig. 7.9** The estimation results of lumped faults $\Theta_{i2}$, $\Theta_{i3}$, and the estimation errors $\|\tilde{\Theta}_{i2}\|$, $\|\tilde{\Theta}_{i3}\|$

## 7.5 Conclusions

This chapter addresses prescribed-time fault-tolerant formation control for multi-6-DOF unmanned helicopters with actuator gain/bias faults. The under-actuated 6-DOF dynamics are transformed into three high-order fully-actuated subsystems, diverging from traditional outer/inner-loop decompositions. A prescribed-time fault estimation observer is proposed to simultaneously estimate subsystem states and lumped faults within a user-specified time, imposing no constraints on initial conditions. Observer parameters are tuned via convergence time requirements. Based on the HOFAS approach and an enhanced APF, a prescribed-time fault-tolerant formation control protocol is developed to mitigate fault impacts and prevent xoy-plane collisions. This ensures prescribed-time convergence and trajectory tracking. Simulations with a virtual leader and four followers validate the FTC scheme, demonstrating effective fault estimation and formation maintenance within predefined times.

## References


1. Cai M, He X, Zhou D (2023) An active fault tolerance framework for uncertain nonlinear high-order fully-actuated systems. Automatica 152:110969
2. Cai M, He X, Zhou D (2023) Low-power fault-tolerant control for nonideal high-order fully actuated systems. IEEE Trans Syst Man Cybern Syst 53(8):4875–4887
3. Chen H, Wang P, Tang G (2023) Prescribed-time control for hypersonic morphing vehicles with state error constraints and uncertainties. Aerospace Sci Technol 108671
4. Cui Y, Duan G, Liu X, Zheng H (2023) Adaptive fuzzy fault-tolerant control of high-order nonlinear systems: a fully actuated system approach. Int J Fuzzy Syst 25(5):1895–1906
5. Dong R, Hua C, Li K, Meng R (2023) Adaptive fault-tolerant control for high-order fully actuated system with full-state constraints. J Franklin Inst 360(12):8062–8074
6. Duan G (2021) High-order fully actuated system approaches: part I. Models and basic procedure. Int J Syst Sci 52(2):422–435
7. Duan G (2023) Robust stabilization of time-varying nonlinear systems with time-varying delays: a fully actuated system approach. IEEE Trans Cybern 53(12):7455–7468
8. Gao Z, Zhang Y, Guo G (2022) Prescribed-time control of vehicular platoons based on a disturbance observer. IEEE Trans Circuits Syst II Exp Briefs 69(9):3789–3793
9. Kuo CW, Tsai CC, Lee CT (2019) Intelligent leader-following consensus formation control using recurrent neural networks for small-size unmanned helicopters. IEEE Trans Syst Man Cybern Syst 51(2):1288–1301
10. Li P, Duan G (2023) High-order fully actuated control approaches of flexible servo systems based on singular perturbation theory. IEEE/ASME Trans Mechatron 28(6):3386–3397
11. Li Y, Feng K, Li K (2024) Finite-time fuzzy adaptive dynamic event-triggered formation tracking control for usvs with actuator faults and multiple constraints. IEEE Trans Industr Inf 20(4):5285–5296
12. Liu C, Jiang B, Zhang K, Ding SX (2021) Hierarchical structure-based fault-tolerant tracking control of multiple 3-DOF laboratory helicopters. IEEE Trans Syst Man Cybern Syst 52(7):4247–4258
13. Liu X, Chen M, Sheng L, Zhou D (2022) Adaptive fault-tolerant control for nonlinear high-order fully-actuated systems. Neurocomputing 495:75–85
14. Lu S, Tsakalis K, Chen Y (2022) Development and application of a novel high-order fully actuated system approach-part i: 3-DOF quadrotor control. IEEE Control Syst Lett 7:1177–1182
15. Wang D, Zong Q, Tian B, Shao S, Zhang X, Zhao X (2018) Neural network disturbance observer-based distributed finite-time formation tracking control for multiple unmanned helicopters. ISA Trans 73:208–226
16. Wang D, Zong Q, Tian B, Wang F, Dou L (2018) Finite-time fully distributed formation reconfiguration control for UAV helicopters. Int J Robust Nonlinear Control 28(18):5943–5961
17. Wang J, Wang D, Wang C, Deng F (2020) Robust formation control for unmanned helicopters with collision avoidance. J Franklin Inst 357(17):11997–12018
18. Wang X, Duan G (2023) High-order fully actuated system approaches: model predictive control with applications to under-actuated systems. J Franklin Inst 360(10):6953–6975
19. Wang X, Wang Y, Zhang Z, Wang X, Patton R (2023) Sensor fault tolerant control for a 3-DOF helicopter considering detectability loss. IEEE Trans Circuits Syst I Regul Pap 70(10):1112–4125
20. Wang Y, Song Y, Hill DJ, Krstic M (2018) Prescribed-time consensus and containment control of networked multiagent systems. IEEE Trans Cybern 49(4):1138–1147
21. Wang Z, Lam H-K, Guo Y, Xiao B, Li Y, Su X, Yeatman EM, Burdet E (2023) Adaptive event-triggered control for nonlinear systems with asymmetric state constraints: a prescribed-time approach. IEEE Trans Autom Control 68(6):3625–3632

22. Wei C, Wu X, Xiao B, Wu J, Zhang C (2022) Adaptive leader-following performance guaranteed formation control for multiple spacecraft with collision avoidance and connectivity assurance. Aerosp Sci Technol 120:107266
23. Wu X, Wei C, Chen T, Dai M (2023) On novel distributed fixed-time formation tracking of multiple hypersonic flight vehicles with collision avoidance. Aerosp Sci Technol 141:108517
24. Xu C, Wu B, Zhang Y (2021) Distributed prescribed-time attitude cooperative control for multiple spacecraft. Aerosp Sci Technol 113:106699
25. Yang H, Jiang B, Liu HH, Yang H, Zhang Q (2019) Attitude synchronization for multiple 3-DOF helicopters with actuator faults. IEEE/ASME Trans Mechatron 24(2):597–608
26. Ye H, Song Y (2023) Prescribed-time tracking control of mimo nonlinear systems under non-vanishing uncertainties. IEEE Trans Autom Control 68(6):3664–3671
27. Yu Y, Liu G-P, Huang Y, Guerrero JM (2023) Coordinated predictive secondary control for dc microgrids based on high-order fully actuated system approaches. IEEE Trans Smart Grid 15(1):19–33
28. Zhao Z, Zhang J, Liu Z, He W, Hong K-S (2023) Adaptive quantized fault-tolerant control of a 2-DOF helicopter system with actuator fault and unknown dead zone. Automatica 148:110792
29. Zou A, Liu Y, Hou Z, Hu Z (2023) Practical predefined-time output-feedback consensus tracking control for multiagent systems. IEEE Trans Cybern 53(8):5311–5322

# Chapter 8
# Adaptive Neural Network Prescribed-Time FTC for UAVs-UGVs

## 8.1 Introduction

Since convergence time is a key performance index, it should be considered in the application of actual formation control. However, although the convergence time is uniformly bounded by the fixed-time formation tracking, the designer cannot define prescribed specifications in advance because the upper bound depends on controller or observer gains. In addition, most of the finite-time and fixed-time control methods inevitably require nonsmooth control action as a result of the use of the signum function. Due to the physical limitation of UAV and UGV, it is of great theoretical and practical importance to study a novel control framework that leads to a smooth control action. Fortunately, a novel concept of "prescribed-time stability" was developed to solve the above considerations [25], where the convergence time can be arbitrarily defined by the designer and not influenced by the initial system conditions or even any control parameters. Moreover, different from the discontinuous or nonsmooth finite-time methods, the prescribed-time algorithm can generate smooth control action everywhere. Based on the above advantages, the prescribed-time control technology for MASs has attracted the interest of researchers in many fields [4, 9–11, 27]. However, the aforementioned results are focused on the prescribed-time control for MASs, where agents have homogeneous single-integrator, second-order integrator dynamics or high-order integrator dynamics. It is obvious that, in practical aerial-ground cooperative system, the models of multiple agents are significantly different. Therefore, it is meaningful and necessary to extend the system dynamics to heterogeneous MASs.

In addition, note that the above-mentioned works suppose that the actuators operate normally. In practice of MASs, the actuator is prone to partial loss of effectiveness and bias faults due to the increasing complexity and automation. As a result, the system performance will degrade and even suffer from catastrophic consequence [22, 31]. Therefore, fault-tolerant capability is required to accommodate the unexpected actuator faults [17]. Over the past decades, plentiful fault-tolerant results

 
K. Zhang et al., *Fixed-Time/Prescribed-Time Fault-Tolerant Control for Multiple Unmanned Systems*, Unmanned System Technologies,
https://doi.org/10.1007/978-981-95-2913-1_8

have been reported for homogeneous or heterogeneous MASs based on asymptotic stability [1, 6, 12, 13, 18, 29], finite-time stability [7, 23, 26, 30], or fixed-time stability [5, 24, 28]. However, for an aerial-ground formation system, despite the influence of the initial conditions of MAS and the algebraic connectivity of the topology network, formation tracking under actuator faults can be implemented within the specified time, which is the most satisfactory. Up to now, no previous result has reported the prescribed-time FTC for MASs. Moreover, it is known that the existing chattering phenomenon produced by a switching finite/fixed-time FTC impedes its implementation. Hence, it is significant to study new fault-tolerant approaches for MASs, especially for heterogeneous MASs, that can ensure the convergence errors to origin within a prescribed time without requiring nonsmooth control inputs.

In view of the above analyses, this chapter investigates the distributed prescribed-time control problem for air-ground systems affected by partial loss of effectiveness and bias actuator faults under directed graphs. Compared with the relevant results, this study possesses the following characteristics.

(1) A novel distributed prescribed-time observer is developed for each follower agent to estimate the state of the leader agent. Different from the existing asymptotic observer [6], finite-time observer [14], and fixed-time observer [2, 5], the settling time of the constructed one can be assigned beforehand under directed graph, thus can be uniformly prespecified.
(2) A novel prescribed-time sliding mode manifold is developed, which can guarantee that the error reaches the desired bounded region within the specified time. Then, a prescribed-time fault-tolerant formation tracking controller is proposed for the considered heterogeneous MASs by utilizing the proposed sliding manifold. The designed FTC method can generate smooth control action due to the use of regular feedback of local system states, which can not achieved in the fixed-time strategies [2, 3, 8, 19, 21].
(3) Two novel adaptive laws are constructed to estimate the unexpected actuator faults and the upper bound of the lumped uncertainties, respectively, such that the adaptive prescribed-time fault-tolerant formation can be achieved. As a result, the safety, reliability and mobility of the system can be effectively ensured, which has the potential of operational aerial-ground formation systems under multivariable environments.

## 8.2 Problem Formulation and Preliminaries

### *8.2.1 Problem Formulation*

Here, a heterogeneous formation MAS involving one leader and $N$ followers is studied. The studied heterogeneous MAS is located in the same coordinate frame, so the longitudinal positions of UAVs and UGVs have the same physical representation.

A UGV is assumed as the physical leader since it experiments with more difficulties and can work in different environments.

Let's consider the existing external disturbance and a model uncertainty. Then, based on the model introduction in Chap. 4, the $i$th ($i \in F_1 \cup F_2$) follower model has the following form

$$\ddot{x}_i(t) = (I - \rho_i(t))u_i(t) + \Delta_i(t) \tag{8.1}$$

where $x_i \in R^{n_i}$ and $u_i \in R^{n_i}$ denote the coordinate state and control signal of the $i$th follower, respectively. In addition, $\rho_i(t) = \text{diag}\left\{\rho_{i1}(t), \rho_{i2}(t), ..., \rho_{in_i}(t)\right\}$; $\Delta_i(t) \in R^{n_i}$ denotes the following lumped uncertainty

$$\Delta_i(t) = K_i\delta_i(t) + f_i(t) + d_i(t) \tag{8.2}$$

where $\delta_i(t)$, $f_i(t)$ and $d_i(t) \in R^{n_i}$. $\rho_i(t)$ and $\delta_i(t)$ represent the partial loss of effectiveness and the bias faults, respectively. The case of $\rho_i(t) = 0$ and $\delta_i(t) = 0$ means that the actuator works properly. If $\rho_i(t) \in (0, 1)$ and $\delta_i(t) \neq 0$, the actuator has partial loss of effectiveness and bias faults but still works; whilts $\rho_i(t) = 1$ implies that the actuator is completely faulty. The diagnoal matrix $K_i = I_2$ if and only if $i \in F_1$, otherwise it is defined as

$$K_i = \begin{pmatrix} \left(\cos\phi_i^q \sin\theta_i^q \cos\psi_i^q + \sin\phi_i^q \sin\psi_i^q\right)/m_i \\ \left(\cos\phi_i^q \sin\theta_i^q \sin\psi_i^q - \sin\phi_i^q \sin\psi_i^q\right)/m_i \\ \cos\theta_i^q \cos\phi_i^q/m_i \end{pmatrix} \in R^3 \tag{8.3}$$

The following assumptions are used to design the prescribed-time observer.

**Assumption 8.1** The directed graph among $N + 1$ agents contains a spanning tree, where the leader is the root node.

**Assumption 8.2** The parameters of the actuator faults are bounded, i.e., there are constants $\rho_{im}$ and $\bar{\delta}_i$ such that $0 \leq \rho_{im} \leq \rho_{ik} < 1$ and $|\delta_{ik}| \leq \bar{\delta}_i$.

**Assumption 8.3** The external disturbance $d_i$ is bounded by an unknown constant $\bar{d}$, i.e., $\|d_i\| \leq \bar{d} < \infty$.

**Assumption 8.4** The input $u_0$ of the leader satisfies the relation $\|\dot{u}_0\| \leq c$, in which $c$ is a known positive constant.

In this chapter, a prescribed-time fault-tolerant formation controller for an aerial-ground heterogeneous MAS is considered. To complete and maintain the expected configuration, a time-varying differentiable vector $h_i(t) \in R^{n_i}$ ($i \in \Pi$) is introduced to describe the desired formation between the $i$th follower and the leader. Moreover, the desired trajectories $h_i(t)$, $\dot{h}_i(t)$, and $\ddot{h}_i(t)$ are bounded.

**Control Objective:** For any given bounded initial conditions, if there exists a convergence time $T_{\max} > 0$ that is a user-assignable finite constant, such that

$$\lim_{t \to T_{\max}} (x_i(t) - x_0(t) - h_i(t)) = 0, i \in \Pi. \tag{8.4}$$

then the heterogeneous MASs model with actuator faults of Eq. (8.1) is assumed to finish the scheduled prescribed-time formation tracking under directed graphs.

**Remark 8.1** Assumption 8.1 is a standard condition to ensure the solvability of the studied control problem, which can be found in [2, 14]. Assumption 8.2 is a standard condition to address unknown actuator faults, and Assumptions 8.3 and 8.4 are generally introduced for addressing the distributed leader-follower tracking problem [18, 29].

**Remark 8.2** In practice, the tracking errors can hardly arrive at origin exactly as a result of the existing of actuator faults and other unknown uncertainties. Therefore, the design objective in this study is to construct robust fault-tolerant tracking control protocols for the heterogeneous MASs, such that the position of each follower $x_i$ can arrive at a sufficiently small neighborhood of the expected formation specified by $h_i$ within a prescribed time, which is acceptable in most practical circumstances. Moreover, the size of the convergence region can be adjusted and the convergence time is independent of controller parameters. In this way, the practically prescribed-time formation strategy developed in this chapter is more satisfactory and realistic than the practically finite/fixed-time control protocols.

### 8.2.2 *Preliminaries*

Radial basis function neural network (RBFNN) is employed to approximate this unknown upper bound. Similar to [13], we proposed to use a RBFNN to handle the time-varying lumped uncertainty in the studied follower system. According to Assumptions 8.2 and 8.3, the unknown lumped uncertainty $\Delta_i$ can be approximated by using the following RBFNN

$$\Delta_i(.) = W_i^{\mathrm{T}} \theta (Z_i) + \varepsilon_i (Z_i) \tag{8.5}$$

where $W_i \in R^{\vartheta}$ denotes the ideal weight vector with $\vartheta > 1$ nodes, $Z_i$ is the input of the NN, $\varepsilon_i$ is the approximation error, and $\theta (Z_i)$ is the following Gaussian function

$$\theta (Z_i) = \exp \left( -\frac{(Z_i - \mu_i)^T (Z_i - \mu_i)}{\nu_i^2} \right) \tag{8.6}$$

where $\mu_i$ and $\nu_i$ denote the center and width of the NN, respectively. It should be pointed that $\varepsilon_i$ and $W_i$ are bounded and satisfy the relations $\|\varepsilon_i\| \leq \bar{\varepsilon}$, $\|W_i\| \leq W_m$ with $\bar{\varepsilon}$ and $W_m$ being the unknown constants.

**Lemma 8.1** ([27]) *For the following system*

$$\dot{x}(t) = f(x(t)), \quad x(0) = x_0 \tag{8.7}$$

*where $x \in R^n$ and $f(.) : R^n \to R^n$ denotes a given nonlinear function. If there is a continuous positive definite function $V(x)$ and a positive scalar $\gamma$ such that*

$$\dot{V}(x) \leq -\gamma V(x) - 2\frac{\dot{\varsigma}(t_0, T)}{\varsigma(t_0, T)} V(x), t \in [t_0, t_0 + T) \tag{8.8}$$

*with*

$$\varsigma(t_0, T) = \begin{cases} \left(\frac{T}{t_0+T-t}\right)^h, t \in [t_0, t_0 + T) \\ 1, t \in [t_0 + T, \infty) \end{cases} \tag{8.9}$$

*where $h > 2$, $T$ is a user-assignable time, and $t_0$ is the start time point, then the equilibrium point of the considered system is assumed to be globally prescribed-time stable within the prescribed time $T$.*

**Remark 8.3** Lemma 8.1 provides a control method where, even if $x_0$ is large enough, the finishing time can still be preassigned independently of the initial system states. Moreover, the settling time $T$ is not subject to certain restrictions and can be uniformly prespecified, which is different from the existing finite-time and fixed-time convergence schemes [5, 14].

## 8.3 Main Results

In this section, we propose a new adaptive prescribed-time formation tracking algorithm for the studied air-ground systems in the presence of actuator faults, based on a distributed prescribed-time observer, to realize the expected formation tracking within an user-defined convergence time.

### *8.3.1 Distributed Prescribed-Time Observer Design for Leader*

For the $i$th ($i \in \Pi$) follower agent, $\hat{x}_i$, $\hat{v}_i$, and $\hat{u}_i$ denote the estimations of $x_0$, $v_0$, and $u_0$, respectively. Let's define $\bar{L}=\frac{1}{2}\left(\Psi H+H^{\mathrm{T}}\Psi\right)$, then a distributed prescribed-time observer has the form of the system of

$$\begin{cases} \dot{\hat{x}}_i = -\left(\alpha+\beta\frac{\dot{\varsigma}}{\varsigma}\right)\left(\sum_{j=1}^{n} a_{ij}(\hat{x}_i - \hat{x}_j) + b_i(\hat{x}_i - x_0)\right) + \hat{v}_i \\ \dot{\hat{v}}_i = -\left(\alpha+\beta\frac{\dot{\varsigma}}{\varsigma}\right)\left(\sum_{j=1}^{n} a_{ij}(\hat{v}_i - \hat{v}_j) + b_i(\hat{v}_i - v_0)\right) + \hat{u}_i \\ \dot{\hat{u}}_i = -\left(\alpha+\beta\frac{\dot{\varsigma}}{\varsigma}\right)\left(\sum_{j=1}^{n} a_{ij}(\hat{u}_i - \hat{u}_j) + b_i(\hat{u}_i - u_0)\right) \\ \quad - \gamma \operatorname{sign}\left(\sum_{j=1}^{n} a_{ij}(\hat{u}_i - \hat{u}_j) + b_i(\hat{u}_i - u_0)\right) \end{cases} \tag{8.10}$$

in which the observer parameters satisfy the relations $\alpha > 0$, $\beta \geq \frac{\max\{\omega_i\}}{\lambda_{\min}(\bar{L})}$ and $\gamma > c$. In addition, $\omega_i$ is given in Lemma 8.1. Let $\tilde{x}_i = \hat{x}_i - x_0$, $\tilde{v}_i = \hat{v}_i - v_0$, and $\tilde{u}_i = \hat{u}_i - u_0$ represent the estimation errors of the corresponding states of the leader.

**Theorem 8.1** *Consider the leader-following dynamics, if Assumptions 8.1 and 8.4 are satisfied, then each follower can obtain the leader states $x_0$, $v_0$, and $u_0$ in a prescribed time by using the proposed observer.*

***Proof*** Using the last relation in Eq. (8.10), one can obtain that

$$\begin{aligned} \dot{\tilde{u}}_i = &-\left(\alpha+\beta\frac{\dot{\varsigma}}{\varsigma}\right)\left(\sum_{j=1}^{n} a_{ij}(\hat{u}_i - \hat{u}_j) + b_i(\hat{u}_i - u_0)\right) \\ &- \gamma \operatorname{sign}\left(\sum_{j=1}^{n} a_{ij}(\hat{u}_i - \hat{u}_j) + b_i(\hat{u}_i - u_0)\right) - \dot{u}_0 \end{aligned} \tag{8.11}$$

Let's define $U_i = \sum_{j=1}^{n} a_{ij}(\hat{u}_i - \hat{u}_j) + b_i(\hat{u}_i - u_0)$, then we have $U = (H \otimes I_2)\,\tilde{u}$, in which $U = [U_1, U_2, ..., U_N]^{\mathrm{T}}$ and $\tilde{u} = \left[\tilde{u}_1, \tilde{u}_2, ..., \tilde{u}_N\right]^{\mathrm{T}}$. Since $H$ is an invertible matrix, $\hat{u}_i$ can provide the estimation of $u_0$ when $U$ goes to zero. Then, we will prove that $U$ achieves prescribed-time convergence.

The time derivative of $U$ has the following compact form

$$\dot{U} = -\left(\alpha+\beta\frac{\dot{\varsigma}}{\varsigma}\right)(H \otimes I_2)\,U - \gamma\,(H \otimes I_2)\operatorname{sign}(U) - (H \otimes I_2)(1_N \otimes \dot{u}_0) \tag{8.12}$$

The Lyapunov function can be defined as

$$V_1 = \sum_{i=1}^{N}\sum_{k=1}^{2} \frac{\varpi_i}{2} |U_{ik}|^2 \tag{8.13}$$

Based on Assumption 8.4, we can obtain that the following inequality holds: $-U\left(\Psi H \otimes I_2\right)\left(\gamma \operatorname{sign}(U)-1_n \otimes \dot{u}_0\right) \leq 0$. Then we can get the following inequality for $\dot{V}_1$

$$\begin{aligned}
\dot{V}_1 &= \sum_{i=1}^{N} \sum_{k=1}^{2} \varpi_i \left|U_{ik}\right| \operatorname{sign}(U_{ik}) \dot{U}_{ik} \\
&\leq \left(\alpha+\beta \frac{\dot{\varsigma}}{\varsigma}\right) U^{\mathrm{T}} \bar{L} U \\
&\leq \lambda_{\min}(\bar{L})\left(\alpha+\beta \frac{\dot{\varsigma}}{\varsigma}\right) \sum_{i=1}^{N} \sum_{k=1}^{2}\left|U_{ik}\right|^2 \\
&\leq -\frac{2 \alpha \lambda_{\min}(\bar{L})}{\max \left\{\varpi_i\right\}} V_1-\frac{2 \beta \lambda_{\min}(\bar{L})}{\max \left\{\varpi_i\right\}} \frac{\dot{\varsigma}}{\varsigma} V_1 \\
&\leq -\frac{2 \alpha \lambda_{\min}(\bar{L})}{\max \left\{\varpi_i\right\}} V_1-2 \frac{\dot{\varsigma}}{\varsigma} V_1
\end{aligned} \tag{8.14}$$

According to the prescribed-time stable theorem, we have that the origin of $V_1=0$ is prescribed-time stable. Therefore, also obtain that each follower can effectively estimate $u_0$ within a prespecified finite time $T_1$. Furthermore, we will prove that the estimation error $\tilde{v}_i=\hat{v}_i-v_0$ meets the prescribed-time convergence within a prescribed time $T_2$.

Let's define $V_i=\sum_{j=1}^n a_{ij}\left(\hat{v}_i-\hat{v}_j\right)+b_i\left(\hat{v}_i-v_0\right)$. Then, we have $V=\left(H \otimes I_2\right) \tilde{v}$, in which $V=\left[V_1, V_2, \ldots, V_N\right]^{\mathrm{T}}$ and $\tilde{v}=\left[\tilde{v}_1, \tilde{v}_2, \ldots, \tilde{v}_N\right]^{\mathrm{T}}$. As for $T_1$, we consider the relation of

$$\dot{V}=-\left(\alpha+\beta \frac{\dot{\varsigma}}{\varsigma}\right)\left(H \otimes I_2\right) V \tag{8.15}$$

The Lyapunov function is $V_2=\sum_{i=1}^N \sum_{k=1}^2 \frac{\varpi_i}{2}\left|V_{ik}\right|^2$. Then, for the derivative of $V_2$, the following inequalities hold

$$\begin{aligned}
\dot{V}_2 &= \sum_{i=1}^{N} \sum_{k=1}^{2} \varpi_i \left|V_{ik}\right| \operatorname{sign}(V_{ik}) \dot{V}_{ik} \\
&\leq -\left(\alpha+\beta \frac{\dot{\varsigma}}{\varsigma}\right) V^{\mathrm{T}} \bar{L} V \\
&\leq -\lambda_{\min}(\bar{L})\left(\alpha+\beta \frac{\dot{\varsigma}}{\varsigma}\right) \sum_{i=1}^{N} \sum_{k=1}^{2}\left|V_{ik}\right|^2 \\
&\leq -\frac{2 \alpha \lambda_{\min}(\bar{L})}{\max \left\{\varpi_i\right\}} V_2-\frac{2 \beta \lambda_{\min}(\bar{L})}{\max \left\{\varpi_i\right\}} \frac{\dot{\varsigma}}{\varsigma} V_2 \\
&\leq -\frac{2 \alpha \lambda_{\min}(\bar{L})}{\max \left\{\varpi_i\right\}} V_2-2 \frac{\dot{\varsigma}}{\varsigma} V_2
\end{aligned} \tag{8.16}$$

Then, we can obtain that $V_2$ will achieve the prescribed-time convergence, which means that $\hat{v}_i$ can provide the estimation of $v_0$ within a prespecified time $T_2$. Then, the following expression of $\tilde{x}_i$ can be easily derived

$$\dot{\tilde{x}}_i = -\left(\alpha+\beta\frac{\dot{\varsigma}}{\varsigma}\right)\left(\sum_{j=1}^{n} a_{ij}(\hat{x}_i - \hat{u}_j) + b_i(\hat{x}_i - x_0)\right) + \hat{v}_i - v_0 \tag{8.17}$$

For $t \geq T_1 + T_2$, one has that $\hat{v}_i = v_0$ and $\hat{u}_i = u_0$. If we define $\tilde{x} = \left[\tilde{x}_1, \tilde{x}_2, ..., \tilde{x}_N\right]^{\mathrm{T}}$, then we can get that $X = (H \otimes I_2)\,\tilde{x}$, where $X = [X_1, X_2, ..., X_N]^{\mathrm{T}}$. Then, the prescribed-time stability of $X$ can be proved.

Let's consider the following Lyapunov function

$$V_3 = \sum_{i=1}^{N}\sum_{k=1}^{2}\frac{\varpi_i}{2}|X_{ik}|^2 \tag{8.18}$$

Following the same procedure as in Eq. (8.15), it can be proved that $X$ can achieve the prescribed-time convergence within a prespecified time $T_3$. Moreover, if the prescribed time is selected as $T_1 = T_2 = T_3$, then all followers can provide the estimation of the leader states based on the proposed observer within $3T_1$.

**Remark 8.4** The solution developed in this chapter is different from the existing finite/fixed-time observer [2, 14, 24], in which the convergence time required to realize the desired observation depends on the controller parameters and even the initial states of the system. The designed prescribed-time observer developed in this study is set by the designer and independent of any control gains and the initial states. Therefore, the settling time of the MASs is always invariant with respect to the initial states, which is more practical and has better design flexibility. However, the selection of the upper bound should be combined with the physical acceptable time range of unmanned vehicles to further improve the actual reference value.

### 8.3.2 *Prescribed-Time Sliding Mode Surface Design*

To deal with the tracking problem, we first define the following formation error for the $i$th ($i \in F$) agent

$$e_i = x_i - \hat{x}_0 - h_i \tag{8.19}$$

Then, a new prescribed-time sliding mode surface has the form of

$$s_i = \dot{e}_i + k_{i1}e_i + k_{i2}e_i + \frac{\dot{\varsigma}_i}{\varsigma_i}e_i \tag{8.20}$$

where $k_{i1}, k_{i2} > 0$ are suitable constants, and $\varsigma_i$ is defined in prescribed-time stability theorem.

**Theorem 8.2** *If the sliding mode surface satisfies the inequality $\|s_i\| \leq \Delta_s$, then the formation error $e_i$ converges to the region $\|e_i\| \leq l_i \Delta_s$ in a prescribed time $T_s$, where $l_i$ is a positive constant.*

***Proof*** Let's define the Lyapunov function $V_s = \frac{1}{2} e_i^{\mathrm{T}} e_i$, and taking the derivative of $V_s$, it yields to the relation of

$$\begin{aligned} e_i^{\mathrm{T}} s_i &\leq \frac{\upsilon_i}{2}\|e_i\|^2 + \frac{1}{2\upsilon_i}\|s_i\|^2 \\ &= e_i^{\mathrm{T}} s_i - k_i e_i^{\mathrm{T}} e_i - k_{i2} e_i^{\mathrm{T}} e_i - \frac{\dot{\varsigma}_i}{\varsigma_i} e_i^{\mathrm{T}} e_i \end{aligned} \tag{8.21}$$

Note that

$$\begin{aligned} e_i^T s_i &\leq \frac{\upsilon_i}{2}\|e_i\|^2 + \frac{1}{2\upsilon_i}\|s_i\|^2 \\ &\leq \frac{\upsilon_i}{2}\|e_i\|^2 + \frac{\Delta_s^2}{2\upsilon_i} \end{aligned} \tag{8.22}$$

where $\upsilon_i > 0$ is a constant. Then, the following inequality holds

$$\begin{aligned} \dot{V}_s &\leq -(2k_i - \upsilon_i) V - 2k_{i2} V - 2\frac{\dot{\varsigma}_i}{\varsigma_i} V + \frac{\Delta_s^2}{2\upsilon_i} \\ &\leq -(2k_i - \upsilon_i)\left(V - \frac{\Delta_s^2}{2\upsilon_i (2k_i - \upsilon_i)}\right) - 2k_{i2} V - 2\frac{\dot{\varsigma}_i}{\varsigma_i} V \end{aligned} \tag{8.23}$$

By appropriately selecting the positive constant $k_i$ such that $2k_i - \upsilon_i > 0$, then, if $\|e_i\| \geq \frac{\Delta_s}{\sqrt{\upsilon_i (2k_i - \upsilon_i)}} \triangleq l_i \Delta_s$, it yields to

$$\dot{V}_s \leq -2k_{i2} V_s - 2\frac{\dot{\varsigma}_i}{\varsigma_i} V_s. \tag{8.24}$$

Based on the prescribed-time stability theorem, $e_i$ converges to the region $\|e_i\| \leq l_i \Delta_s$ in an user-defined time $T_s$.

### 8.3.3 Prescribed-Time FTC Design

Usually, the information of the lumped uncertainty $\Delta_i$ in follower system is demanded to design the controller parameters for obtaining a robust controller. However, the upper bound of the uncertainty cannot be easily obtained for the practical operation of unmanned vehicles. Moreover, to reduce the number of parameters that have to be updated with the adaptive algorithm, an effective indirect approach exploiting the RBFNN is considered, as highlighted by

$$\|\Delta_i\| \le \|K_i\delta_i\| + \|f_i\| + \|d_i\| = \Gamma_i(Z_i) \tag{8.25}$$

Although $\Gamma_i(Z_i)$ is completely unknown, it is a nonlinear and scalar function and relies on $Z_i = \left[x_i, \dot{x}_i, h_i, \dot{h}_i\right]$. Therefore, the following inequality holds

$$\|\Gamma_i(Z_i)\| \le \|W_i\| \, \|\theta\left(Z_i\right)\| + \bar{\varepsilon} \le a_i\Phi_i(Z_i) \tag{8.26}$$

where $a_i = \max\left\{\|W_i\|, \bar{\varepsilon}\right\}$ is an unknown constant, and $\Phi_i(Z_i) = 1 + \|\theta\left(Z_i\right)\| > 0$ is a known scalar function, which can be directly utilized in the controller construction.

**Theorem 8.3** *Let's consider the heterogeneous MASs described by Eq. (8.1) with prescribed-time sliding manifold of Eq. (8.20), while Assumptions 8.1–8.4 hold. If the proposed adaptive fault-tolerant formation tracking algorithms of Eqs. (8.27) and (8.28) with adaptive laws of Eqs. (8.29) and (8.30), $s_i$ can converge to the designed area $D_s = \left\{s_i \,|\, \|s_i\| \le \sqrt{\Delta_{i1} + \Delta_{i2}} \overset{\Delta}{=} \Delta_s\right\}$, and the formation error $e_i$ converges to the region $D_e = \{e_i \,|\, \|e_i\| \le l_i\Delta_s\}$ within the prescribed time $3T_1 + T_s + T_{cs}$, in which $\Delta_{i1}$ and $\Delta_{i2}$ are several positive assignable constants. Moreover*

$$u_i = \left[A_i\left(I - \hat{\rho}_i\right)\right]^{-1}\left(\ddot{x}_0 + \ddot{h}_i - k_{i1}\dot{e}_i - k_{i2}\dot{e}_i - \frac{\dot{\varsigma}_i}{\varsigma_i}\dot{e}_i - \dot{\Upsilon}_i e_i\right) + u_{im} \tag{8.27}$$

$$u_{im} = \left[A_i\left(I - \hat{\rho}_i\right)\right]^{-1}\left(-\hat{a}_i\Phi_i\frac{s_i}{\|s_i\|} - \omega_{i1}s_i - \omega_{i2}s_i - \frac{\dot{\varsigma}_i}{\varsigma_i}s_i\right) \tag{8.28}$$

*where $\Upsilon_i = \frac{\dot{\varsigma}_i}{\varsigma_i}$, and $\omega_{i1}, \omega_{i2}$ are designed positive parameters. Then, the adaptive compensation laws have the following forms*

$$\dot{\hat{\theta}}_i = -\lambda_i\hat{\theta}_i - \mu_i\mathrm{diag}(u_i)^{\mathrm{T}}A_i^{\mathrm{T}}s_i \tag{8.29}$$

$$\dot{\hat{a}}_i = -\vartheta_i\hat{a}_i + \kappa_i\Phi_i\|s_i\| \tag{8.30}$$

*where $\lambda_i, \mu_i, \vartheta_i$ and $\kappa_i$ are design positive parameters. Moreover, the adaptive parameters $\hat{a}_i$ and $\hat{\theta}_i$ represent the online estimation of the unknown gains $a_i$ and $\theta_i$. In addition, $\hat{\theta}_i$ satisfies the relation $\hat{\rho}_i u_i = \mathrm{diag}(u_i)\hat{\theta}_i$, in which $\hat{\rho}_i$ is the estimation of the actuator fault parameter $\rho_i$.*

***Proof*** For the considered heterogeneous MASs, the candidate Lyapunov function is defined as

$$V_c = \frac{1}{2}s_i^{\mathrm{T}}s_i + \frac{1}{2\kappa_i}\tilde{a}_i^{\mathrm{T}}\tilde{a}_i + \frac{1}{2\mu_i}\tilde{\theta}_i^{\mathrm{T}}\tilde{\theta}_i \tag{8.31}$$

where $\tilde{\theta}_i = \theta_i - \hat{\theta}_i$, and $\tilde{a}_i = a_i - \hat{a}_i$.

Firstly, based on proposed control laws, the derivative of $s_i$ has

$$\begin{aligned}\dot{s}_i =&\ddot{x}_i - \ddot{\hat{x}}_0 - \ddot{h}_i + k_{i1}\dot{e}_i + k_{i2}\dot{e}_i + \frac{\dot{\varsigma}_i}{\varsigma_i}\dot{e}_i + \dot{\Upsilon}_i e_i \\ =&A_i\left(I-\rho_i\right)u_i + a_i\Phi_i - \ddot{\hat{x}}_0 - \ddot{h}_i + k_{i1}\dot{e}_i + k_{i2}\dot{e}_i + \frac{\dot{\varsigma}_i}{\varsigma_i}\dot{e}_i + \dot{\Upsilon}_i e_i\end{aligned} \tag{8.32}$$

Then, by using Eqs. (8.27) and (8.28), we can obtain the following inequalities for the derivative of Lyapunov function

$$\begin{aligned}\dot{V}_c =&s_i^{\mathrm{T}}\dot{s}_i - \frac{1}{\kappa_i}\tilde{a}_i^{\mathrm{T}}\dot{\hat{a}}_i - \frac{1}{\mu_i}\tilde{\theta}_i^{\mathrm{T}}\dot{\hat{\theta}}_i \\ \le&s_i^{\mathrm{T}}\Big(A_i\left(I-\hat{\rho}_i\right)u_i - A_i\tilde{\rho}_i u_i + a_i\Phi_i - \ddot{\hat{x}}_0 - \ddot{h}_i + k_{i1}\dot{e}_i \\ &+k_{i2}\dot{e}_i + \frac{\dot{\varsigma}_i}{\varsigma_i}\dot{e}_i + \dot{\Upsilon}_i e_i - \frac{1}{\kappa_i}\tilde{a}_i^{\mathrm{T}}\dot{\hat{a}}_i - \frac{1}{\mu_i}\tilde{\theta}_i^{\mathrm{T}}\dot{\hat{\theta}}_i \\ \le&s_i^{\mathrm{T}}\Big(-A_i\tilde{\rho}_i u_i + a_i\Phi_i - \hat{a}_i\Phi_i\frac{s_i}{\|s_i\|} - \omega_{i1}s_i \\ &-\omega_{i2}s_i - \frac{\dot{\varsigma}_i}{\varsigma_i}s_i - \frac{1}{\kappa_i}\tilde{a}_i^{\mathrm{T}}\dot{\hat{a}}_i - \frac{1}{\mu_i}\tilde{\theta}_i^{\mathrm{T}}\dot{\hat{\theta}}_i \\ \le&-\omega_{i1}s_i^{\mathrm{T}}s_i - \omega_{i2}s_i^{\mathrm{T}}s_i - \frac{\dot{\varsigma}_i}{\varsigma_i}s_i^{\mathrm{T}}s_i - s_i^{\mathrm{T}}A_i\tilde{\rho}_i u_i \\ &+\tilde{a}_i\Phi_i\|s_i\| - \frac{1}{\kappa_i}\tilde{a}_i^{\mathrm{T}}\dot{\hat{a}}_i - \frac{1}{\mu_i}\tilde{\theta}_i^{\mathrm{T}}\dot{\hat{\theta}}_i\end{aligned} \tag{8.33}$$

Substituting the laws of Eqs. (8.29) and (8.30) into the above inequality, it yields to

$$\begin{aligned}\dot{V}_c \le&-\omega_{i1}s_i^{\mathrm{T}}s_i - \omega_{i2}s_i^{\mathrm{T}}s_i - \frac{\dot{\varsigma}_i}{\varsigma_i}s_i^{\mathrm{T}}s_i + \frac{\vartheta_i}{\kappa_i}\tilde{a}_i^{\mathrm{T}}\hat{a}_i + \frac{\lambda_i}{\mu_i}\tilde{\theta}_i^{\mathrm{T}}\hat{\theta}_i \\ \le&-\omega_{i1}s_i^{\mathrm{T}}s_i - \omega_{i2}s_i^{\mathrm{T}}s_i - \frac{\dot{\varsigma}_i}{\varsigma_i}s_i^{\mathrm{T}}s_i \\ &+\frac{\vartheta_i}{2\kappa_i}\left(\|a_i\|^2 - \|\tilde{a}_i\|^2\right) + \frac{\lambda_i}{2\mu_i}\left(\|\theta_i\|^2 - \left\|\tilde{\theta}_i\right\|^2\right) \\ \le&-\omega_{i1}s_i^{\mathrm{T}}s_i - \omega_{i2}s_i^{\mathrm{T}}s_i - \frac{\dot{\varsigma}_i}{\varsigma_i}s_i^{\mathrm{T}}s_i \\ &-\frac{\vartheta_i}{2\kappa_i}\|\tilde{a}_i\|^2 - \frac{\lambda_i}{2\mu_i}\left\|\tilde{\theta}_i\right\|^2 + \frac{\vartheta_i}{2\kappa_i}\|a_i\|^2 + \frac{\lambda_i}{2\mu_i}\|\theta_i\|^2 \\ \le&-\omega_{i1}\left(\|s_i\|^2 - \frac{\vartheta_i}{2\omega_{i1}\kappa_i}\|a_i\|^2 - \frac{\lambda_i}{2\omega_{i1}\mu_i}\|\theta_i\|^2\right) \\ &-\omega_{i2}s_i^{\mathrm{T}}s_i - \frac{\dot{\varsigma}_i}{\varsigma_i}s_i^{\mathrm{T}}s_i - \frac{\vartheta_i}{2\kappa_i}\|\tilde{a}_i\|^2 - \frac{\lambda_i}{2\mu_i}\left\|\tilde{\theta}_i\right\|^2\end{aligned} \tag{8.34}$$

Let's define $\Delta_{i1} = \frac{\vartheta_i}{2\omega_{i1}\kappa_i}\|a_i\|^2$ and $\Delta_{i2} = \frac{\lambda_i}{2\omega_{i1}\mu_i}\|\theta_i\|^2$. It follows that $s_i$ is bounded and satisfies $\|s_i\| \leq \sqrt{\Delta_{i1} + \Delta_{i2}}$ as the time goes to infinity. Hence, it can be concluded that $V_i$ is bounded during the control interval, which implies that $u_i$ , $\tilde{a}_i$ and $\tilde{\theta}_i$ are also bounded at any finite time interval. Moreover, we will prove that the formation error can achieve prescribed-time stability.

Let define $V_{cs} = \frac{1}{2}s_i^{\mathrm{T}} s_i$. Then, when $t \in [3T_1, \infty)$, the derivative of $V_{cs}$ can be computed as

$$
\begin{aligned}
\dot{V}_{cs} &= s_i^{\mathrm{T}} \dot{s}_i \\
&= s_i^{\mathrm{T}} \left( A_i \left( I - \hat{\rho}_i \right) u_i - A_i \tilde{\rho}_i u_i + a_i \Phi_i - \ddot{\hat{x}}_0 - \ddot{h}_i + k_{i1} \dot{e}_i \right. \\
&\quad \left. + k_{i2} \dot{e}_i + \frac{\dot{\varsigma}_i}{\varsigma_i} \dot{e}_i + \dot{\Upsilon}_i e_i \right) \\
&= -\omega_{i1} s_i^{\mathrm{T}} s_i - \omega_{i2} s_i^{\mathrm{T}} s_i - \frac{\dot{\varsigma}_i}{\varsigma_i} s_i^{\mathrm{T}} s_i - s_i^{\mathrm{T}} A_i \tilde{\rho}_i u_i + \tilde{a}_i \Phi_i \|s_i\| \\
&\leq -\omega_{i1} \|s_i\|^2 + \|s_i\|^2 + \frac{1}{4} \|\tilde{a}_i\|^2 \|\Phi_i\|^2 + A_i^2 \|s_i\|^2 \\
&\quad + \frac{1}{4} \|\tilde{\rho}_i\|^2 \|u_i\|^2 - \omega_{i2} s_i^{\mathrm{T}} s_i - \frac{\dot{\varsigma}_i}{\varsigma_i} s_i^{\mathrm{T}} s_i \\
&\leq \left(1 - \omega_{i1} + A_i^2\right) \|s_i\|^2 + \frac{1}{4} \|\tilde{\rho}_i\|^2 \|u_i\|^2 + \frac{1}{4} \|\tilde{a}_i\|^2 \|\Phi_i\|^2 \\
&\quad - \omega_{i2} s_i^{\mathrm{T}} s_i - \frac{\dot{\varsigma}_i}{\varsigma_i} s_i^{\mathrm{T}} s_i \\
&\leq \left(1 - \omega_{i1} + A_i^2\right) \left( \|s_i\|^2 - \frac{\|\tilde{\rho}_i\|^2 \|u_i\|^2}{4\left(1 - \omega_{i1} + A_i^2\right)} - \frac{\|\tilde{a}_i\|^2 \|\Phi_i\|^2}{4\left(1 - \omega_{i1} + A_i^2\right)} \right) \\
&\quad - \omega_{i2} s_i^{\mathrm{T}} s_i - \frac{\dot{\varsigma}_i}{\varsigma_i} s_i^{\mathrm{T}} s_i
\end{aligned}
\tag{8.35}
$$

It is always possible to find a $\omega_{i1}$, such that $1 - \omega_{i1} + A_i^2 < 0$ holds. Then, the following inequality holds

$$
\begin{aligned}
\dot{V}_{cs} &\leq -\omega_{i2} s_i^{\mathrm{T}} s_i - \frac{\dot{\varsigma}_i}{\varsigma_i} s_i^{\mathrm{T}} s_i \\
&\leq -2\omega_{i2} V_{cs} - 2\frac{\dot{\varsigma}_i}{\varsigma_i} V_{cs}.
\end{aligned}
\tag{8.36}
$$

Based on Theorem 8.2, there exist a finite time $T_{cs}$, such that $s_i$ and $e_i$ can converge to the region $\Delta_s$ and $l_i \Delta_s$ in the time $3T_1 + T_s + T_{cs}$, respectively.

By considering the previous results, we have that the bounded region $\Delta_{cs}$ of $s_i$ is always globally prescribed-time stable. Then, based on Theorem 8.2, it can be concluded that the formation system error converges to the region $\|e_i\| \leq l_i \Delta_{cs}$ within the prescribed time $3T_1 + T_s + T_{cs}$.

**Remark 8.5** This chapter uses a RBFNN to estimate the upper bound of lumped uncertainty $\Delta_i$. As it is a complete self-optimization process, there is no need to conduct data sampling learning in advance. However, different from the existing results [16, 23], the adaptively undated parameter is just a scalar $a_i$ rather than the vector $W_i$, which significantly reduces the computational complexity. Furthermore, it ensures that the calculation process occurs within the sampling rate range of the system. Therefore, the designed control protocol provides a more affordable solution to formation tracking of aerial-ground cooperative systems.

**Remark 8.6** In this chapter, the proposed formation tracking scheme is based on the implicit separation principle, that is, the design control process is based on the premise that the leader observer and the formation controller can achieve convergence within a specified time, respectively. Based on this strategy, it can be proven that the formation error is uniformly bounded within a prescribed time, which is acceptable in practical engineering. The size of the error convergence domain can be effectively adjusted by tuning the controller gains. With the continuous increase in the scale of modern unmanned clusters, the use of the separation principle can effectively reduce the information transmission burden of interactive networks, and has been adopted by many researchers to achieve different tracking tasks [2, 3, 24].

**Remark 8.7** Compared with the relevant asymptotical fault-tolerant results for heterogeneous MASs [6, 29], an adaptive updating law is designed in this chapter to estimate the fault factors $\rho_i$, which features the characteristics of the active fault-tolerant strategy to achieve a more effective estimation of the fault values. Then, by using the estimated value, an adaptive prescribed-time fault-tolerant controller is developed, which can better compensate the influence of actuator faults within the specified time. Therefore, the designed formation strategy can ensure not only the fast and high-precision control, but also the strong fault tolerance of the heterogeneous MASs. Moreover, the existing finite/fixed-time fault-tolerant controllers for MASs, such as [7, 24, 30], can not produce smooth control action due to the use of signum function. However, the proposed method in this chapter renders the control action to be continuous and smooth everywhere. The above user-friendly features bring great convenience to the users, so have important practical significance.

**Remark 8.8** Taking into account the constraints of the model dynamics (position and velocity), actuation, and safety (minimum position separation), the APF function can be further incorporated into the prescribed-time controller design to achieve faster fault-tolerant collision avoidance formation control. Then, during the formation process, when packet loss occurs, it is necessary to adjust the topology connection between agents to ensure that the entire system can maintain control performance in the new topology environment. In addition, time delay often occurs in realistic network systems. In the face of this phenomenon, we can learn from the fixed-time lag tracking control scheme, which can simultaneously manage convergence and stability issues [20].

## 8.4 Simulation Experiments

An aerial-ground formation MAS is considered in the simulation tests. Specifically, this cooperative system includes a UGV leader marked as $i = 0$, two UGV followers denoted as $i = 1, 2$, and two UAV followers marked as $i = 3, 4$, which is used to assess the effectiveness of the proposed prescribed-time algorithms. The model parameters of the studied heterogeneous vehicles are same as Chap. 4. Figure 8.1 shows the directed topology network. The initial conditions of four followers are designed as $x_0(0) = [0, 0]^{\mathrm{T}}, x_1(0) = [-0.3, 0.4]^{\mathrm{T}}, x_2(0) = [1.2, -0.6]^{\mathrm{T}}, x_3(0) = [2, 1.5, 0]^{\mathrm{T}}$, and $x_4(0) = [-2, 3, 0]^{\mathrm{T}}$. The expected formation is specified by $h_1(t) = [1, -1]^{\mathrm{T}}, h_2(t) = [2, -2]^{\mathrm{T}}, h_3(t) = [3, -3, 2t]^{\mathrm{T}}$, and $h_4(t) = [4, -4, 5t]^{\mathrm{T}}$. The control input signal of the leader-UGV is chosen as $u_0(t) = [\sin(t), \cos(t)]^{\mathrm{T}}$, which is unknown to all followers and satisfies Assumption 8.1.

For the proposed prescribed-time observer, the parameters are selected as $\alpha = 1$, and $\beta = 3$ for each follower, and the user-assignable time is designed as $T_1 = T_2 = T_3 = 2s$. Then, the controller parameters are $k_{i1} = 3, k_{i2} = 2, \omega_{i1} = 3, \omega_{i2} = 5, \vartheta_i = 0.5, \kappa_i = 1, \lambda_i = 0.001$, and $\mu_i = 0$ that are fixed in the simulations. In order to better simulate the impact of real working conditions on the UAV system, we add the model uncertainties (20% variation) in the aerodynamic damping parameters and the moments of inertia. Furthermore, the disturbances are set to $d_1(t) = [0.1\cos 2t, 0.3\sin t]^{\mathrm{T}}, d_2(t) = [0.1\sin 0.5t, 0.16\sin 1.7t]^{\mathrm{T}}$ for UGV followers, and the disturbances of UAV followers are $d_3(t) = [0.2\cos 0.3t, -0.18\sin 1.2t, -0.2\cos t]^{\mathrm{T}}$, and $d_4(t) = [0.2\sin 1.8t, 0.3\cos 1.2t, 0.25\sin t]^{\mathrm{T}}$. In the simulation, the actuators of agent 1 and 3 lose their effectiveness of $\rho_1(t) = [0.3, 0.4]^{\mathrm{T}}$ and $\rho_3(t) = [0.15, 0.2, 0.1]^{\mathrm{T}}$, respectively. The biased faults are set to be $\delta_1 = [-0.2\sin t, 0.1\cos 0.2t]^{\mathrm{T}}$, and $\delta_3 = [0.12\sin 1.3t, 0.1\cos 2t, 0.23\sin 2t]^{\mathrm{T}}$. The prescribed time are selected as $T_s = T_{cs} = 2s$.

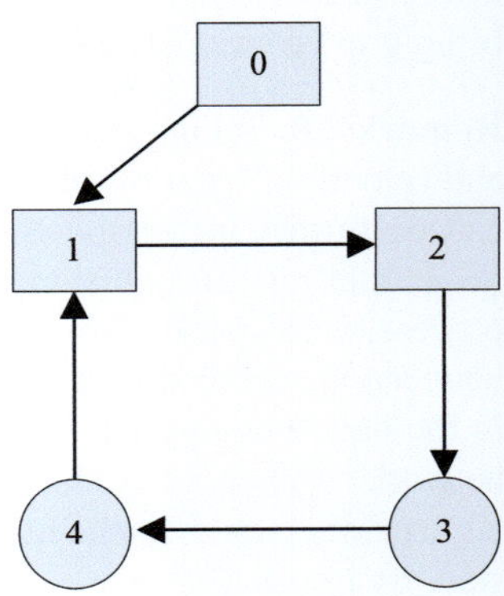


**Fig. 8.1** Directed communication topology

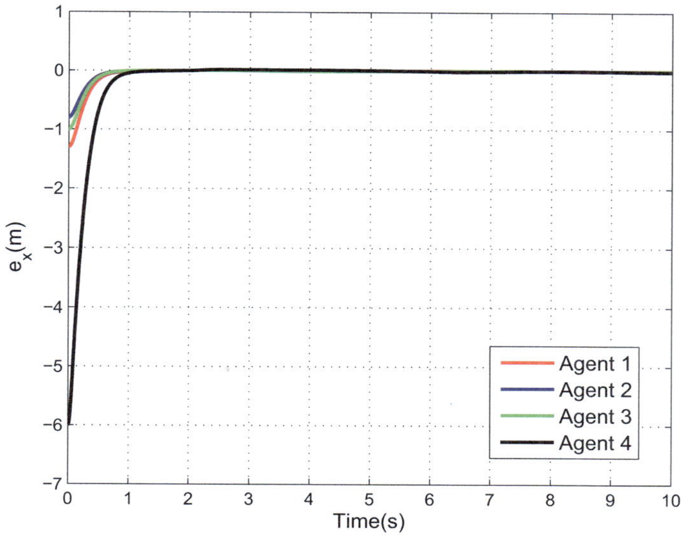


**Fig. 8.2** Tracking error for each follower in $x$ axis

The achieved results summarized in Figs. 8.2, 8.3 and 8.4 depict that by using the proposed approach, the position tracking errors of each follower can arrive at a satisfactory region of zero very fast, which implies the effectiveness of the developed prescribed-time observer. In addition, it can be observed that the formation tracking can be realized after 2 s, which verifies the prescribed-time stability of the developed controller. Figures 8.5, 8.6 and 8.7 show that the control inputs are smooth and uniformly bounded. As the proposed method is based on regular feedback of local error system and the sliding-mode based controller does not contain the sign function, it can generate smooth control action. However, due to the use of the prescribed-time stability relation, transient chattering may occur when the convergence time is met, which is clearly found in Fig. 8.6. We can adjust the control quantity at this moment by reasonably selecting the controller parameters. Furthermore, Fig. 8.8 depicts the convergence process of the designed adaptive parameter $\hat{a}_i$. Figure 8.9 presents the evolution of the fault factors, which reveals that the proposed adaptive laws can realize the satisfactory estimation of the unknown actuator fault parameters.

From the achieved simulation results, it can be concluded that the proposed formation tracking scheme does not generate high gain control inputs. Then, through the recursive adjustment of adaptive parameter $\hat{a}_i$, the controller achieves effective compensation for the uncertainty $\Delta_i$ composed of the biased faults, external disturbances, and model uncertainties, which also avoids significant fluctuations of the tracking errors.

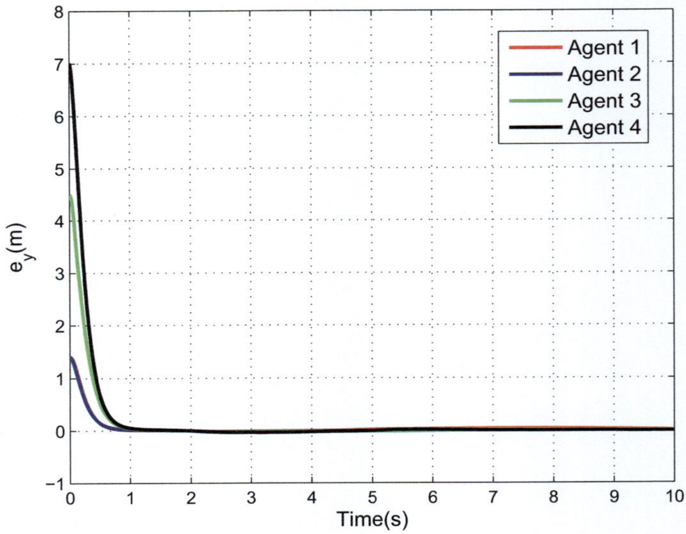


**Fig. 8.3** Tracking error for each follower in $y$ axis

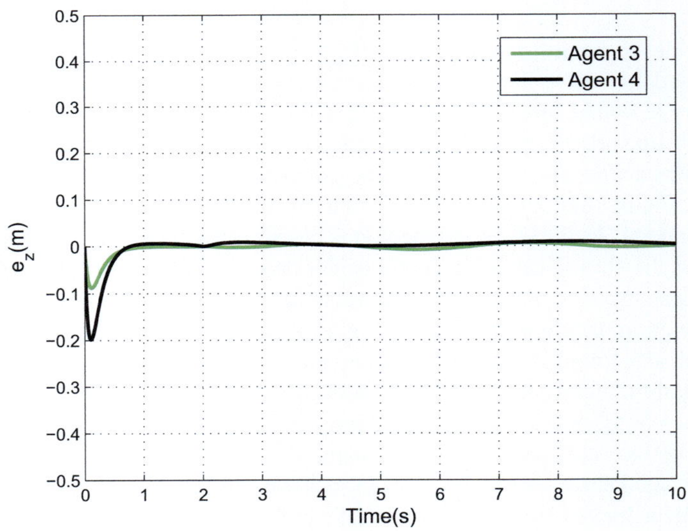


**Fig. 8.4** Tracking error for each follower in $z$ axis

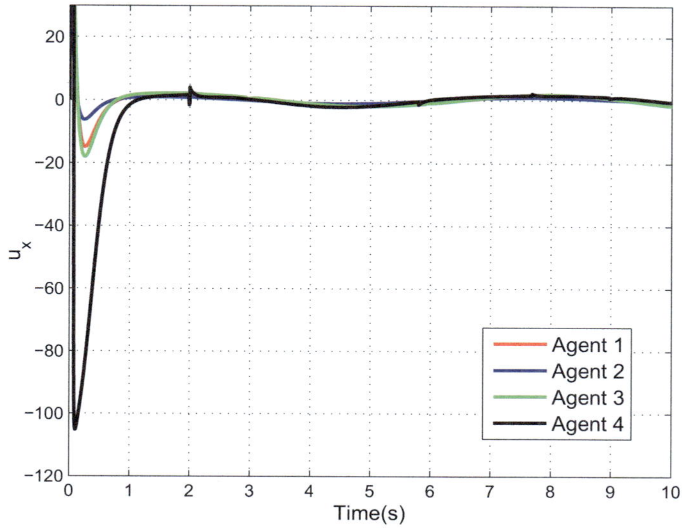


**Fig. 8.5** Control input for each follower in $x$ axis

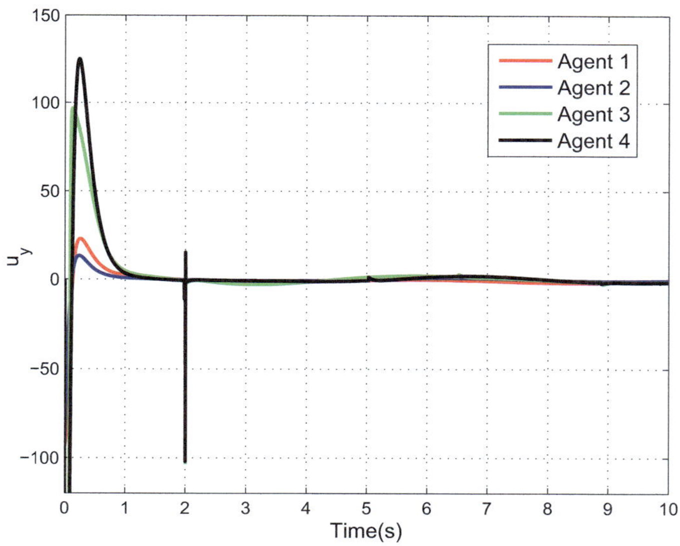


**Fig. 8.6** Control input for each follower in $y$ axis

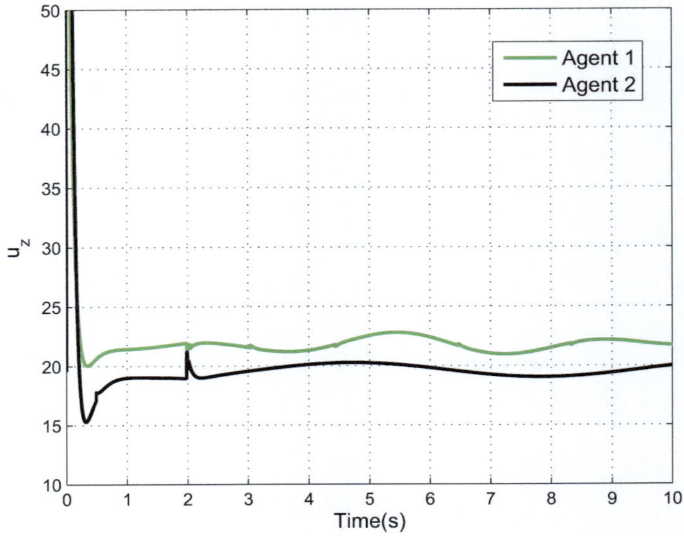


**Fig. 8.7** Control input for each follower in $z$ axis

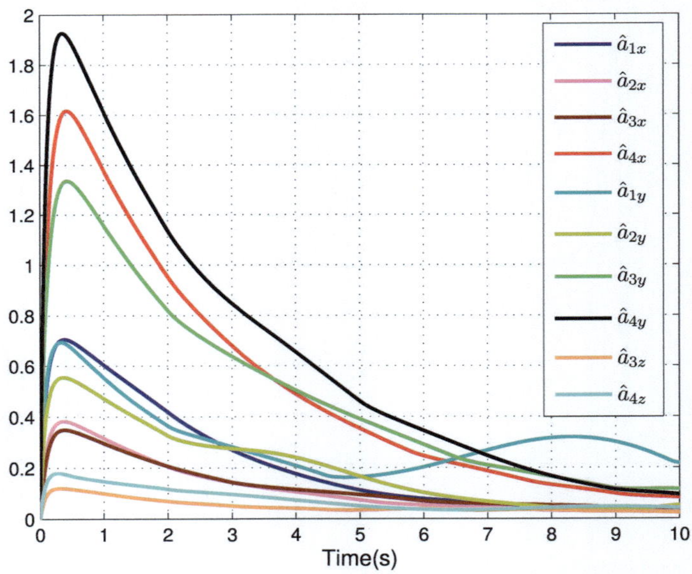


**Fig. 8.8** Profiles of the estimated parameter $\hat{a}_i$

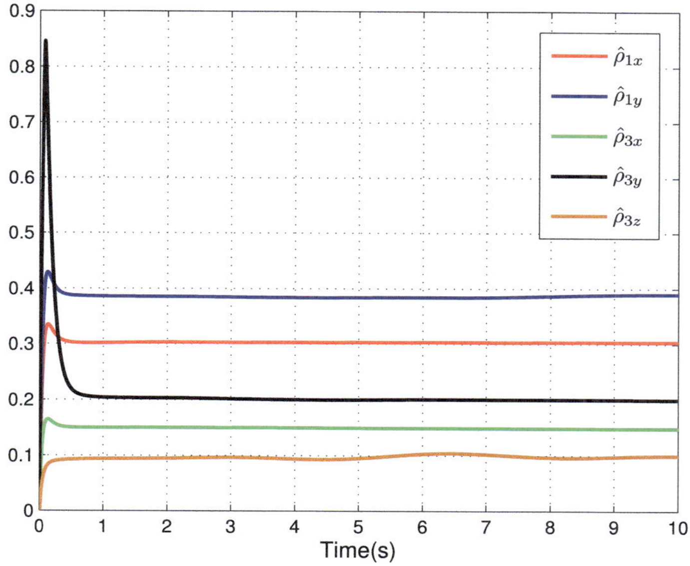


**Fig. 8.9** Profiles of the estimated parameter $\hat{\rho}_i$

## 8.5 Conclusions

This chapter researches the issue of prescribed-time formation control for air-ground heterogeneous systems with actuator faults. Based on prescribed-time stability theory, a leader observer is developed firstly. Then, we construct a prescribed-time sliding mode surface. By using RBFNN and adaptive technique, a novel neural adaptive prescribed-time formation controller is proposed under actuator faults. Furthermore, simulations prove that the proposed formation controller has fast convergence speed and satisfactory fault tolerance performance.

## References

1. Cai Y, Zhang H, Li W, Mu Y, He Q (2022) Distributed bipartite adaptive event-triggered fault-tolerant consensus tracking for linear multiagent systems under actuator faults. IEEE Trans Cybern 52(11):11313–11324
2. Cai Y, Zhang H, Wang Y, Gao Z, He Q (2022) Adaptive bipartite fixed-time time-varying output formation-containment tracking of heterogeneous linear multiagent systems. IEEE Trans Neural Netw Learn Syst 33(9):4688–4698
3. Chang S, Wang Y, Zuo Z, Yang H (2022) Fixed-time formation control for wheeled mobile robots with prescribed performance. IEEE Trans Control Syst Technol 30(2):844–851
4. Chen X, Yu H, Hao F (2022) Prescribed-time event-triggered bipartite consensus of multiagent systems. IEEE Trans Cybern 52(4):2589–2598

5. Cheng W, Zhang K, Jiang B, Ding SX (2021) Fixed-time fault-tolerant formation control for heterogeneous multi-agent systems with parameter uncertainties and disturbances. IEEE Trans Circuits Syst I Regul Pap 68(5):2121–2133
6. Deng C, Wen C (2020) Distributed resilient observer-based fault-tolerant control for heterogeneous multiagent systems under actuator faults and DoS attacks. IEEE Trans Control Netw Syst 7(3):1308–1318
7. Dong G, Li H, Ma H, Lu R (2021) Finite-time consensus tracking neural network FTC of multi-agent systems. IEEE Trans Neural Netw Learn Syst 32(2):653–662
8. Gao K, Liu Y, Zhou Y, Zhao Y, Huang P (2022) Practical fixed-time affine formation for multi-agent systems with time-based generators. IEEE Trans Circuits Syst II Express Briefs 69(11):4433–4437
9. Gong X, Cui Y, Shen J, Shu Z, Huang T (2021) Distributed prescribed-time interval bipartite consensus of multi-agent systems on directed graphs: theory and experiment. IEEE Trans Netw Sci Eng 8(1):613–624
10. Gong X, Cui Y, Shen J, Xiong J, Huang T (2022) Distributed optimization in prescribed-time: theory and experiment. IEEE Trans Netw Sci Eng 9(2):564–576
11. Gong X, Cui Y, Wang T, Shen J, Huang T (2022) Distributed prescribed-time consensus observer for high-order integrator multi-agent systems on directed graphs. IEEE Trans Circuits Syst II Express Briefs 69(4):2216–2220
12. Guan Y, Yang H, Jiang B, Tang Q, Zhang X (2022) Fault-tolerant control of swarm deployment: a PDE approach. IEEE Trans Control Netw Syst 10(2):741–751
13. Guo X-G, Xu W-D, Wang J-L, Park JH, Yan H (2022) BLF-based neuroadaptive fault-tolerant control for nonlinear vehicular platoon with time-varying fault directions and distance restrictions. IEEE Trans Intell Transp Syst 23(8):12388–12398
14. Hua Y, Dong X, Han L, Li Q, Ren Z (2020) Finite-time time-varying formation tracking for high-order multiagent systems with mismatched disturbances. IEEE Trans Syst Man Cybern Syst 50(10):3795–3803
15. Hua Y, Dong X, Hu G, Li Q, Ren Z (2019) Distributed time-varying output formation tracking for heterogeneous linear multiagent systems with a nonautonomous leader of unknown input. IEEE Trans Autom Control 64(10):4292–4299
16. Hwang C-L, Abebe HB (2022) Generalized and heterogeneous nonlinear dynamic multiagent systems using online RNN-based finite-time formation tracking control and application to transportation systems. IEEE Trans Intell Transp Syst 23(8):13708–13720
17. Liu C, Jiang B, Patton RJ, Zhang K (2019) Hierarchical-structure-based fault estimation and fault-tolerant control for multiagent systems. IEEE Trans Control Netw Syst 6(2):586–597
18. Liu Y, Dong X, Shi P, Ren Z, Liu J (2022) Distributed fault-tolerant formation tracking control for multiagent systems with multiple leaders and constrained actuators. IEEE Trans Cybern 53(6):3738–3747
19. Lv Y, Wen G, Huang T (2020) Adaptive protocol design for distributed tracking with relative output information: a distributed fixed-time observer approach. IEEE Trans Control Netw Syst 7(1):118–128
20. Ni J, Liu L, Liu C, Liu J (2017) Fixed-time leader-following consensus for second-order multiagent systems with input delay. IEEE Trans Ind Electron 64(11):8635–8646
21. Ning B, Han Q-L, Lu Q (2020) Fixed-time leader-following consensus for multiple wheeled mobile robots. IEEE Trans Cybern 50(10):4381–4392
22. Qin J, Zhang G, Zheng WX, Kang Y (2019) Neural network-based adaptive consensus control for a class of nonaffine nonlinear multiagent systems with actuator faults. IEEE Trans Neural Netw Learn Syst 30(12):3633–3644
23. Shang L, Cai M, Wang B, Yu J (2022) Adaptive fault-tolerant fast finite-time consensus protocols for multiple mechanical systems with output constraints. IEEE Trans Syst Man Cybern Syst 52(9):6028–6035
24. Song G, Shi P, Lim CP (2022) Distributed fault-tolerant cooperative output regulation for multiagent networks via fixed-time observer and adaptive control. IEEE Trans Control Netw Syst 9(2):845–855

25. Song YD, Wang YJ, Holloway J, Krstic M (2017) Time-varying feedback for regulation of normal-form nonlinear systems in prescribed finite time. Automatica 83:243–251
26. Wang Q, Dong X, Wang B, Hua Y, Ren Z (2023) Finite-time observer-based $H_\infty$ fault-tolerant output formation tracking control for heterogeneous nonlinear multi-agent systems. IEEE Trans Netw Sci Eng 10(4):1822–1834
27. Wang Y, Song Y, Hill DJ, Krstic M (2019) Prescribed-time consensus and containment control of networked multiagent systems. IEEE Trans Cybern 49(4):1138–1147
28. Wu Y, Wang Z (2021) Fuzzy adaptive practical fixed-time consensus for second-order nonlinear multiagent systems under actuator faults. IEEE Trans Cybern 51(3):1150–1162
29. Xiao S, Dong J (2022) Distributed adaptive fuzzy fault-tolerant containment control for heterogeneous nonlinear multiagent systems. IEEE Trans Syst Man Cybern Syst 52(2):954–965
30. Yin Y, Wang F, Liu Z, Chen Z (2022) Finite-time leader-following consensus of multiagent systems with actuator faults and input saturation. IEEE Trans Syst Man Cybern Syst 52(5):3314–3325
31. Zhang K, Jiang B, Shi P (2017) Adjustable parameter-based distributed fault estimation observer design for multiagent systems with directed graphs. IEEE Trans Cybern 47(2):306–314

# Chapter 9
# Game Theory-Based Prescribed-Time Optimal FTC for UAVs-UGVs

## 9.1 Introduction

Due to the increasing demand for control performance, the research of MASs cannot be limited only to the formation tracking. Specifically, in many practical systems, the MASs not only passively receive instructions, but also need to optimize certain performance indicators in the process of interacting with other individuals, in order to form a rational behavior that reflects the individual and MASs [12, 14, 17, 24]. Therefore, optimization control has attracted much attention [1, 20, 23, 29]. However, relevant achievements in the field of air-ground collaborative heterogeneous MASs are extremely rare. Recently, reinforcement learning-based game control is an impactful method that acquires training information from the external circumstance, to learn optimal algorithms without knowledge of the game MASs [10, 16, 21, 27]. In the existing classic reinforcement learning-based optimal algorithms, the actor-critic framework is wildly utilized, in which the actor neural network generates executor input signal and the critic RBFNN is designed to evaluate the execution cost in order to decide whether the current condition is better or worse than the desired and to determine a signal to be measured [14, 28]. After that, the actor revises the produced control signal by using the measurement. Then, the performance function will arrive at the global optimization. However, the above results can only achieve the objective with asymptotic convergence for homogeneous MASs, and cannot realized the desired tasks within a fixed time or a prescribed time for heterogeneous MASs.

In addition, note that the above-mentioned works suppose that the actuators operate normally. In practice of MASs, the actuator is prone to partial loss of effectiveness and bias faults due to the increasing complexity and automation [4, 11, 13]. As a result, the system performance will degrade and even suffer from catastrophic consequence [2, 31]. Therefore, fault-tolerant capability is required to accommodate the unexpected actuator faults. Recently, for the increasing requirement of safety and reliability in safety-critical systems, some effective cooperative fault-tolerant algorithms for MASs have been proposed [2, 7, 8, 18, 26, 30]. However, the existing



K. Zhang et al., *Fixed-Time/Prescribed-Time Fault-Tolerant Control for Multiple Unmanned Systems*, Unmanned System Technologies,
https://doi.org/10.1007/978-981-95-2913-1_9

results focus on the consensus tracking and output regulation problems. The extension to the formation control of heterogeneous MASs, especially to the UAVs-UGVs system, is of practical significance to enhance the system control performance of fast fault accommodation for the practical systems. In the previous works, conventional finite/fixed-time formation algorithms were proposed to deal with the tracking issue of UAVs-UGVs heterogeneous formation system under actuator faults [5]. However, because of the theoretical bound of the convergence time relies on the controller parameters, the convergence time of the fixed-time protocol cannot be preassigned. Recently, a novel approach of "prescribed-time control" was developed in [19], which ensures that the system can finish the finite stability and the convergence time can be set by the user and independent of any other control parameters. However, the existing results are focused on the cooperation control for MASs, where agents have homogeneous single-integrator, second-order integrator dynamics or high-order integrator dynamics [6, 9, 15, 25]. It is obvious that, the models of multiple agents in practical UAVs-UGVs formation system are significantly different. Therefore, it is meaningful and necessary to extend the system dynamics to heterogeneous MASs. Due to the obvious heterogeneity between agents, a completely different control framework should be redesigned to ensure the predefined-time tracking, which is very challenging. Up to now, there is no attention paid to intelligent control of UAVs-UGVs heterogeneous formation systems with prescribed-time convergence property.

Purposed by the above analysis, this chapter investigates the hierarchical structure-based optimal formation control problem of a heterogeneous formation system of UAVs and UGVs. The chief works of this chapter are as follows

(1) This study focuses on the optimization control with fault-tolerant performance and fast convergence speed, and constructs an prescribed-time optimal fault-tolerant control algorithm based on hierarchical control, graphical game theory, and prescribed-time control method. The effectiveness of the constructed control strategy is verified through simulation experiment.
(2) In virtual layer, an quadratic performance index function based on consistency tracking error and energy consumption is designed, and approximate optimal control strategy is obtained by using coupled HJB equation and reinforcement learning, which achieves synchronous optimal control and interactive Nash equilibrium of MASs.
(3) In actual control layer, a prescribed-time fault-tolerant tracking controller is designed based on the optimal signal, sliding-mode and adaptive technologies, which realizes the finite-time tracking of the optimal formation trajectory. The proposed method ensures that the global convergence time is completely independent of the initial states of the system and controller parameters, while also effectively approximating the actuator fault parameters.

## 9.2 Problem Formulation and Preliminaries

### 9.2.1 Problem Formulation

The studied heterogeneous MASs consists of $N$ agents, including $N_1$ UGVs with two degrees of freedom and $N_2$ UAVs with six degrees of freedom. The simplified structural diagram is shown in Fig. 6.1. In order to construct a globally unified description model, the ideal model of the $i$th agent is unified into the following expression based on the model introduction in Chap. 4.

$$\ddot{\chi}_{vi}(t) = u_{vi}(t) + f_{vi}(t) \tag{9.1}$$

where $\chi_{vi}(t) \in R^{p_i}, u_{vi}(t) \in R^{p_i}$ and $f_{vi}(t) \in R^{p_i}$ represent the position coordinates, control inputs, and system nonlinearity of the agent, respectively, and $p_i$ represent the state dimension of the $i$th agent.

When the $k$th actuator of the $i$th follower fails, the input $u_{ik}$ and output $u_{ik}^F$ of the actuator exhibit inconsistency and have the following relationship

$$u_{ik}^F = (1 - \rho_{ik}(t))u_{ik} + \tau_{ik}(t) \tag{9.2}$$

where $\rho_{ik}(t)$ is the efficiency loss factor of the actuator and is an unknown time-varying parameter, while $\tau_{ik}(t)$ is the additive fault parameter of the actuator.

Then, unify the actual control model of the $i$th agent as follows

$$\ddot{\chi}_{ai}(t) = (I_i - \Lambda_i)\,u_{ai}(t) + \tau_i(t) + f_{ai}(t) + d_i(t) \tag{9.3}$$

where $\chi_{ai}(t)$, $u_{ai}(t)$ and $f_{ai}(t)$ represents the position coordinates, control inputs, and system nonlinearity terms in complex environments, respectively.

**Control Objective**: This study considers a special heterogeneous formation scenario, where a heterogeneous MASs consisting of UAVs and UGVs executes distributed formation tasks in 3D space. Assuming the expected formation of agent $i$ is $h_i(t)$, and the globally consistent optimal position signal is $\chi_{vi}(t) = [\chi_{v1}(t), \chi_{v2}(t), ..., \chi_{vN}(t)]^{\mathrm{T}}$. Then, when there exists a prescribed time $T_{\max}$ that is independent of the initial state of the system and any other controller parameters, such that the following equation holds

$$\lim_{t \to T_{\max}} (\chi_{ai}(t) - \chi_{vi}(t) - h_i(t)) = 0 \tag{9.4}$$

then it is said that the heterogeneous MASs can complete the prescribed-time fault-tolerant optimal formation control.

To obtain the desired prescribed-time controller, the following assumptions and Lemma are required.

**Assumption 9.1** The actuator fault parameters $\rho_{ik}$ and $\tau_{ik}$ are unknown, may vary rapidly and be unmeasurable but bounded, i.e. there are unknown constants $\rho_m$ and $\tau_m$ such that $0 \leq \rho_{ik} \leq \rho_m < 1$ and $\tau_{ik} \leq \tau_m$ hold.

**Assumption 9.2** The external disturbance $d_i(t)$ experienced by the agent is bounded, that is, there exists an unknown constant $d_m$ such that $0 \leq \|d_i\| \leq d_m$ holds.

**Remark 9.1** The heterogeneous characteristics of UGVs and UAVs in dynamic space and system models affect the level of collaborative formation, and also prevent the distributed formation strategy of MASs from effectively playing a role in the field of air-ground collaboration. Compared to the existing control strategy designed solely based on accurate system models, this chapter unifies the heterogeneous multiagent model while fully considering external disturbances, model uncertainties, and unknown faults. This strategy provides a relatively unified research model for the unified management of multiple types of mobile unmanned vehicles in complex environments and facilitates the design of subsequent controller.

## 9.2.2 Preliminaries

**Definition 9.1** For agent $i$, let $u_{-vi}$ represent the set of control strategies for all neighbors of the $i$th agent. If

$$V_{vi}^* \triangleq V_{vi}\left(u_{vi}^*, u_{-vi}^*\right) \leq V_{vi}\left(u_{vi}, u_{-vi}^*\right) \tag{9.5}$$

then it is called that the optimal performance indicators $\left\{V_{v1}^*, V_{v2}^*, \ldots, V_{vN}^*\right\}$ reach the Nash equilibrium of the graph game.

**Lemma 9.1** ([22]) *For the following system*

$$\dot{x}(t) = f(t, x(t)), \quad x(0) = x_0 \tag{9.6}$$

*If there exists a continuous positive definite function $V(x)$ and a positive definite scalar $\gamma$ that satisfy*

$$\dot{V}(x) \leq -\gamma V(x) - 2\frac{\left|\dot{\varsigma}\left(t_0, T_\varsigma\right)\right|}{\varsigma\left(t_0, T_\varsigma\right)} V(x) + \frac{\eta}{\varsigma\left(t_0, T_\varsigma\right)} + c \tag{9.7}$$

*and*

$$\varsigma\left(t_0, T_\varsigma\right) = \begin{cases} \exp\left(\alpha\left(t_0 + T_\varsigma - t\right)\right) - 1, t \in \left[t_0, t_0 + T_\varsigma\right) \\ \varepsilon - \varepsilon\tanh\left(\alpha\left(t - t_0 - T_\varsigma\right)\right), t \in \left[t_0 + T_\varsigma, \infty\right) \end{cases} \tag{9.8}$$

*where $\gamma > 0$, $\eta \geq 0$, $c \geq 0$, $\alpha > 0$. $\varepsilon$ is a positive definite constant, $T_\varsigma$ is an artificially set constant, and $t_0$ is the start time. Then, the system is globally stable at the equilibrium point with a convergence time of $T_\varsigma$.*

## 9.3 Main Results

### 9.3.1 *Prescribed-Time Optimal Controller Design for Virtual Layer*

The consensus optimal control problem of MASs can be regarded as a multi-objective optimization problem. Therefore, based on graph game theory, an Nash equilibrium search rule is designed in the virtual layer to achieve global interactive Nash equilibrium, providing a reference path for the optimal fault-tolerant tracking at the lower layer.

Firstly, design a consensus error based on local interaction information

$$e_{vi}(t) = \sum_{j=1}^{N} a_{ij} \left( \chi_{vi} - \chi_{vj} \right) \tag{9.9}$$

To achieve controller design, the following error variable is designed

$$s_{vi} = \dot{e}_{vi} + \lambda e_{vi} \tag{9.10}$$

By taking its derivative, one can obtain

$$\dot{s}_{vi} = \sum_{j=1}^{N} a_{ij} A_i u_{vi} - \sum_{j=1}^{N} a_{ij} A_j u_{vj} + \sum_{j=1}^{N} a_{ij} \left( f_{vi} - f_{vj} \right) + \lambda \dot{e}_{vi} \tag{9.11}$$

The goal of the virtual layer is to design a set of optimal control strategies $\{u_{v1}^*, u_{v1}^*, ..., u_{v1}^*\}$ that enable the error dynamic system of Eq. (9.10) to achieve uniformly bounded convergence while minimizing the following local performance indicators

$$V_{vi} = \int_0^{\infty} \left( s_{vi}^{\mathrm{T}} Q_i s_{vi} + u_{vi}^{\mathrm{T}} R_i u_{vi} \right) \mathrm{d}t \tag{9.12}$$

where $Q_i$ and $R_i$ are positive definite matrices, and $R_i$ is a symmetric matrix.

Assuming $V_{vi}$ is continuously differentiable, then the Hamiltonian function of Eq. (9.12) is

$$H_i \left( s_{vi}, u_{vi}, u_{vj}, \nabla V_{vi} \right) = \nabla V_{vi}^T \left( \sum_{j=1}^{N} a_{ij} A_i u_{vi} - \sum_{j=1}^{N} a_{ij} A_j u_{vj} + \sum_{j=1}^{N} a_{ij} \left( f_i - f_j \right) + \lambda \dot{e}_{vi} \right) + s_{vi}^{\mathrm{T}} Q_i s_{vi} + u_{vi}^{\mathrm{T}} R_i u_{vi} \tag{9.13}$$

According to the balance condition $\frac{\partial H_i}{\partial u_{vi}} = 0$, let $g_i = \sum_{j=1}^{N} a_{ij} A_i$, then the optimal control strategy can be obtained as

$$u_{vi}^* = -\frac{1}{2} R_i^{-1} g_i^{\mathrm{T}} \nabla V_{vi}^* \tag{9.14}$$

By substituting the optimal control strategy Eq. (9.14) into the Hamiltonian function of Eq. (9.13), the coupled HJB equation can be obtained

$$\begin{aligned} 0 = & \left(\nabla V_{vi}^*\right)^{\mathrm{T}} \left( \sum_{j=1}^{N} a_{ij} \left(f_i - f_j\right) + \lambda \dot{e}_{vi} \right) \\ & - \sum_{j=1}^{N} \tfrac{1}{2} a_{ij} A_j \left(\nabla V_{vj}^*\right)^{\mathrm{T}} R_j^{-1} g_j^{\mathrm{T}} \nabla V_{vj}^* - \tfrac{1}{4} \left(\nabla V_{vi}^*\right)^{\mathrm{T}} g_i R_i^{-1} g_i^{\mathrm{T}} \nabla V_{vi}^* + s_{vi}^{\mathrm{T}} Q_i s_{vi} \end{aligned} \tag{9.15}$$

Due to the fact that the coupled HJB equation is a nonlinear partial differential equation, it is difficult to obtain the local optimal performance index $V_{vi}^*$ and the control input $u_{vi}^*$. Therefore, a neural network-based adaptive evaluation mechanism is designed to approximate the optimal performance index and obtain an approximate optimal control strategy. Firstly, the optimal performance metric $V_{vi}^*$ of the intelligent agent $i$ is represented as

$$V_{vi}^* = W_{vi}^{\mathrm{T}} \sigma_{vi}(Z_{vi}) + \varepsilon_{vi} \tag{9.16}$$

where, $W_{vi}$ is the ideal weight vector, $\sigma_{vi}$ is the activation function, $Z_{vi}$ is the input vector, and $\varepsilon_{vi}$ is the approximation error. Furthermore, the gradient of the following optimal performance indicators can be obtained

$$\nabla V_{vi}^* = \nabla \sigma_{vi}^{\mathrm{T}} W_{vi} + \nabla \varepsilon_{vi} \tag{9.17}$$

Substituting the gradient expression Eq. (9.17) of the performance indicators into the optimal control input of Eq. (9.14), then Eq. (9.13) can be reorganized as

$$\begin{aligned} & W_{vi}^{\mathrm{T}} \nabla \sigma_{vi} \left( \sum_{j=1}^{N} a_{ij} \left(f_i - f_j\right) + \lambda \dot{e}_{vi} \right) + s_{vi}^{\mathrm{T}} Q_i s_{vi} - \tfrac{1}{4} W_{vi}^{\mathrm{T}} \nabla \sigma_{vi} g_i R_i^{-1} g_i^{\mathrm{T}} \nabla \sigma_{vi}^{\mathrm{T}} W_{vi} \\ & - \sum_{j=1}^{N} \tfrac{1}{2} a_{ij} A_j W_{vj}^{\mathrm{T}} \nabla \sigma_{vj} R_j^{-1} g_j^{\mathrm{T}} \nabla \sigma_{vj}^{\mathrm{T}} W_{vj} + \nabla \varepsilon_{vi} \left( \sum_{j=1}^{N} a_{ij} \left(f_i - f_j\right) + \lambda \dot{e}_{vi} \right) \\ & - \sum_{j=1}^{N} \tfrac{1}{2} a_{ij} A_j \nabla \varepsilon_{vi}^{\mathrm{T}} R_j^{-1} g_j^{\mathrm{T}} \nabla \varepsilon_{vj}^{\mathrm{T}} - \tfrac{1}{4} \nabla \varepsilon_{vi}^{\mathrm{T}} g_i R_i^{-1} g_i^{\mathrm{T}} \nabla \varepsilon_{vi} = 0 \end{aligned} \tag{9.18}$$

It is worth noting that the $W_{vi}$ in Eq. (9.18) is unknown, so $u_{vi}^*$ cannot be obtained. Therefore, we design a weight adaptive law

$$\hat{V}_{vi} = \hat{W}_{vi}^{\mathrm{T}} \sigma_{vi} \tag{9.19}$$

Then, the approximate optimal control law is

$$\hat{u}_{vi} = -\frac{1}{2} R_i^{-1} g_i^{\mathrm{T}} \nabla \sigma_{vi}^{\mathrm{T}} \hat{W}_{vi} \tag{9.20}$$

Next, design a weight training rule based on performance metric of Eq. (9.12) and create the following error terms

$$\eta_i = c_{vi} - \frac{1}{\psi} \hat{V}_{vi} + \dot{\hat{V}}_{vi} \tag{9.21}$$

where $c_{vi} = s_{vi}^T Q_i s_{vi} + u_{vi}^T R_i u_{vi}$, $\psi$ is a gain parameter greater than zero. By adjusting this parameter, the degree of influence of the error term $\eta_i$ on the weight adaptive adjustment and controller can be adjusted. Then, using gradient descent, the weight adaptive update law is designed as

$$\dot{\hat{W}}_{vi} = -\lambda_{vi} \frac{\partial E_{vi}}{\partial \hat{W}_{vi}} \tag{9.22}$$

where $\lambda_{vi}$ is an artificially designed positive definite learning rate, $E_{vi} = \frac{1}{2} \eta_{vi}^{\mathrm{T}} \eta_{vi}$. Substituting Eq. (9.21) into Eq. (9.22) yields

$$\begin{aligned} \dot{\hat{W}}_{vi} &= -\lambda_{vi} \eta_i \frac{\partial \left[ c_{vi} - \frac{1}{\psi} \hat{V}_{vi} + \dot{\hat{V}}_{vi} \right]}{\partial \hat{W}_{vi}} \\ &= -\lambda_{vi} \eta_i \left[ -\frac{1}{\psi} \frac{\partial \hat{V}_{vi}}{\partial \hat{W}_{vi}} + \frac{\partial}{\partial \hat{W}_{vi}} \left( \frac{\partial \hat{V}_{vi}}{\partial Z_{vi}} \right) \dot{Z}_{vi} \right] \\ &= -\lambda_{vi} \eta_i \left[ -\frac{\sigma_{vi}}{\psi} + \nabla \sigma_{vi} \dot{Z}_{vi} \right] \\ &= -\lambda_{vi} \left[ c_{vi} - \frac{\hat{W}_{vi}^{\mathrm{T}} \sigma_{vi}}{\psi} + \dot{\hat{V}}_{vi} \right] * \left[ -\frac{\sigma_{vi}}{\psi} + \nabla \sigma_{vi} \dot{Z}_{vi} \right] \\ &= -\lambda_{vi} \left( c_{vi} + \hat{W}_{vi}^{\mathrm{T}} \Lambda_{vi} \right) \Lambda_{vi} \end{aligned} \tag{9.23}$$

where $\Lambda_{vi} = -\frac{\sigma_{vi}}{\psi} + \nabla \sigma_{vi} \dot{Z}_{vi}$. To ensure the boundedness of the weights, the adaptive law is redesigned as

$$\dot{\hat{W}}_{vi} = \begin{cases} -\lambda_{vi}\left(c_{vi} + \hat{W}_{vi}^{\mathrm{T}}\Lambda_{vi}\right)\Lambda_{vi}, \text{if } \left\|\hat{W}_{vi}\right\| \le \|\bar{W}_{vi}\| \text{ or } \left\|\hat{W}_{vi}\right\| = \|\bar{W}_{vi}\|, \\ \hat{W}_{vi}^{\mathrm{T}}\left(c_{vi} + \hat{W}_{vi}^{\mathrm{T}}\Lambda_{vi}\right)\Lambda_{vi} > 0. \\ -\lambda_{vi}\left(c_{vi} + \hat{W}_{vi}^{\mathrm{T}}\Lambda_{vi}\right)\Lambda_{vi} + \lambda_{vi}\phi_{vi}, \text{if } \left\|\hat{W}_{vi}\right\| = \|\bar{W}_{vi}\|, \\ \hat{W}_{vi}^{\mathrm{T}}\left(c_{vi} + \hat{W}_{vi}^{\mathrm{T}}\Lambda_{vi}\right)\Lambda_{vi} \le 0. \end{cases} \tag{9.24}$$

where $\|\bar{W}_{vi}\|$ is the upper bound, and $\phi_{vi} = \frac{\hat{W}_{vi}^{\mathrm{T}}\left(c_{vi}+\hat{W}_{vi}^{\mathrm{T}}\Lambda_{vi}\right)\Lambda_{vi}}{\left\|\hat{W}_{vi}\right\|^2}\hat{W}_{vi}$.

**Theorem 9.1** *Considering the optimal consensus tracking problem of heterogeneous MASs of Eq. (9.1) in the virtual layer, if the neural network weight adaptation law is designed as Eq. (9.24), then the convergence process of the adaptive weight* $\hat{W}_{vi}$ *is bounded.*

***Proof*** If $\left\|\hat{W}_{vi}\right\| \le \|\bar{W}_{vi}\|$ or $\left\|\hat{W}_{vi}\right\| = \|\bar{W}_{vi}\|$, $\hat{W}_{vi}\left(c_{vi} + \hat{W}_{vi}^{\mathrm{T}}\Lambda_{vi}\right)\Lambda_{vi} > 0$, one has

$$\dot{V}_{vi} = \frac{1}{\lambda_{vi}}\hat{W}_{vi}^{\mathrm{T}}\dot{\hat{W}}_{vi} = -\hat{W}_{vi}^{\mathrm{T}}\left(c_{vi} + \hat{W}_{vi}^{\mathrm{T}}\Lambda_{vi}\right)\Lambda_{vi} < 0. \tag{9.25}$$

If $\left\|\hat{W}_{vi}\right\| = \|\bar{W}_{vi}\|$, $\hat{W}_{vi}^{\mathrm{T}}\left(c_{vi} + \hat{W}_{vi}^{\mathrm{T}}\Lambda_{vi}\right)\Lambda_{vi} \le 0$, one has

$$\dot{V}_{vi} = \frac{1}{\lambda_{vi}}\hat{W}_{vi}^{\mathrm{T}}\dot{\hat{W}}_{vi} = -\hat{W}_{vi}^{\mathrm{T}}\left(c_{vi} + \hat{W}_{vi}^{\mathrm{T}}\Lambda_{vi}\right)\Lambda_{vi} + \frac{\hat{W}_{vi}^{\mathrm{T}}\left(c_{vi} + \hat{W}_{vi}^{\mathrm{T}}\Lambda_{vi}\right)\Lambda_{vi}}{\left\|\hat{W}_{vi}\right\|^2}\hat{W}_{vi}^{\mathrm{T}}\hat{W}_{vi} = 0 \tag{9.26}$$

By designing $\left\|\hat{W}_{vi}(0)\right\| \le \|\bar{W}_{vi}\|$, then one can obtain $\left\|\hat{W}_{vi}\right\| \le \|\bar{W}_{vi}\|$.

**Remark 9.2** Equation (9.10) has a sliding mode surface form, which is used to construct the tracking error of the agent. On the one hand, it helps to achieve synchronous convergence of distributed position tracking error $e_{vi}$ and distributed velocity tracking error $\dot{e}_{vi}$. On the other hand, by designing $\dot{e}_{vi}$, an equation containing the explicit control input $u_{vi}$ can be derived to facilitate direct design of the control input. Secondly, the design of Eq. (9.21) integrates the instantaneous performance index $c_{vi}$ and the long-term performance index $V_{vi}$, thus taking into account both current and long-term system optimization problems, and maximizing system performance. The weight adaptive law of Eq. (9.22) searches for the optimal weight values along the gradient descent direction of the coupling performance index of Eq. (9.21), which can better ensure the efficiency of the search and the optimization of the search path.

**Remark 9.3** In some applications of MASs, in order to meet the design requirements of resource conservation, it is necessary to consider the optimal control problem aimed at minimizing performance indicators. The performance indicator of Eq. (9.12) in this chapter balances the relationship between the required control objectives and control resources, and along the gradient descent direction of the performance indicator function, it can independently generate approximate optimal control signals for each agent to achieve the effect of global Nash equilibrium.

### 9.3.2 Prescribed-Time Sliding Mode Surface Design

Based on the above results, design the following prescribed-time sliding mode surface

$$s_{ai} = \dot{e}_{ai} + \lambda_{ai} e_{ai} + 2\frac{\dot{\mu}\,(t_0, T_i)}{\mu\,(t_0, T_i)} e_{ai} \tag{9.27}$$

The structure of $\mu\,(t_0, T_i)$ is shown in Eq. (9.8) of Lemma 9.1. $\lambda_{ai}$ is a positive definite adjustable parameter, $e_{ai} = \chi_i - \chi_{vi} - h_i$ is the tracking error. Let $\Omega_i = 2\frac{\dot{\mu}(t_0, T_i)}{\mu(t_0, T_i)} e_{ai}$ and $u_{aie} = \lambda_{ai}\dot{e}_{ai} + \dot{\Omega}_i$. Then taking the derivative of the sliding mode surface, one can obtain

$$\begin{aligned} \dot{s}_{ai} &= \ddot{e}_{ai} - u_{aie} \\ &= (I_i - \Lambda_i)\, u_{ai} + \tau_i + f_{ai} + d_i - \ddot{\chi}_{vi} - \ddot{h}_i - u_{aie} \end{aligned} \tag{9.28}$$

To estimate and compensate for additive faults in actuators, external disturbances, and nonlinear terms in uncertain systems, the following inequalities are designed using the boundedness of each term

$$\|\tau_i + f_{ai} + d_i\| \le \|\tau_{mi}\| + \|f_{mi}\| + \|d_{mi}\| \le \zeta_i \tag{9.29}$$

Therefore, it can be concluded that $\zeta_i$ is a positive definite constant to be estimated.

### 9.3.3 Adaptive Prescribed-Time FTC Design for Actual Layer

**Theorem 9.2** *Considering the fault-tolerant tracking control problem of actual heterogeneous MASs of Eq. (9.3) under actuator faults that satisfy Assumptions 9.1 and 9.2, if the following control law is adopted based on the prescribed-time sliding mode surface of Eq. (9.27)*

$$u_i = \left(I_i - \hat{\Lambda}_i\right)^{-1} \left(\ddot{h}_i + \ddot{\chi}_{vi} + u_{aie} - \mu_i s_{ai} + \frac{b}{\varsigma_i s_{ai}} - \frac{|\dot{\varsigma}_i|}{\varsigma_i} s_{ai} - \hat{\zeta}_i \frac{s_{ai}}{\|s_{ai}\|}\right) \tag{9.30}$$

*and the following parameter adaptive estimation laws*

$$\dot{\hat{\theta}}_i = -\Gamma_i \hat{\theta}_i - \Upsilon_i \mathrm{diag}(u_i)^{\mathrm{T}} s_{ai} \tag{9.31}$$

*and*

$$\dot{\hat{\zeta}}_i = -\mathrm{M}_i \hat{\zeta}_i + \mathrm{N}_i \left\| s_{ai} \right\| \tag{9.32}$$

*The parameters meet the following conditions*

$$\Gamma_i = \nu_i + 2\Upsilon_i \frac{|\dot{\varsigma}_i|}{\varsigma_i} \tag{9.33}$$

$$\mathrm{M}_i = \upsilon_i + 2\mathrm{N}_i \frac{|\dot{\varsigma}_i|}{\varsigma_i} \tag{9.34}$$

*where $\hat{\theta}_i$ and $\hat{\zeta}_i$ are adaptive estimates of the efficiency loss factor $\theta_i$ and the complex uncertainty $\zeta_i$, and $\hat{\theta}_i$ satisfies $\hat{\theta}_i u_i = \mathrm{diag}(u_i)\,\hat{\Lambda}_i$. In addition, $\Upsilon_i, \mathrm{N}_i, \nu_i$ and $\upsilon_i$ are all artificially set positive definite constants. Then, the formation error of heterogeneous MASs in 3D space can achieve actual prescribed-time convergence.*

***Proof*** Based on sliding mode surface and parameter adaptive estimation error, design the following Lyapunov function

$$V_{ai} = \frac{1}{2} s_{ai}^{\mathrm{T}} s_{ai} + \frac{1}{2\Upsilon_i} \tilde{\theta}_i^{\mathrm{T}} \tilde{\theta}_i + \frac{1}{2\mathrm{N}_i} \tilde{\zeta}_i^{\mathrm{T}} \tilde{\zeta}_i \tag{9.35}$$

By taking the derivative of Eq. (9.28) from Eq. (9.35) and substituting it into control law of Eq. (9.30), one can obtain

$$\begin{aligned}
\dot{V}_{ai} &= s_{ai}^{\mathrm{T}} \dot{s}_{ai} + \frac{1}{\Upsilon_i} \tilde{\theta}_i^{\mathrm{T}} \dot{\tilde{\theta}}_i + \frac{1}{\mathrm{N}_i} \tilde{\zeta}_i^{\mathrm{T}} \dot{\tilde{\theta}}_i \\
&\leq s_{ai}^{\mathrm{T}} \left[ (I_i - \Lambda_i) u_i + \zeta_i - \ddot{\chi}_{vi} - \ddot{h}_i - u_{aie} \right] - \frac{1}{\Upsilon_i} \tilde{\theta}_i^{\mathrm{T}} \dot{\hat{\theta}}_i - \frac{1}{\mathrm{N}_i} \tilde{\zeta}_i^{\mathrm{T}} \dot{\hat{\zeta}}_i \\
&\leq s_{ai}^{\mathrm{T}} \left[ \left( I_i - \hat{\Lambda}_i \right) u_i - \tilde{\Lambda}_i u_i + \zeta_i - \ddot{\chi}_{vi} - \ddot{h}_i \right] - \frac{1}{\Upsilon_i} \tilde{\theta}_i^{\mathrm{T}} \dot{\hat{\theta}}_i - \frac{1}{\mathrm{N}_i} \tilde{\zeta}_i^{\mathrm{T}} \dot{\hat{\zeta}}_i \\
&\leq s_{ai}^{\mathrm{T}} \left( -\tilde{\Lambda}_i u_i - \mu_i s_{ai} + \zeta_i + \frac{b}{\varsigma_i s_{ai}} - \frac{|\dot{\varsigma}_i|}{\varsigma_i} s_{ai} - \hat{\zeta}_i \frac{s_{ai}}{\|s_{ai}\|} \right) - \frac{1}{\Upsilon_i} \tilde{\theta}_i^{\mathrm{T}} \dot{\hat{\theta}}_i - \frac{1}{\mathrm{N}_i} \tilde{\zeta}_i^{\mathrm{T}} \dot{\hat{\zeta}}_i
\end{aligned} \tag{9.36}$$

Then, substitute the adaptive laws of Eq. (9.31) and of Eq. (9.32) to obtain

$$\begin{aligned}
\dot{V}_{ai} \leq & - \mu_i s_{ai}^{\mathrm{T}} s_{ai} + \frac{b}{\varsigma_i} + \tilde{\zeta}_i \left\| s_{ai} \right\| - \frac{|\dot{\varsigma}_i|}{\varsigma_i} s_{ai}^{\mathrm{T}} s_{ai} - s_{ai}^{\mathrm{T}} \tilde{\Lambda}_i u_i - \frac{1}{\Gamma_i} \tilde{\theta}_i^{\mathrm{T}} \dot{\hat{\theta}}_i - \frac{1}{\mathrm{N}_i} \tilde{\zeta}_i^{\mathrm{T}} \dot{\hat{\zeta}}_i \\
\leq & - \left( \mu_i + \frac{|\dot{\varsigma}_i|}{\varsigma_i} \right) s_{ai}^{\mathrm{T}} s_{ai} + \frac{b}{\varsigma_i} + \frac{\Gamma_i}{\Upsilon_i} \tilde{\theta}_i^{\mathrm{T}} \hat{\theta}_i + \frac{\mathrm{M}_i}{\mathrm{N}_i} \tilde{\zeta}_i^{\mathrm{T}} \hat{\zeta}_i \\
\leq & - \left( \mu_i + \frac{|\dot{\varsigma}_i|}{\varsigma_i} \right) s_{ai}^{\mathrm{T}} s_{ai} + \frac{b}{\varsigma_i} + \frac{\Gamma_i}{\Upsilon_i} \left( \tilde{\theta}_i^{\mathrm{T}} \theta_i - \left\| \tilde{\theta}_i \right\|^2 \right) + \frac{\mathrm{M}_i}{\mathrm{N}_i} \left( \tilde{\zeta}_i^{\mathrm{T}} \zeta_i - \left\| \tilde{\zeta}_i \right\|^2 \right) \\
\leq & - \left( \mu_i + \frac{|\dot{\varsigma}_i|}{\varsigma_i} \right) s_{ai}^{\mathrm{T}} s_{ai} + \frac{b}{\varsigma_i} + \frac{\Gamma_i}{2\Upsilon_i} \left( \left\| \theta_i \right\|^2 - \left\| \tilde{\theta}_i \right\|^2 \right) + \frac{\mathrm{M}_i}{2\mathrm{N}_i} \left( \left\| \zeta_i \right\|^2 - \left\| \tilde{\zeta}_i \right\|^2 \right)
\end{aligned} \tag{9.37}$$

Using Eqs. (9.33) and (9.34), one has

$$\begin{aligned}
\dot{V}_{ai} \leq & - \left( \mu_i + \frac{|\dot{\varsigma}_i|}{\varsigma_i} \right) s_{ai}^{T} s_{ai} + \frac{b}{\varsigma_i} - \frac{\Gamma_i}{2\Upsilon_i} \left( \left\| \tilde{\theta}_i \right\| \right)^2 - \frac{\mathrm{M}_i}{2\mathrm{N}_i} \left( \left\| \tilde{\zeta}_i \right\| \right)^2 + \frac{\Gamma_i}{2\Upsilon_i} \left\| \theta_i \right\|^2 + \frac{\mathrm{M}_i}{2\mathrm{N}_i} \left\| \zeta_i \right\|^2 \\
\leq & - \left( \mu_i + \frac{|\dot{\varsigma}_i|}{\varsigma_i} \right) \left( \left\| s_{ai} \right\|^2 \right) - \frac{\Gamma_i}{2\Upsilon_i} \left( \left\| \tilde{\theta}_i \right\|^2 \right) - \frac{\mathrm{M}_i}{2\mathrm{N}_i} \left( \left\| \tilde{\zeta}_i \right\|^2 \right) + \frac{b}{\varsigma_i} + c_i \\
\leq & - \left( \mu_i + \frac{|\dot{\varsigma}_i|}{\varsigma_i} \right) \left( \left\| s_{ai} \right\|^2 \right) - \left( \frac{\nu_i}{2\Upsilon_i} + \frac{|\dot{\varsigma}_i|}{\varsigma_i} \right) \left( \left\| \tilde{\theta}_i \right\|^2 \right) \\
& - \left( \frac{\upsilon_i}{2\mathrm{N}_i} + \frac{|\dot{\varsigma}_i|}{\varsigma_i} \right) \left( \left\| \tilde{\zeta}_i \right\|^2 \right) + \frac{b}{\varsigma_i} + c_i \\
\leq & - \left( \delta_i + 2 \frac{|\dot{\varsigma}_i|}{\varsigma_i} \right) V_{ai} + \frac{b}{\varsigma_i} + c_i
\end{aligned} \tag{9.38}$$

where $\delta_i = \min \left\{ 2\mu_i, \frac{\nu_i}{\Upsilon_i}, \frac{\upsilon_i}{\mathrm{N}_i} \right\}, c_i = \frac{\Gamma_i}{2\Upsilon_i} \left\| \theta_i \right\|^2 + \frac{\mathrm{M}_i}{2\mathrm{N}_i} \left\| \zeta_i \right\|^2$. According to Lemma 9.1, the system can achieve global actual prescribed-time fault-tolerant tracking control, and the errors $\tilde{\theta}_i$, $\tilde{\zeta}_i$ and $e_{ai}$ are uniformly bounded within $[0, T_i]$.

**Remark 9.4** In classical reinforcement learning based control problems, the adaptive law of neural network weights is often related to tracking error, with the aim of achieving uniform asymptotic bounded convergence of estimation error. The design goal of this chapter is to achieve prescribed-time stability as the algorithm. If traditional adaptive law design methods are used, more adjustment parameters and switching functions such as Eq. (9.8) need to be introduced, which sharply increases the complexity of the overall algorithm and the complexity of the proof process. This chapter assumes that the upper bound of weights exists and is known, with the aim of proving that weights are bounded. By using gradient descent method in performance indicators to adaptively adjust the weights of neural networks, on the one hand, it ensures that the weight adjustment is aimed at performance optimization, and on the other hand, it avoids coupling effects with fault-tolerant algorithms in the actual control layer.

**Remark 9.5** This chapter aims to solve the problem of prescribed-time formation tracking under actuator faults, disturbances, and model uncertainties. In order to ensure the fault tolerance and anti-interference ability of the system, a complex uncertainty term $\zeta_i(t)$ to be estimated is designed. Then, adaptive laws of Eqs. (9.31) and (9.32) are designed to effectively estimate the efficiency loss fault $\Lambda_i(t)$ and the complex uncertainty term $\zeta_i(t)$, respectively. Secondly, a prescribed-time FTC scheme is designed based on the estimated values to achieve finite time convergence of global formation error and adaptive parameter estimation error. This scheme simultaneously improves the fault tolerance performance, system response speed, and anti-interference ability of the entire formation system. The introduction of parameter adaptive technology improves the system's ability to detect interference and faults, the proactivity of fault-tolerant controller design, and the robustness of the entire formation system. Moreover, by appropriately adjusting the adaptive parameters to narrow the convergence region, the sensitivity of the system's steady-state error to interference, faults, and model uncertainty can be further improved.

**Remark 9.6** The adjustable parameter $R_i$ in the control algorithm of Eq. (9.20) of the virtual layer ensures the positive definiteness of the quadratic performance index function, which can be set as an identity matrix for convenience. In the actual control layer, the parameters related to Eq. (9.8) in the prescribed-time theorem play a critical role in the system response performance. The setting of other parameters, based on meeting the basic rules, does not have a significant impact on the system performance whether they are the optimal values. Therefore, after the initial adjustment is completed, there is no need for tedious repeated adjustments in later use. In order to reduce the spatial complexity of the algorithm, the preset time parameters can be adjusted to ensure that the running cycle is acceptable to industrial computers, avoiding the algorithm from occupying processor space for a long time. Through subsequent simulation verification, it is found that the proposed algorithm can achieve a running cycle of less than 0.0001 s. According to Shannon's sampling theorem, to recover a given signal from a sampled signal, the frequency of a given harmonic signal can be greater than 5KHZ. Considering the intelligence level of drones and unmanned vehicles, the acceptable sampling time for general flight control is 0.01 s, so the acceptable harmonic signal frequency is much less than 5 KHZ. This also indicates that the proposed algorithm can be effectively implemented in drone unmanned vehicle formation systems.

## 9.4 Simulation Experiments

In order to verify the effectiveness of the theoretical derivation results, this section designed a simulation experiment of an air-ground heterogeneous formation MAS consisting of two UGVs ($i = 1, 2$) and two UAVs ($i = 3, 4$). The agent model parameters used in the simulation are same as Chap. 8. In addition, its network topology structure is shown in Fig. 9.1, with connection weights of 0 or 1.

The parameters $\lambda_{ai}$ and $\mu_i$ in the controller affect the robustness of the controller and the convergence speed of the error reaching the sliding mode surface.

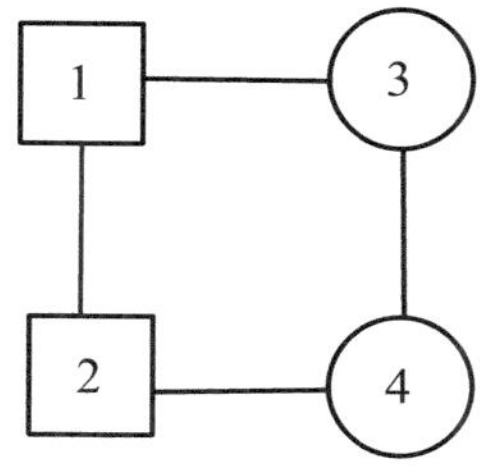


**Fig. 9.1** Communication topology

When the initial values of disturbance and error are large, the value of this parameter can be increased. The parameter design in adaptive law should be combined with practical application background, and relatively more reasonable values should be selected through trial and error combined with control effects. The controller parameters in this experiment are set to $\lambda_{ai} = 5$, $\mu_i = 30$, $b = 8$, $\nu_i = 0.55$, $\Upsilon_i = 1$, $\upsilon_i = 0.05$, $\mathrm{N}_i = 2$. The external time-varying disturbances are designed as

$$d_1 = [-0.1\cos(0.1t), 0.2\sin(0.5t)]^{\mathrm{T}},$$

$$d_2 = [-0.1\cos(0.2t), -0.2cos(0.5t)]^{\mathrm{T}},$$

$$d_3 = [-0.2\cos(0.4t), 0.1\cos(0.3t), 0.15\cos(t)]^{\mathrm{T}},$$

$$d_4 = [0.05\cos(1.5t), 0.15\sin(0.5t), 0.1\sin(0.2t)]^{\mathrm{T}}.$$

To fully simulate the actual operating environment of UAVs, a random fluctuation of 20% amplitude is added to the system damping coefficient. In addition, to verify the fault tolerance of the algorithm, actuator faults of varying degrees are designed for each intelligent agent. The specific fault parameters are designed as follows

$$\rho_{1x} = \begin{cases} 0, t < 5\,s \\ 0.2, t \geq 5\,s \end{cases}, \rho_{2x} = \begin{cases} 0, t < 7\,s \\ 0.15, t \geq 7\,s \end{cases},$$
$$\rho_{3x} = \begin{cases} 0, t < 10\,s \\ 0.18, t \geq 10\,s \end{cases}, \rho_{4x} = \begin{cases} 0, t < 13\,s \\ 0.1, t \geq 13\,s. \end{cases}$$

Consider the occurrence of the following additive faults in the actuator

$$\tau_1 = [0.1\sin(0.1t), 0.15\cos(0.2t)]^{\mathrm{T}},$$

$$\tau_2 = [0.2\cos(0.3t), 0.3\sin(0.1t)]^{\mathrm{T}},$$

$$\tau_3 = [-0.2\sin(t), 0.25\sin(0.2t), 0.17\sin(t)]^{\mathrm{T}},$$

$$\tau_4 = [0.1\cos(t), 0.2\sin(t), 0.12\sin(0.5t)]^{\mathrm{T}}.$$

To verify the superiority of the algorithm, the following comparative experiments are designed.

Simulation Experiment 1: Use the method designed in this article to complete the formation tracking task. The initial state of each agent system is $x_1(0) = [1, 1]^{\mathrm{T}}, x_2(0) = [0.2, 0.1]^{\mathrm{T}}, x_3(0) = [2, 1, 5]^{\mathrm{T}}, x_4(0) = [1, 1, 5]^{\mathrm{T}}$. The expected formation for each agent to be completed is $h_1 = [-2, 2]^{\mathrm{T}}, h_2 = [-4, 4]^{\mathrm{T}}, h_3 = [-6, 6, 10]^{\mathrm{T}}$, and $h_4 = [-8, 8, -10]^{\mathrm{T}}$. The prescribed-time for the formation to complete is $T_i = 7s$.

Simulation Experiment 2: Using the robust backstepping sliding mode tracking control method proposed in reference [3] to complete the formation tracking task. The initial state of each agent is designed as $x_1(0) = [4, 0]^{\mathrm{T}}, x_2(0) = [0.8, 0.2]^{\mathrm{T}}, x_3(0) = [0.6, 2, 0]^{\mathrm{T}}, x_4(0) = [1.5, 3, 0]^{\mathrm{T}}$. The formation trajectory that each agent needs to track is $h_i = [i \sin(i * 0.5t), i \cos(i * 0.5t)]^{\mathrm{T}}$.

From Figs. 9.2, 9.3 and 9.4, it can be concluded that the designed distributed optimal mechanism aimed at optimizing global performance indicators can achieve consensus optimal control of the agent in three-dimensional space, and achieve uniformly bounded convergence of errors. The tracking error in Fig. 9.2 dose not converge to 0, mainly due to disturbance, model uncertainty, and actuator fault. In general, when the formation error converges to a sufficiently small area under the control protocol, it can be considered that the system has achieved the expected formation task, which is acceptable in practical situations. Secondly, the controller in this chapter is designed based on the theory of actual prescribed time. Although it can ensure that the convergence time is completely independent of any other parameters, there may also be a phenomenon where the error converges to a bounded adjustable neighborhood near

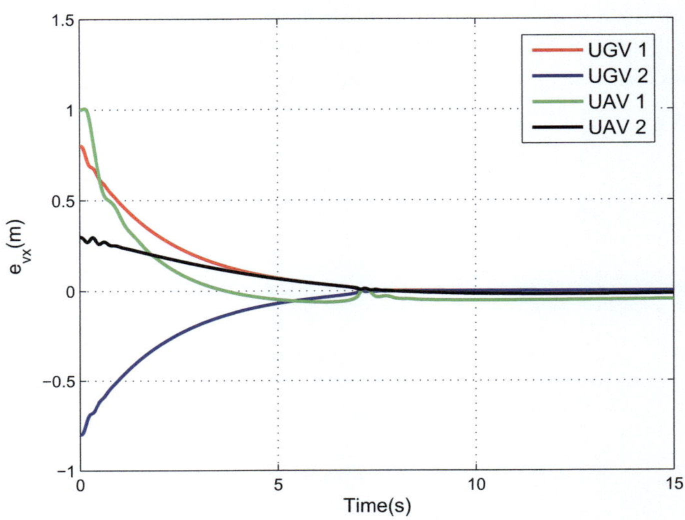


**Fig. 9.2** Consensus tracking errors for agents in $x$ axis of virtual layer

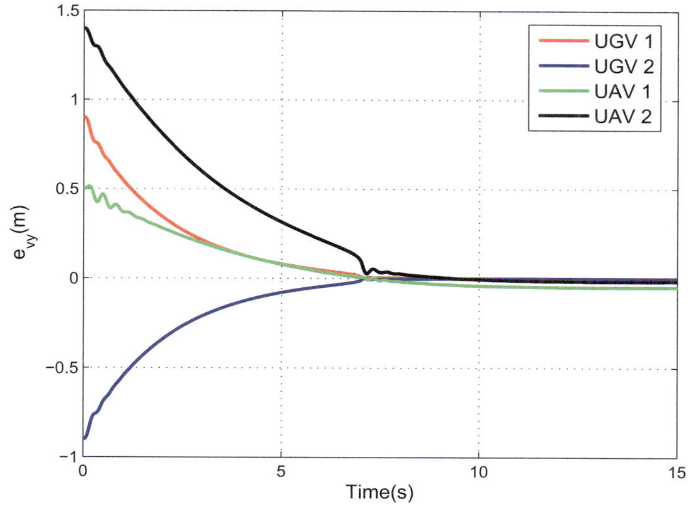


**Fig. 9.3** Consensus tracking errors for agents in $y$ axis of virtual layer

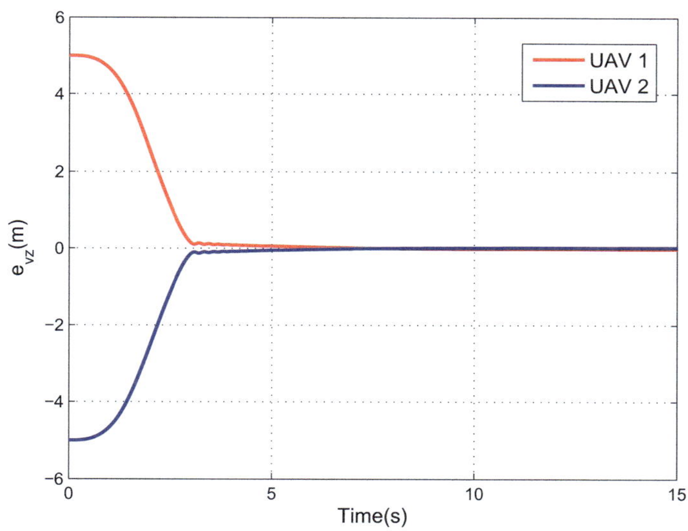


**Fig. 9.4** Consensus tracking errors for agents in $z$ axis of virtual layer

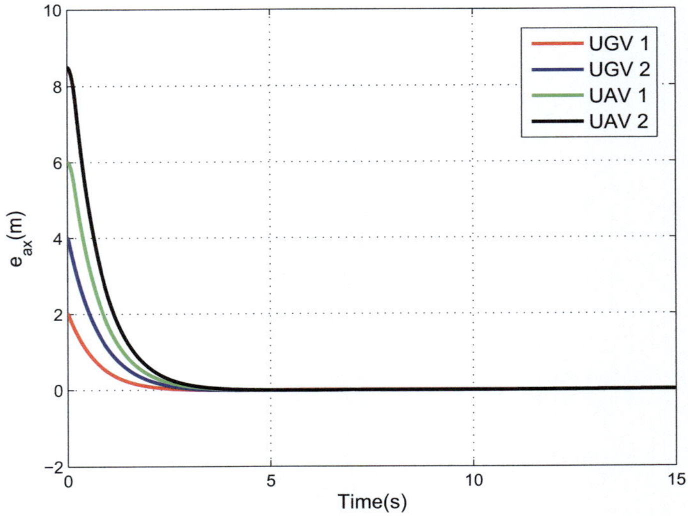


**Fig. 9.5** Formation tracking errors for agents in $x$ axis of actual layer

the origin. This is an inherent characteristic of the theory of actual prescribed-time control, which is inevitable and reasonable.

From Figs. 9.5, 9.6 and 9.7, it can be seen that each intelligent agent can still quickly complete the expected formation tracking task under the conditions of actuator fault, disturbance, and model uncertainty, and the formation completion time is controllable. Although the dynamic characteristics of different agents and the different formation tasks they need to perform can lead to different actual convergence performance of each controlled agent, the convergence completion time meets the maximum time limit of 7 s specified.

As shown in Figs. 9.8, 9.9 and 9.10, although the robust backstepping sliding mode control method proposed in reference [3] can achieve rapid error convergence, the steady-state error is not satisfactory, and therefore cannot ensure that the formation system always maintains good formation tracking performance under complex uncertainties such as actuator faults. The main reason is that the control performance of the backstepping sliding mode control method overly relies on accurate model parameters, and the controller is highly sensitive to actuator faults and uncertainties without prior knowledge. At the same time, the selection of control gains in this method often results in the controller's conservatism, leading to energy loss of the unmanned vehicle in orbit.

In contrast, the adaptive method proposed in this chapter does not rely on prior knowledge and has satisfactory stability and anti-interference ability. Figure 9.11 shows the estimation effect of the designed adaptive law on actuator multiplicative faults, verifying the effectiveness of the adaptive law. Therefore, the proposed solu-

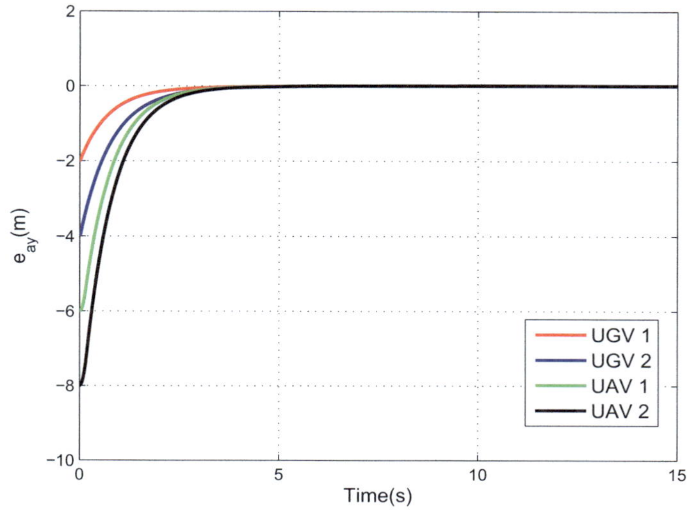


**Fig. 9.6** Formation tracking errors for agents in $y$ axis of actual layer

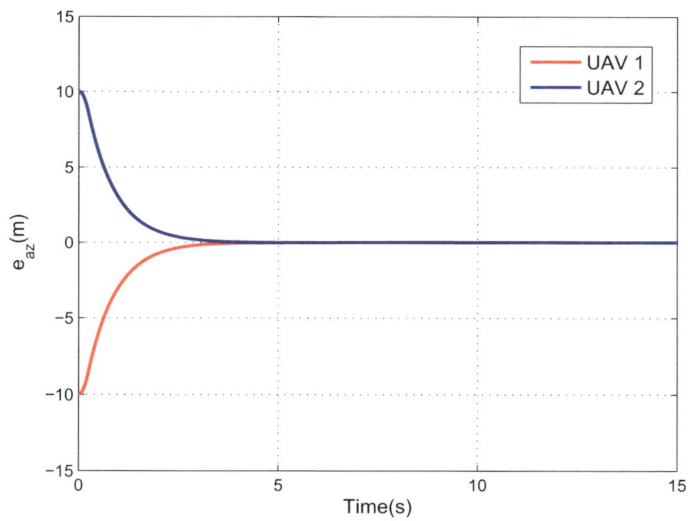


**Fig. 9.7** Formation tracking errors for agents in $z$ axis of actual layer

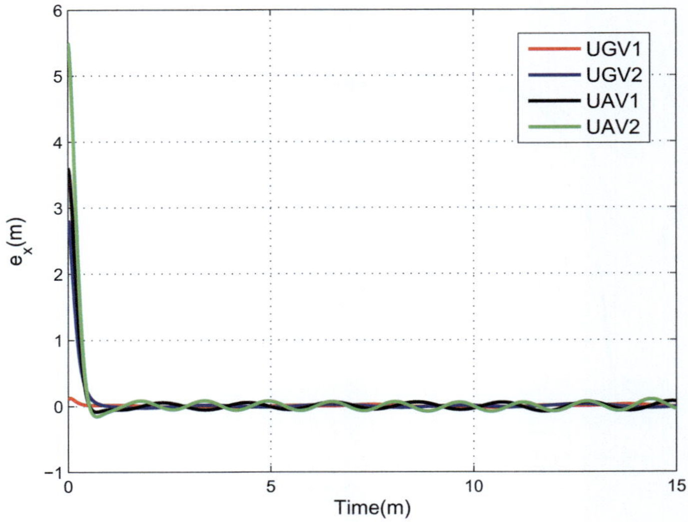


**Fig. 9.8** Formation tracking errors for agents in $x$ axis based on the method proposed in [3]

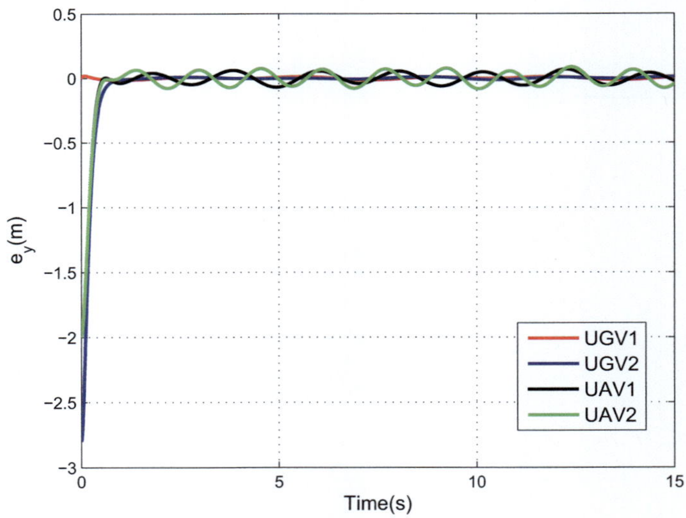


**Fig. 9.9** Formation tracking errors for agents in $y$ axis based on the method proposed in [3]

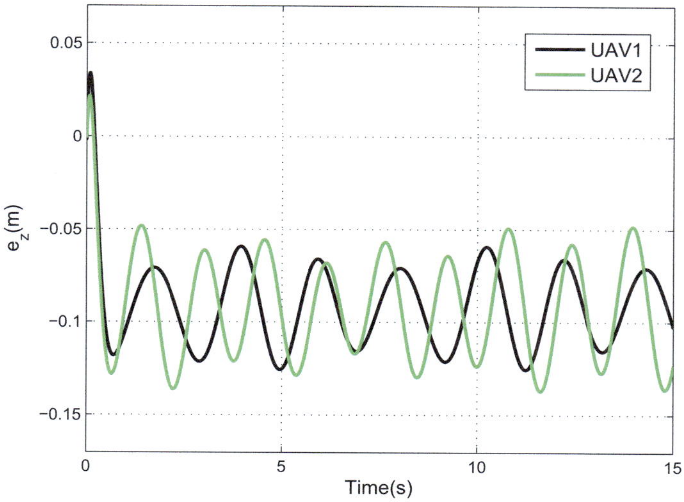


**Fig. 9.10** Formation tracking errors for agents in $z$ axis based on the method proposed in [3]

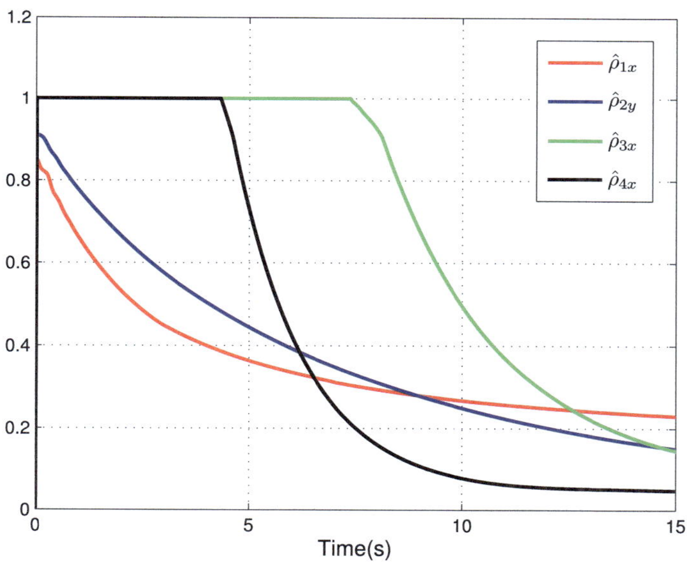


**Fig. 9.11** Profiles of the estimated fault parameters

tion in this chapter has good maneuverability, robustness, and fault tolerance, and can provide a favorable solution for the safe, stable, and fast execution of expected tasks in heterogeneous MASs.

However, from the simulation results, the theoretical value of the convergence time of the prescribed-time controller still has a certain degree of conservatism, mainly because the theoretical value is only an upper bound of the convergence time. With different engineering environments and initial states, the actual convergence time will exhibit different numerical effects, which is an inherent characteristic of this control method. Moreover, the payload capacity, task configuration capability, and control performance of heterogeneous UAVs are different, which leads to differences in response speed when executing local control instructions, ultimately resulting in differences in convergence time.

In order to demonstrate the impact of faults on the system and the response changes of the controller after faults, we take agent 2 as an example and present its control input change chart after being affected by faults. As shown in Fig. 9.12, since agent 2 is affected by the fault after 7 s, the controller needs to increase the amplitude of the control command at 7 s to effectively compensate for the losses caused by the actuator fault. Although the controller may experience brief non smooth phenomena when a fault occurs, it will quickly recover to a smooth state after adjustment. This is because the prescribed-time controller is designed based on the state feedback method, so the entire formation control system will show a stable tracking trend and smooth control input, and the closed-loop behavior of the system will also exhibit good response performance.

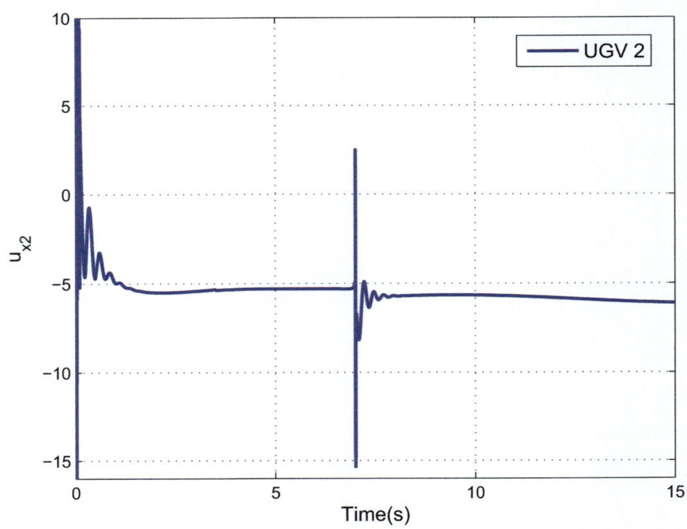


**Fig. 9.12** Control input of agent 2

## 9.5 Conclusions

Due to the large number of functional units in the airborne system of air-ground heterogeneous MASs, each agent is prone to sudden actuator faults, and it is difficult to accurately describe their state. Traditional fault-tolerant methods are difficult to achieve fast and effective processing. This chapter proposes a multi-level FTC scheme to address this issue. The virtual layer constructs a distributed optimal consistency tracking controller, while the actual control layer designs a decentralized adaptive prescribed-time fault-tolerant controller. While ensuring the convergence time of global formation error states, it effectively estimates actuator fault parameters, achieves safe and rapid adjustment of equipment functions, maintains overall system health, and ensures the smooth execution of expected formation tasks. The final simulation experiment verified the effectiveness of the proposed algorithm.

## References


1. Bai N, Wang Q, Duan Z, Chen G (2024) Distributed optimal consensus control of constrained multiagent systems: a nonseparable optimization perspective. IEEE Trans Autom Control 69(10):6880–6894
2. Cai Y, Zhang H, Li W, Mu Y, He Q (2022) Distributed bipartite adaptive event-triggered fault-tolerant consensus tracking for linear multiagent systems under actuator faults. IEEE Trans Cybern 52(11):11313–11324
3. Chen F, Jiang R, Zhang K, Jiang B, Tao G (2016) Robust backstepping sliding-mode control and observer-based fault estimation for a quadrotor UAV. IEEE Trans Ind Electron 63(8):5044–5056
4. Cheng S, Xin B, Chen J, Deng F (2025) Predefined-time distributed fault-tolerant control for nonlinear multiagent systems suffering from nonaffine faults. IEEE Trans Syst Man Cybern Syst 55(5):3300–3311
5. Cheng W, Zhang K, Jiang B, Ding SX (2021) Fixed-time fault-tolerant formation control for heterogeneous multi-agent systems with parameter uncertainties and disturbances. IEEE Trans Circuits Syst I Regul Pap 68(5):2121–2133
6. Cui Q, Song Y, Wen C (2024) Prescribed time consensus control of multiagent systems with minimized time-varying cost function. IEEE Trans Autom Control 69(5):3381–3388
7. Deng C, Wen C (2020) Distributed resilient observer-based fault-tolerant control for heterogeneous multiagent systems under actuator faults and dos attacks. IEEE Trans Control Netw Syst 7(3):1308–1318
8. Fu J, Chai T, Jin Y, Su C-Y (2016) Fault-tolerant control of a class of switched nonlinear systems with structural uncertainties. IEEE Trans Circuits Syst II Express Briefs 63(2):201–205
9. Jiang Y, Lv J, Wang C, Kao Y, Wang F (2024) Prescribed-time disturbance observer-based fully distributed prescribed-time containment control of multiagent systems. IEEE Trans Circuits Syst II Express Briefs 71(4):2044–2048
10. Li J, Yuan L, Cheng W, Chai T, Lewis FL (2024) Reinforcement learning for synchronization of heterogeneous multiagent systems by improved q-functions. IEEE Trans Cybern 54(11):6545–6558
11. Li K, Ding SX, Zheng WX, Hua C-C (2024) Global distributed fault-tolerant consensus control of nonlinear delayed multiagent systems with hybrid faults. IEEE Trans Autom Control 69(3):1967–1974

12. Lin L, Huang J (2025) Refined algorithms for adaptive optimal output regulation and adaptive optimal cooperative output regulation problems. IEEE Trans Control Netw Syst 12(1):241–250
13. Liu C, Jiang B, Patton RJ, Zhang K (2019) Hierarchical-structure-based fault estimation and fault-tolerant control for multiagent systems. IEEE Trans Control Netw Syst 6(2):586–597
14. Liu Q, Yan H, Wang M, Li Z, Liu S (2024) Data-driven optimal bipartite consensus control for second-order multiagent systems via policy gradient reinforcement learning. IEEE Trans Cybern 54(6):3468–3478
15. Lv Y, Wu Z, Zhang H, Wang Z, Yan H (2024) Local-bearing-based prescribed-time distributed localization of multiagent systems with noisy measurement. IEEE Trans Ind Inform 20(7):9518–9526
16. Golmisheh FM, Shamaghdari S (2025) Data-driven inverse reinforcement learning for heterogeneous optimal robust formation control. IEEE Trans Cybern 55(5):2024–2037
17. Rao K, Yan H, Liu Q, Dang Q, Shi K (2025) Optimal tracking control of second-order multiagent systems with input delay via data-driven forward reward q-learning framework. IEEE Trans Syst Man Cybern Syst 55(3):1858–1869
18. Song G, Shi P, Lim CP (2022) Distributed fault-tolerant cooperative output regulation for multiagent networks via fixed-time observer and adaptive control. IEEE Trans Control Netw Syst 9(2):845–855
19. Song YD, Wang YJ, Holloway J, Krstic M (2017) Time-varying feedback for regulation of normal-form nonlinear systems in prescribed finite time. Automatica 83:243–251
20. Wang J, Zhang Z, Tian B, Zong Q (2024) Event-based robust optimal consensus control for nonlinear multiagent system with local adaptive dynamic programming. IEEE Trans Neural Netw Learn Syst 35(1):1073–1086
21. Wang Y, Wang Z, Zhang H, Yan H (2025) Optimal group consensus of multiagent systems in graphical games using reinforcement learning. IEEE Trans Syst Man Cybern Syst 55(3):2343–2353
22. Wang Z, Liang B, Sun Y, Zhang T (2020) Adaptive fault-tolerant prescribed-time control for teleoperation systems with position error constraints. IEEE Trans Ind Inform 16(7):4889–4899
23. Wu W, Tong S (2025) Adaptive fuzzy distributed optimal FTC for nonlinear multiagent systems based multiplayer differential game. IEEE Trans Fuzzy Syst 33(2):657–668
24. Xu Y, Li T, Yang Y, Tong S, Chen CLP (2023) Simplified ADP for event-triggered control of multiagent systems against FDI attacks. IEEE Trans Syst Man Cybern Syst 53(8):4672–4683
25. Yan Y, Li T, Liang H (2025) Event-based prescribed-time output regulation of uncertain nonlinear multiagent systems. IEEE Trans Cybern 55(3):1070–1082
26. Yang T, Dong J (2023) Predefined-time adaptive fault-tolerant control for switched odd-rational-power multi-agent systems. IEEE Trans Autom Sci Eng 20(4):2423–2434
27. Yang X, Zhang H, Wang Z (2022) Data-based optimal consensus control for multiagent systems with policy gradient reinforcement learning. IEEE Trans Neural Netw Learn Syst 33(8):3872–3883
28. Zhang J, Fu Y, Fu J (2023) Adaptive finite-time optimal formation control for second-order nonlinear multiagent systems. IEEE Trans Syst Man Cybern Syst 53(10):6132–6144
29. Zhang Q, Chen F, Yuan J, Liu Z, Feng G (2025) Distributed optimal control of multiagent systems under control variation cost. IEEE Trans Control Netw Syst 12(1):104–113
30. Zhang S, Wu Y, He X, Liu Z (2022) Cooperative fault-tolerant control for a mobile dual flexible manipulator with output constraints. IEEE Trans Autom Sci Eng 19(4):2689–2698
31. Zhang Y, Zhao B, Liu D, Zhang S (2024) Distributed fault tolerant consensus control of nonlinear multiagent systems via adaptive dynamic programming. IEEE Trans Neural Netw Learn Syst 35(7):9041–9053

# Chapter 10
# Conclusions and Future Directions

## 10.1 Conclusions

For multiple unmanned systems, especially quadrotor UAVs, fixed-wing UAVs, unmanned helicopters, and UAVs-UGVs system, the existence of multiple agents and communication topologies bring demanding requirements of safety and reliability for realizing cooperative control under component faults and unreliable interconnections. Control system with the capability of accommodating unexpected faults in order to maintain certain pre-specified safety performance are called FTC systems. This monograph focuses on the FTC design and the stability analysis for multiple unmanned systems. Specifically, the contributions and achievements of this monograph are summarized as the following aspects.

(1) A command filter-based fixed-time FTC strategy is proposed for multiple quadrotor UAVs under actuator faults and undirected communication topologies. A robust integral sliding-mode control scheme is designed for the position loop, and a sliding mode observer-based fixed-time FTC is designed for the attitude loop. The constructed algorithms can meet the practical requirements on convergence time and ensure the fixed-time stability of the system.
(2) A fixed-time FTC protocol with collision avoidance is proposed for quadrotor UAVs for the safety of the formation, especially when the intensive formation is required to pass through narrow spaces. To estimate and compensate actuator faults, an augmented fixed-time observer is designed and a fixed-time FTC is proposed for position tracking bansed on sliding-mode technology. Numerical simulations and pixhawk platform are chosen to implement the proposed control scheme. The experiments reveal that the proposed scheme effectively can achieve leader-follower formation control and collision avoidance.
(3) An adaptive fixed-time fault-tolerant formation control problem of UAVs-UGVs is proposed with finite-time prescribed performance. The online update protocols are designed to estimate the unknown fault factors and the upper bound of



K. Zhang et al., *Fixed-Time/Prescribed-Time Fault-Tolerant Control for Multiple Unmanned Systems*, Unmanned System Technologies,
https://doi.org/10.1007/978-981-95-2913-1_10

lumped uncertainties. Based on the simulation experiments, the effectiveness of the proposed scheme is proved. The proposed scheme not only can have fast convergence and fault tolerance, but also can effectively estimate the parameter of actuator faults.

(4) A hierarchical-based fixed-time FTC is proposed for fixed-wing UAVs with actuator faults and input saturation under switching topologies. In the upper layer, an ITSM FTC is designed and a fixed-time auxiliary system is introduced in the lower layer to solve the control input saturation problem. A fixed-time disturbance observer is designed to estimate the aggregate information of model uncertainties and actuator faults. Finally, the effectiveness of the designed FTC scheme is verified by simulation experiments.

(5) A continuous fixed-time FTC protocol is proposed for UAVs-UGVs formation systems with actuator faults and disturbances under fixed and switching topologies. A fixed-time observer is constructed for each follower to obtain the estimation of actuator faults. Then, a novel FTC is proposed based on the integral fixed-time sliding mode technology and backstepping method, which provides many great features such as chattering elimination, fault tolerance, fixed-time convergence.

(6) A prescribed-time fault-tolerant formation control problem for collision-avoidance multi-unmanned helicopters via the HOFAS approach. A prescribed-time fault estimation observer is thus presented to possess the capacity of approximating the lumped faults. Then, a prescribed-time fault-tolerant formation control protocol is further developed, which can not only compensate the actuator faulty redundancy and assure a higher formation tracking accuracy with collision-avoidance in a predefined time, but require fewer control design steps.

(7) A distributed neural network prescribed-time FTC is proposed for air-ground systems affected by partial loss of effectiveness and bias actuator faults under directed graphs. A novel prescribed-time sliding manifold is developed, and then a prescribed-time fault-tolerant formation tracking controller is proposed based on the neural network and adaptive parameter estimation, which can generate smooth control action due to the use of regular feedback of local system states.

(8) A hierarchical structure-based optimal FTC scheme is proposed of a heterogeneous formation system of UAVs and UGVs by using graphical game theory, and prescribed-time control method. The virtual layer constructs a distributed optimal consistency tracking controller, and then a prescribed-time fault-tolerant tracking controller is designed for the actual control layer, which achieves safe and rapid adjustment of equipment functions, maintains overall system health, and ensures the smooth execution of expected formation tasks.

## 10.2 Future Directions

This monograph presents a series of control strategies of the fixed-time/prescribed-time FTC issues for quadrotor UAVs, fixed-wing UAVs, unmanned helicopters, and UAVs-UGVs system. However, there still exist some open and challenging problems to be addressed for the advanced FTC design. Some possible future research directions are listed as follows.

(1) In recent years, with the continuous emergence of new technologies such as game theory, hybrid driving, and big data processing, a new generation of advanced intelligent control algorithms has emerged. Combining the new generation of artificial intelligence algorithms with FTC theory can ensure safer and more efficient operation of highly complex industrial systems in the event of faults. Therefore, it is necessary to combine new generation artificial intelligence technology to design highly intelligent fault-tolerant controllers to ensure the safety of multiple unmanned systems in the future.
(2) The future research direction should further focus on the cross-linking characteristics and fault mechanisms of heterogeneous MASs, develop models and multi-source data-driven health maintenance mechanisms and methods, expand and deepen the integral fault diagnosis and FTC mechanisms of the system from single source to multiple sources, from single to complex, from decentralized to comprehensive, and further provide technical support for the safe operation of modern airborne systems.
(3) For heterogeneous/homogeneous MASs that perform complex tasks in highly uncertain open environments, further research should be conducted on effective human-machine cross cognitive behavior models and optimization evolution mechanisms, while balancing the dual advantages of intelligence and data. Implementing an intelligent joint control strategy based on human-machine hybrid enhancement, promoting collaborative solutions based on the combination of human intelligence, artificial intelligence, and machine intelligence as an effective way to solve complex task problems.

MIX
Papier aus verantwortungsvollen Quellen
Paper from responsible sources
FSC® C105338

If you have any concerns about our products,
you can contact us on
**ProductSafety@springernature.com**

In case Publisher is established outside the EU,
the EU authorized representative is:
**Springer Nature Customer Service Center GmbH**
**Europaplatz 3, 69115 Heidelberg, Germany**

Printed by Libri Plureos GmbH
in Hamburg, Germany